LEGAL PRACTICE COURSE

KU-050-534

BLACKSTONE'S

ROAD TRAFFIC LAW
INDEX

CASE PRECEDENTS 1900-1997

WITHDRAWN

WP 2141642 7

BLACKSTONE'S

ROAD TRAFFIC LAW INDEX

CASE PRECEDENTS 1900-1997

Dr Maxwell Barrett

UNIVERSITY OF WOLVERHAMPTON LIBRARY

Acc No. 2141642   CLASS

CONTROL
1854318527

DATE
25. AUG. 1998   SITE LW

LPC
Room
343.
094
BLA

LPc

BLACKSTONE
PRESS LIMITED

First published in Great Britain 1998 by Blackstone Press Limited,
Aldine Place, London W12 8AA. Telephone: 0181-740 2277

© Max Barrett, 1998

ISBN: 1 85431 852 7

British Library Cataloguing in Publication Data
A CIP catalogue record for this book is available from the British Library

Typeset by Style Photosetting Limited, Mayfield, East Sussex
Printed by Bell & Bain Limited, Glasgow

All rights reserved. No part of this book may be reproduced or transmitted in any form or by any
means, electronic or mechanical including photocopying, recording, or any information storage or
retrieval system without prior permission from the publisher.

CONTENTS

INTRODUCTION

'The decisions . . . of courts are held in the highest regard, and
are not only preserved as authentic records in the treasuries of the
several courts, but are handed out to public view in the numerous
volumes of *reports* which furnish the lawyer's library.[1]

So wrote Blackstone in 1765. Since then those 'numerous volumes of reports' to which he makes reference have multiplied in number to a level that is surely beyond anything he ever imagined. And it is not just that the total number of reports being published has increased in the past two and a half centuries. Mirroring the present tendency among lawyers towards ever greater specialisation into ever more specific fields of legal expertise has been a growing tendency among publishers in recent years to bring out law reports that concentrate on particular subject areas.

All of this has placed lawyers in something of a predicament. Even if it were economically feasible to subscribe to each of the many series of law reports and legal periodicals that have been or become available during this century alone (and even the most generously funded library would surely baulk at such an enormous expense) the amount of overlap in coverage between the different sources could well mean that those cases to which particular lawyers need access are already handsomely covered in the publications to which they or their chambers or firm presently subscribe and that outside reference is so occasional it could be met by a reasonably priced publication which would not only index the reports to which those lawyers or their chambers or firms now subscribe but contain additional comprehensive references to alternative reports and journals that feature relevant case-law and to which occasional recourse might be necessary.

Economic feasibility aside, there is the matter of convenience. However well prepared they are on entering court every barrister and many solicitors have on occasion found themselves on their feet addressing or pressing a perhaps unanticipated argument which they wished they could support by reference to a convincing precedent. That precedent may well exist somewhere among the many sources of case-law now available. Indeed oftentimes a barrister or solicitor may have a very good notion of when a relevant case was decided or with what subject-matter the relevant case was concerned but simply cannot remember its name. Standing before the court, away from law reports and legal periodicals practitioners would at such times undoubtedly find a one-stop, single-volume index relevant to the subject area being litigated, detailing the location and content of many thousands of cases and drawn from a very wide array of sources to be of real usefulness.

Of course such an index would also be of enormous benefit back in the office. It would in a single self-contained book provide lawyers with a cost-effective, hassle-free, easy-to-use means of accessing in their particular subject area many thousands of cases in many thousands of law report and law journal volumes and so open up a whole new world of precedents for them to call in aid when seeking to buttress whatever case they are called upon to make. And with lawyers working ever longer hours, many of them at home, such a reference work that gave immediate access to thousands of cases would be of considerable help there too.

---

[1] Blackstone, *Commentaries on the Laws of England* (Dawsons of Pall Mall, 1966), Volume I, p.71.

Such a reference work now exists. The publication of the first eight volumes of the *Blackstone's Index of Case Precedents*, each volume being an entirely self-contained index to cases in one of eight subject areas means that there is now a reasonably priced, convenient and readily usable one-stop reference source for legal practitioners, academic lawyers and legal students who are seeking to locate a specific twentieth century precedent or trying to discover whether there is a relevant and helpful precedent in a particular area of the law.

Embracing Child Law, Criminal Law, Evidence, Family Law, Marriage Breakdown, Road Traffic Law, Sentencing, and Tort these first eight volumes of *Blackstone's Index* contain over 120,000 references to more than 40,000 cases decided by the English and Welsh courts (and by the Privy Council) throughout the twentieth century. Within the various volumes cases are listed alphabetically as well as by subject. The court which decided each case is also identified. In addition case-entries are followed by brief one or two sentence 'pointers' which seek to indicate in a little greater detail the content of the many decisions that have been indexed. These pointers do not, it should be noted, attempt to summarise the *ratio decidendi* of each case or indeed to encapsulate the entire array of issues that have been considered in a case. They are intended merely to provide a modicum of assistance to the reader in deciding whether it is worth moving on from having unearthed a case in the *Blackstone's Index* to actually reading it in its entirety in a case report.

Blackstone's Index is uniquely comprehensive in indexing not only law reports but legal journals and periodicals as well. Cases have been extracted from the *All England Law Reports, Cox's Criminal Cases*, the *Criminal Appeal Reports, Criminal Appeal Reports (Sentencing), Criminal Law Review (Case and Comment), Family Court Reporter, Family Law Reports, Justice of the Peace Reports, Law Journal, Law Journal County Court Reporter, Law Journal Newspaper County Court Appeals, Law Journal Newspaper County Court Reports, Law Journal Reports, Law Reports (Appeal Cases, Chancery Division, Family Division, King's/Queen's Bench Division, Probate, Divorce* and *Admiralty Division), Law Times Reports, New Law Journal, New Law Journal Reports, Road Traffic Reports, Solicitors Journal, Times Law Reports* (old and new series), *Weekly Notes, Weekly Reporter* and *Weekly Law Reports*.

In short *Blackstone's Index of Case Precedents* represents a uniquely powerful low-cost reference tool through which the legal wisdom of the twentieth century is made readily available to lawyers heading into the twenty-first. In addition to publishing each volume of the work as a separate self-contained and easy-to-use book all eight volumes have been published on CD-ROM.

HOW TO USE BLACKSTONE'S ROAD TRAFFIC LAW INDEX

This volume contains references to those Road Traffic cases decided by the English and Welsh courts (and the Privy Council) during the twentieth century which have been reported in the wide array of reports and periodicals that form the basis of the *Blackstone's Index of Case Precedents*.

Like each other volume of a *Blackstone's Index* this volume is a self-contained reference source. In other words users of *Blackstone's Road Traffic Law Index* will not be referred to any other volume in their quest for a particular *Road Traffic* Law case.

On the whole *Blackstone's Road Traffic Law Index* embraces those areas of law that would feature in a typical Road Traffic law treatise. As with all volumes of the Index cases primarily concerned with points of practice have not been included.

Blackstone's Road Traffic Law Index is divided into two Parts. In Part I cases are grouped into subject categories. In Part II cases are listed alphabetically. Readers looking for cases by topic are recommended to consult Part I. Readers with a rough (or indeed exact) idea of a case name and who wish merely to discover where a specific case is located ought to consult Part II.

Each case is succeeded by a brief 'pointer'. The 'pointer' gives a better idea of what exactly each case is concerned with. It cannot be overemphasised that the 'pointer' does not represent an attempt to summarise the entirety of a case nor does it seek to give a one or two line *ratio decidendi* for each case. It is meant merely to provide a flavour of what has been decided in each case and to thereby enable the reader to make a somewhat more informed choice as to whether it is worth consulting a comprehensive account of the case in the relevant source or sources to which the reader is referred.

The Subject Index Cases in the subject index are grouped into generic categories such as DRUGGED/ DRUNK DRIVING, FAILURE TO PROVIDE SPECIMEN and UNAUTHORISED TAKING OF VEHICLE. Cases are then listed under a variety of highly specific sub-headings. This has the double advantage that readers can either look up all the cases in a loosely defined area of interest or alternatively can very quickly zone in on the few cases decided in a much more tightly defined area of interest.

Either way the subject index is perfectly easy to use. It is preceded by two Tables of Contents. The first (entitled 'Generic Headings') lists the broad categories into which cases have been divided (DRUGGED/DRUNK DRIVING, FAILURE TO PROVIDE SPECIMEN and UNAUTHORISED TAKING OF VEHICLE). The second (entitled 'Generic Headings with Sub-Headings') lists the much more specific headings under which cases are listed within each generic category ('DRUGGED/DRUNK DRIVING (general)', 'FAILURE TO PROVIDE SPECIMEN (general)', 'UNAUTHORISED TAKING OF VEHICLE (general)'). The reader can turn to the 'Generic Headings' page, see in what generic category the case or type of case being looked for is most likely to be found and turn to the appropriate part of the subject index. Alternatively the reader can move on to (if the reader has not in fact gone straight to) the 'Generic Headings with Sub-Headings' page, find out the sub-headings under which cases are listed, decide which sub-heading most closely matches the reader's subject of inquiry and turn to the page indicated. As with any dictionary or index a degree of creative thinking may very occasionally be required of the reader in judging which category or sub-heading best matches the case or genre of case the reader is seeking to locate.

The Alphabetical Index The alphabetical index generally follows the traditional format adopted in the contents of case reports. Hence case names are on the whole listed both in forward and in reverse order. However, there is one important exception. Cases beginning *R* v . . . are listed only in reverse order. Thus users consulting the alphabetical index for, say, *R* v *Adames* should look under *Adames, R* v. Users seeking *R* v *Peters* should look under *Peters, R* v and so on.

Users should note that while every reasonable care has been taken to ensure that the case names and citations in the text of this work are accurate and that the text of the work is correct insofar as it indicates what was decided in each case mentioned the author accepts no responsibility for loss occasioned to any person acting or refraining from acting as a result of material contained in this publication.

ABBREVIATIONS USED

The following abbreviations are used in the text:

AC	*Law Reports (Appeal Cases)*
All ER	*All England Reports*
CA	Court of Appeal
CCA	Court of Criminal Appeal
CCR	Crown Cases Reserved
CC Rep	*Law Journal County Court Reports*
Ch	*Law Reports (Chancery Division)*
Cox CC	*Cox's Criminal Cases*
Cr App R	*Criminal Appeal Reports*
Cr App R (S)	*Criminal Appeal Reports (Sentencing)*
CrCt	Crown Court
Crim LR	*Criminal Law Review*
CyCt	County Court
FamD	*Law Reports (Family Division)*
FCR	*Family Court Reporter*
FLR	*Family Law Reports*
HC ChD	High Court (Chancery Division)
HC FamD	High Court (Family Division)
HC KBD	High Court (King's Bench Division)
HC PDAD	High Court (Probate, Divorce and Admiralty Division)
HC QBD	High Court (Queen's Bench Division)
HL	House of Lords
JP	*Justice of the Peace Reports*
KB	*Law Reports (King's Bench Division)*
LJ	*Law Journal/Law Journal Reports*
LJNCCR	*Law Journal Newspaper County Court Reports*
LJNCCA	*Law Journal Newspaper County Court Appeals*
LTR	*Law Times Reports*
Mag	Magistrates' Court
NLJ	*New Law Journal/New Law Journal Reports*
PC	Privy Council
Police Ct	Police Court
PDAD	*Law Reports (Probate, Divorce and Admiralty Division)*
QB	*Law Reports (Queen's Bench Division)*
RTR	*Road Traffic Reports*
SJ	*Solicitors Journal*
TrTb	Transport Tribunal
TLR	*Times Law Reports*
WN	*Weekly Notes*
WR	*Weekly Reporter*
WLR	*Weekly Law Reports*

ABBREVIATIONS USED

In addition the following letters sometimes appear after page numbers:

1. dt, g, i, t

These letters when they appear immediately before a page number indicate that the source to which the reader is being referred contains an additional reference to a British newspaper report of the case concerned.

dt = *Daily Telegraph*
g   = *The Guardian*
i   = *The Independent*
t   = *The Times*

2. ccr

Not all County Court cases noted by the *Law Journal* were published separately from the main body of the *Journal*. Those notes of County Court reports contained within the *Journal* itself are indicated in the text of this *Index* by prefacing the page number from the relevant volume of the *Law Journal* with the abbreviation 'ccr' to show that the case is contained in the County Court Reporter section of the *Journal*.

3. LB

These letters may appear before the page number in a reference to the *Solicitors Journal* and indicate that the reference is to a page number in the *Lawyer's Brief* section of the relevant *Solicitors Journal*.

ACKNOWLEDGMENTS

Writing the first eight volumes of *Blackstone's Index of Case Precedents* has been a challenging task rendered all the easier by the consistent kindness of family and friends and the very real generosity of many other people who have freely given their advice and assistance throughout the period in which the volumes were prepared.

My parents, Michael and Della Barrett, yet again provided me with that unqualified encouragement and support which they have afforded me in the past. My two brothers, Conor and Dr Gavin Barrett were similarly helpful.

In addition to my parents and brothers I was privileged throughout the period of writing the various volumes of the *Index* to have a close coterie of generous people around me who were eager to provide me with whatever assistance I needed. My very good friend Jennifer Powell proved herself to be a remarkable bastion of support to whom I owe a particular debt of gratitude. I am also deeply grateful to Sue Bate, the Law Librarian at Manchester University whose unfailing and undeserved kindness towards me has immeasurably facilitated and speeded the completion of this work. Professor Frank B. Wright, my onetime doctoral and later post-doctoral research supervisor at the European Occupational Health and Safety Legal Research Centre willingly provided useful and much-appreciated advice whenever it was solicited. My old friend Dr Jonathan Rush, not only undertook the onerous technical task of formatting *Blackstone's Index* for publication both in book and CD-form but was a constant bedrock of support in many other ways. Bryan and Josie Hallows provided me with a warm and welcome retreat in the Cheshire countryside and so much more besides. So far as places to retreat to with my work were concerned I was in fact rather spoiled for choice with — regrettably as yet unavailed of — offers coming in from Duncan Lennox in San Francisco and from Anna Retoula, Christos Retoulas and Constantina Scholidi in Greece. Others among this 'support team' of friends included Fr Des Doyle, Fr Ian Kelly, Fr John McMahon, Stephano Pistillo and Martin Wai-Chung Leung. I must also acknowledge the very great assistance given to me by the library staff at Cambridge University, Manchester University, Salford University and Trinity College Dublin.

Last but far from least I would like to record my special thanks to Agapi Kapeloni, who when I began this work was my girlfriend, who somewhere along the way found herself changed in status to that of honorary research assistant and whose latest and most agreeable change in status was to become my wife.

Finally, I have taken reasonable care to ensure that the case names and citations in the text of this work are accurate and that the text of the work is correct insofar as it indicates what was decided in each case mentioned. However, I accept no responsibility for loss occasioned to any person acting or refraining from acting as a result of material contained in this publication.

Blackstone's Index of Case Precedents embraces cases from the turn of the twentieth century through to the early Autumn of 1997.

Dr Maxwell Barrett,
Dublin,
Feast of the Immaculate Conception, 1997.

GENERIC HEADINGS

AGGRAVATED DRIVING
AGGRAVATED VEHICLE TAKING
ATTEMPT TO TAKE AND DRIVE AWAY
ATTEMPTING TO DRIVE
BREATH TEST
CARELESS DRIVING
CARRIAGE OF GOODS FOR
   HIRE/REWARD
CAUSING BODILY HARM WHEN IN
   CHARGE OF VEHICLE
CAUSING DEATH BY DRIVING
CLAMPING
CONSTRUCTION AND USE
CROSSING
DANGEROUS DRIVING
DANGEROUS VEHICLE
DEFENCES
DOCUMENTS
DRIVER'S HOURS
DRIVING CAR ON PAVEMENT
DRIVING INSTRUCTION
DRIVING TEST
DRIVING WHILE DISQUALIFIED
DRIVING WITH EXCESS ALCOHOL
DRIVING WITHOUT AUTHORITY
DRIVING/ATTEMPTING TO DRIVE
DRUGGED/DRUNK DRIVING
DUTY TO STOP/REPORT
EMERGENCY VEHICLE
EMISSION OF OIL FROM LORRY
EXPRESS CARRIAGE
FURIOUS DRIVING
GENERAL
GOODS VEHICLE
HEALTH AND SAFETY AT WORK
HEAVY MOTOR VEHICLE
IDENTITY
INCONSIDERATE DRIVING
INTERFERENCE WITH FREE PASSAGE
INTERFERENCE WITH MOTOR
   VEHICLE

LICENCE
LIGHTING
LOAD
LOCOMOTIVE
MAKING FALSE STATEMENT
MOTOR MANSLAUGHTER
MOTORWAY
NEGLIGENT DRIVING
NOTICE OF INTENDED
   PROSECUTION
OBSTRUCTION OF HIGHWAY
OVERTAKING
PARKING
PASSENGER VEHICLE
PETROL
PROTECTIVE HEADGEAR
'PUBLIC PLACE'
PUBLIC SERVICE VEHICLE
RECKLESS DRIVING
RECOVERY VEHICLE
REGISTRATION/LICENSING
ROAD
SEAT BELT
SENTENCE
SIGN
SPECIAL VEHICLE
SPECIMEN
SPEEDING
STAGE CARRIAGE
TAXI
TRACTOR
TRAFFIC EXAMINER
TRAFFIC REGULATION
TRAFFIC WARDEN
UNAUTHORISED TAKING OF
   VEHICLE
UNDER-AGE DRIVING
UNINSURED DRIVING
UNLAWFUL USE OF VEHICLE
UNLICENSED DRIVING
WANTON DRIVING

GENERIC HEADINGS WITH SUB-HEADINGS

AGGRAVATED DRIVING (general)
AGGRAVATED VEHICLE TAKING
(general)
ATTEMPT TO TAKE AND DRIVE AWAY
(general)
ATTEMPTING TO DRIVE (general)
BREATH TEST (general)
BREATH TEST (hospital)
BREATH TEST (printout)
CARELESS DRIVING (aiding and abetting)
CARELESS DRIVING (general)
CARRIAGE OF GOODS FOR
HIRE/REWARD (general)
CAUSING BODILY HARM WHEN IN
CHARGE OF VEHICLE (general)
CAUSING DEATH BY DRIVING (general)
CLAMPING (general)
CONSTRUCTION AND USE (agricultural
machine)
CONSTRUCTION AND USE (articulated
vehicle)
CONSTRUCTION AND USE (axle)
CONSTRUCTION AND USE (brakes)
CONSTRUCTION AND USE (defective
tyres)
CONSTRUCTION AND USE (general)
CONSTRUCTION AND USE (land
implement)
CONSTRUCTION AND USE (latent defect)
CONSTRUCTION AND USE (mechanically
propelled vehicle)
CONSTRUCTION AND USE (mirrors)
CONSTRUCTION AND USE (moped)
CONSTRUCTION AND USE (motor
vehicle)
CONSTRUCTION AND USE (motorcycle)
CONSTRUCTION AND USE (special
vehicle)
CONSTRUCTION AND USE (steering)
CONSTRUCTION AND USE (trailer)
CONSTRUCTION AND USE (tyres)
CONSTRUCTION AND USE
(unroadworthy vehicle)
CONSTRUCTION AND USE (use)

CONSTRUCTION AND USE (vehicle)
CONSTRUCTION AND USE (vehicle
testing)
CONSTRUCTION AND USE (weight)
CONSTRUCTION AND USE (works truck)
CROSSING (general)
DANGEROUS DRIVING (general)
DANGEROUS VEHICLE (general)
DEFENCES (alcoholism)
DEFENCES (automatism)
DEFENCES (duress)
DEFENCES (necessity)
DEFENCES (sleep)
DOCUMENTS (general)
DRIVER'S HOURS (general)
DRIVING CAR ON PAVEMENT (general)
DRIVING INSTRUCTION (general)
DRIVING TEST (general)
DRIVING WHILE DISQUALIFIED (aiding
and abetting)
DRIVING WHILE DISQUALIFIED
(general)
DRIVING WITH EXCESS ALCOHOL
(general)
DRIVING WITHOUT AUTHORITY
(general)
DRIVING/ATTEMPTING TO DRIVE
(general)
DRUGGED/DRUNK DRIVING (general)
DUTY TO STOP/REPORT (general)
EMERGENCY VEHICLE (general)
EMISSION OF OIL FROM LORRY
(general)
EXPRESS CARRIAGE (general)
FURIOUS DRIVING (general)
GENERAL (miscellaneous)
GOODS VEHICLE (access permit)
GOODS VEHICLE (driver)
GOODS VEHICLE (driver's records)
GOODS VEHICLE (farmer's vehicle)
GOODS VEHICLE (general)
GOODS VEHICLE (licence)
HEALTH AND SAFETY AT WORK
(general)

HEAVY MOTOR VEHICLE (general)
IDENTITY (general)
INCONSIDERATE DRIVING (general)
INTERFERENCE WITH FREE PASSAGE (general)
INTERFERENCE WITH MOTOR VEHICLE (general)
LICENCE (articulated vehicle)
LICENCE (disabled person)
LICENCE (epileptic)
LICENCE (excise)
LICENCE (express carriage)
LICENCE (false)
LICENCE (false declaration)
LICENCE (farmer's goods vehicle)
LICENCE (forgery)
LICENCE (general)
LICENCE (general trade)
LICENCE (goods vehicle)
LICENCE (identity mark)
LICENCE (locomotive)
LICENCE (motor cycle)
LICENCE (penalty)
LICENCE (provisional)
LICENCE (restricted)
LICENCE (road haulage)
LICENCE (road service)
LICENCE (suspension)
LICENCE (tractor)
LICENCE (trade)
LIGHTING (general)
LOAD (general)
LOCOMOTIVE (general)
MAKING FALSE STATEMENT (general)
MOTOR MANSLAUGHTER (general)
MOTORWAY (general)
NEGLIGENT DRIVING (general)
NOTICE OF INTENDED PROSECUTION (general)
OBSTRUCTION OF HIGHWAY (general)
OVERTAKING (general)
PARKING (general)
PASSENGER VEHICLE (general)
PETROL (general)
PROTECTIVE HEADGEAR (general)
'PUBLIC PLACE' (general)
PUBLIC SERVICE VEHICLE (general)
RECKLESS DRIVING (general)
RECOVERY VEHICLE (general)
REGISTRATION (general)
REGISTRATION/LICENSING (general)
ROAD (general)
SEAT BELT (general)
SENTENCE ('totting up')
SENTENCE (aggravated vehicle-taking)

SENTENCE (careless driving)
SENTENCE (causing bodily harm by wanton driving)
SENTENCE (causing death by dangerous driving)
SENTENCE (causing death by reckless driving)
SENTENCE (concurrent/consecutive sentences)
SENTENCE (conspiracy to obstruct/pervert course of justice)
SENTENCE (criminal damage)
SENTENCE (dangerous driving)
SENTENCE (disqualification)
SENTENCE (driving while disqualified)
SENTENCE (drugged/drunk driving)
SENTENCE (endorsement of licence)
SENTENCE (failure to provide specimen)
SENTENCE (fines)
SENTENCE (fixed penalties)
SENTENCE (general)
SENTENCE (motor manslaughter)
SENTENCE (parking)
SENTENCE (penalty points)
SENTENCE (probation)
SENTENCE (reckless driving)
SENTENCE (speeding)
SENTENCE (suspended sentence)
SENTENCE (tachograph)
SENTENCE (theft of car)
SENTENCE (unauthorised taking of vehicle)
SENTENCE (uninsured vehicle)
SENTENCE (vehicle taking)
SENTENCE (violence)
SENTENCE (wanton driving)
SIGN (general)
SPECIAL VEHICLE (fire engine)
SPECIAL VEHICLE (general)
SPECIAL VEHICLE (mobile crane)
SPECIAL VEHICLE (road tanker)
SPECIMEN (admissibility)
SPECIMEN (blood)
SPECIMEN (container)
SPECIMEN (failure to provide)
SPECIMEN (general)
SPECIMEN (hospital)
SPECIMEN (portion)
SPECIMEN (urine)
SPEEDING (general)
STAGE CARRIAGE (general)
TAXI (advertisement)
TAXI (general)
TRACTOR (general)
TRAFFIC EXAMINER (general)
TRAFFIC REGULATION (general)

SUBJECT INDEX

AGGRAVATED DRIVING (general)

Dawes v Director of Public Prosecutions [1995] 1 Cr App R 65; [1994] Crim LR 604; [1994] RTR 209; [1994] TLR 112 HC QBD Valid arrest for aggravated driving.

AGGRAVATED VEHICLE TAKING (general)

R v Bradshaw and another [1994] TLR 693 CA Extended driving test after disqualification period expired unmerited by person found guilty of aggravated vehicle-taking but who was said to have been passenger in relevant vehicle.

R v Gostkowski [1995] RTR 324 CA Disqualification imposed on passenger party to aggravated vehicle-taking reduced in light of disqualification imposed on driver.

R v Marsh (William) [1997] 1 Cr App R 67; [1997] Crim LR 205; [1997] RTR 195; (1996) 140 SJ LB 157 CA Was aggravated vehicle taking where accident arising in connection with car taken without authority was not driver's responsibility.

R v Timothy (Stephen Brian) (1995) 16 Cr App R (S) 1028 CA Nine months' imprisonment for aggravated vehicle taking.

ATTEMPT TO TAKE AND DRIVE AWAY (general)

R v Fussell [1951] 2 All ER 761; (1951-52) 35 Cr App R 135; (1951) 115 JP 562; (1951) 101 LJ 582 CCA Could try attempted (taking and driving away) indictable offence summarily.

ATTEMPTING TO DRIVE (general)

Stevens v Thornborrow [1969] 3 All ER 1487; [1970] Crim LR 41t; (1970) 134 JP 99; [1970] RTR 31; (1969) 113 SJ 961; [1970] 1 WLR 23 HC QBD Person is driving after vehicle stops if what is doing is connected with driving.

BREATH TEST (general)

Adams v Valentine [1975] Crim LR 238; [1975] RTR 563 HC QBD Constable having validly stopped car for other reason could then form opinion that driver had been drinking and seek to administer breath test (driver still deemed to be driving).

Anderton v Lythgoe [1985] Crim LR 158; [1985] RTR 395; (1984) 128 SJ 856; [1984] TLR 596; [1985] 1 WLR 222 HC QBD Police could not rely on breath specimen as definitive evidence of accused's guilt where failed to inform accused when arrested him of his right to give blood/urine sample.

Anderton v Waring [1986] RTR 74 HC QBD Was not reasonable excuse to non-provision of breath specimen charge that had tried one's best but machine did not record sample as being provided.

Arnold v Chief Constable of Kingston-upon-Hull [1969] 3 All ER 646; [1969] Crim LR 442; (1969) 133 JP 694; (1969) 113 SJ 409; [1969] 1 WLR 1499 HC QBD On taking of tests 'there or nearby'.

Askew v Director of Public Prosecutions [1988] RTR 303 HC QBD Depends on individual circumstances whether person still a hospital patient: here was not.

Atkinson v Walker [1976] Crim LR 138; (1976) 126 NLJ 64t; [1976] RTR 117 HC QBD Police officer to make clear that breath specimen request must be complied with but need not specify statutory basis of demand (test still valid if mistaken justification for it given but valid alternative reason exists).

1

Attorney-General's Reference No 2 of 1974 [1975] 1 All ER 658; (1975) 139 JP 267; [1975] RTR 142 CA Breathalyser test invalid if smoke immediately beforehand.

Badkin v Director of Public Prosecutions [1987] Crim LR 830; [1988] RTR 401 HC QBD Evidence showed breathalyser evidence here to be unreliable but in any event police officer (in belief of device's unreliability) had sought blood specimens and prosecution had to proceed on that evidence thereafter.

Blake v Bickmore [1982] RTR 167 HC QBD Breath test valid where nothing to show manufacturer's instructions not complied with or that if had not been results would be prejudicial towards defendant.

Blake v Pope [1986] 3 All ER 185; [1986] Crim LR 749; [1987] RTR 77; (1970) 130 SJ 731; [1986] 1 WLR 1152 HC QBD Police need not see person driving before forming that driving with excess alcohol.

Braham v Director of Public Prosecutions (1995) 159 JP 527; [1996] RTR 30; [1994] TLR 684 HC QBD Breath specimen obtained when police after tip-off went to house of person and arrested her when she refused to take breath test (later taken at police station) was admissible.

Bunyard v Hayes [1985] RTR 348; [1984] TLR 603 HC QBD Issue of lawfulness of arrest did not not affect culpability for failure to provide breath specimen.

Burditt v Rogers [1986] Crim LR 636; (1986) 150 JP 344; [1986] RTR 391 HC QBD Breathalyser not unreliable where printout merely contained spelling mistakes that were human in provenance.

Burridge v East [1986] Crim LR 632; (1986) 150 JP 347; [1986] RTR 328 HC QBD Single breath specimen admissible in evidence (whether so depends on circumstances of case).

Butcher v Catterall [1975] Crim LR 463; [1975] RTR 436 HC QBD Non-compliance with manufacturer's instructions did not per se render breathalyser evidence useless.

Butler v Easton [1970] Crim LR 45t; [1970] RTR 109; (1969) 113 SJ 906 HC QBD Breath test/specimen request/specimen provision to occur at same police station: otherwise laboratory certificate inadmissible.

Campbell v Tormey [1969] 1 All ER 961; (1969) 53 Cr App R 99; [1969] Crim LR 150t; (1969) 133 JP 267; (1968) 118 NLJ 1196t; (1968) 112 SJ 1023; [1969] 1 WLR 189 HC QBD Cannot ask person who has fully completed drive to take breath test; must arrest person before can request breath test/laboratory specimen.

Cannings v Houghton [1977] RTR 55; [1976] Crim LR 748 HC QBD Constable who when dealing with driver with car on private land reasonably believed driver/car to have been in accident on road could validly require breath test of driver.

Castle v Cross [1985] 1 All ER 87; [1984] Crim LR 682; [1985] RTR 62; (1984) 128 SJ 855; [1984] 1 WLR 1372 HC QBD Breath test computer printout admissible in evidence as was evidence of machine operator based thereon.

Chief Constable of Avon and Somerset Constabulary v Creech [1986] Crim LR 62; [1986] RTR 87 HC QBD Two breath specimens allowed and that obtained during different cycles of breathalyser not improper.

Chief Constable of Avon and Somerset Constabulary v O'Brien [1987] RTR 182 HC QBD That had sought to meet with doctor/solicitor before giving specimen not reasonable excuse defence to charge of failure to provide specimen.

Chief Constable of Gwent v Dash [1985] Crim LR 674; [1986] RTR 41 HC QBD Random breath tests impermissible but police could randomly stop cars, form opinion that person had been drinking and require breath test.

Chief Constable of Northumbria v Browne [1986] RTR 113 HC QBD Secretary of State's mistake in reference to breathalyser manufacturer in breathalyser approval order did not invalidate approval of relevant breathalyser.

Chief Constable of Staffordshire v Lees (1981) 145 JP 208; [1981] RTR 506 HC QBD On what constitutes 'accident' for purposes of taking breath test (Road Traffic Act 1972, s 8(2)).

Chief Constable of Surrey v Wickens (1985) 149 JP 333; [1985] RTR 277; [1984] TLR 637 HC QBD No requirement that copy of breathalyser printout be signed.

Chief Constable of West Midlands Police v Billingham [1979] 2 All ER 182; [1979] Crim LR 256; (1979) 129 (1) NLJ 149; [1979] RTR 446; (1979) 123 SJ 98; [1979] 1 WLR 747 HC QBD 'Accident' in Road Traffic Act 1972 bears common meaning.

Clements v Dams [1978] Crim LR 96; [1978] RTR 206 HC QBD Failed attempt to justify breath test by reference to constable's suspicion of alcohol where requested test in mistaken belief that person driving car had been involved in accident.

Clowser v Chaplin (1981) 72 Cr App R 342; [1981] Crim LR 412; (1981) 125 SJ 221 HC QBD Power of arrest on requesting breath test does not give police officer right to enter property without invitation.

Cooper v Rowlands [1972] Crim LR 53; [1971] RTR 291 HC QBD Could presume from circumstances/absent alternative evidence that police constable in uniform/approved breathalyser used.

Corp v Dalton [1982] Crim LR 756; [1983] RTR 160; [1982] TLR 419 HC QBD Good faith deviation from manufacturer's/police officer's instructions when providing breath test did not render defendant guilty of failure to provide specimen.

Cotgrove v Cooney [1987] Crim LR 272; (1987) 151 JP 736; [1987] RTR 124 HC QBD On finding there to be reasonable cause why specimen not provided.

Darnell v Portal [1972] Crim LR 511; [1972] RTR 483 HC QBD Non-compliance with manufacturer's directions inadequate per se to render breath test invalid (here was not invalid).

DeFreitas v Director of Public Prosecutions [1992] Crim LR 894; (1993) 157 JP 413; [1993] RTR 98; [1992] TLR 337 HC QBD Genuine (though baseless) fear that would contract AIDS by using breathalyser a reasonable excuse for refusing breath specimen.

Denny v Director of Public Prosecutions (1990) 154 JP 460; [1990] RTR 417 HC QBD Not improper for police officer to seek breath specimens on properly working breathalyser after failed attempts on defective breathalyser.

Director of Public Prosecutions v Berry [1995] TLR 571 HC QBD Breath test evidence available where could have given blood/urine sample instead but were incapable of understanding this in light of alcohol had taken.

Director of Public Prosecutions v Carey [1969] 3 All ER 1662; [1970] AC 1072; (1970) 54 Cr App R 119; [1970] Crim LR 107t; (1970) 134 JP 59; [1970] RTR 14; (1969) 113 SJ 962; [1969] 3 WLR 1169 HL On breathalyser and specimen testing.

Director of Public Prosecutions v Charles; Director of Public Prosecutions v Kukadia; Ruxton v Director of Public Prosecutions; Reaveley v Director of Public Prosecutions; Healy v Director of Public Prosecutions; McKean v Director of Public Prosecutions; Edge (James) v Director of Public Prosecutions [1996] RTR 247 HC QBD Slight procedural errors did not render specimens obtained inadmissible.

Director of Public Prosecutions v Coyle [1996] RTR 287; [1995] TLR 427 HC QBD That were unaware/uninformed of mechanical time limit on breathing into breathalyser not reasonable excuse justifying non-provision of specimen inside time limit.

Director of Public Prosecutions v Curtis (1993) 157 JP 899; [1993] RTR 72; [1992] TLR 432 HC QBD Inadequate evidence of asthma at time of being breathalysed to provide reasonable excuse for not providing second breath specimen.

Director of Public Prosecutions v Dixon [1993] RTR 22 HC QBD Matter for subjective adjudication whether police officer has reasonable basis for belief that breathalyser malfunctioning (and so to require blood (as here) or urine specimen).

Director of Public Prosecutions v Frost [1989] Crim LR 154; (1989) 153 JP 405; [1989] RTR 11 HC QBD On nature of offences of being in charge of vehicle while unfit to drive (matter which ordinary person could decide)/in charge with excess alcohol (provable by reference to expert evidence).

Director of Public Prosecutions v Godwin (1992) 156 JP 643; [1991] RTR 303 HC QBD Evidence as to excess alcohol in defendant's breath validly excluded on basis that obtained on foot of unlawful arrest.

Director of Public Prosecutions v Hill (1992) 156 JP 197; [1991] RTR 351 HC QBD Can challenge presumption that breath-alcohol concentration not lower than that shown by breathalyser but was inadequate evidence here that latter not functioning properly.

Director of Public Prosecutions v Hill-Brookes [1996] RTR 279 HC QBD Must tell person consequences of providing/not providing optional blood/urine specimen so that may properly exercise option.

Director of Public Prosecutions v Hutchings (1992) 156 JP 702; [1991] RTR 380 HC QBD Duplicate print-outs from breathalyser admissible where original print-outs lost/police officer could have given oral evidence as to original readings by reference to duplicate print-outs.

Director of Public Prosecutions v Jones [1990] RTR 33 HC QBD Person using car to escape attack by another could not succeed on defence of necessity insofar as entire journey was concerned.

Director of Public Prosecutions v Magill [1989] Crim LR 155; [1988] RTR 337 HC QBD Upon breath reading indicating less than 50mg alcohol in 100ml of breath police officer to specifically inform defendant that may substitute either breath or urine sample instead.

Director of Public Prosecutions v McGladrigan [1991] Crim LR 851; (1991) 155 JP 785; [1991] RTR 297; [1991] TLR 216 HC QBD On effect of unlawful arrest on admissibility of breath test evidence.

Director of Public Prosecutions v McKeown; Director of Public Prosecutions v Jones [1997] 1 All ER 737; [1997] 2 Cr App R 155; [1997] 2 Cr App R (S) 289; [1997] Crim LR 522; (1997) 161 JP 356; (1997) 147 NLJ 289; [1997] RTR 162; [1997] TLR 88; [1997] 1 WLR 295 HL Malfunction of clock on otherwise properly functioning breathalyser did not render print-out consequent upon breath test inadmissible.

Director of Public Prosecutions v Pearman (1993) 157 JP 883; [1992] RTR 407; [1992] TLR 154 HC QBD Was open to justices to find that very shocked and distressed woman had been incapable of providing proper breath specimen.

Director of Public Prosecutions v Rose [1993] Crim LR 407; (1992) 156 JP 733 HC QBD Tiny distance which travelled/begrudgery over apprehension not special reasons justifying non-disquali-fication for failure to provide breath specimens.

Director of Public Prosecutions v Skinner and Director of Public Prosecutions v Cornell (1989) 153 JP 605 HC QBD On non-access to solicitor as a reasonable excuse for refusing breath specimen.

Director of Public Prosecutions v Thomas (Elwyn Kenneth) (1993) 157 JP 480; [1996] RTR 293; [1992] TLR 343 HC QBD Accused afforded proper opportunity of providing second breath sample where had argued over providing same (as first over the limit) until machine could no longer process second sample.

Director of Public Prosecutions v Whalley [1991] Crim LR 211; (1992) 156 JP 661; [1991] RTR 161 HC QBD Refusal to provide breath specimen only reasonable if to give specimen would for mental/physical reasons be adverse to donor's health.

Director of Public Prosecutions v Winstanley (1994) 158 JP 1062; [1993] RTR 222; [1993] TLR 68 HC QBD Breath specimen relied upon even though failure to provide alternative specimen was not through fault of defendant.

Dixon v Director of Public Prosecutions (1994) 158 JP 430 HC QBD Police were wrong to obtain blood samples on basis that breathalyser defective where had no reasonable ground for believing breathalyser to be defective.

Donegani v Ward [1969] 3 All ER 636; [1969] Crim LR 493t; (1969) 133 JP 693; (1969) 119 NLJ 649t; (1969) 113 SJ 588; [1969] 1 WLR 1502 HC QBD Issue of fact for justices whether test in a place 'nearby'.

Endean v Evens [1973] Crim LR 448 HC QBD Person did not still have to be driving when breath testing equipment being assembled for testing to be valid.

Erskine v Hollin [1971] Crim LR 243t; (1971) 121 NLJ 154t; [1971] RTR 199; (1971) 115 SJ 207 HC QBD Improper to arrest person and then wait for breathalyser to be brought to scene.

Evans v Director of Public Prosecutions (1996) TLR 28/5/96 HC QBD Breath test evidence did not prove person guilty of offence of driving with excess alcohol in urine.

Everitt v Trevorrow [1972] Crim LR 566 HC QBD Was reasonable cause to suspect had alcohol in body where driving not especially bad but was smell of alcohol from driver's breath when police stopped him.

Farrow v Bosomworth [1969] Crim LR 320t; (1969) 119 NLJ 390; (1969) 113 SJ 368 HC QBD Home Office circular could not be relied on to establish that breathalyser an approved device.

Fawcett v Gasparics [1987] Crim LR 53; [1986] RTR 375 HC QBD Printout showing wrong day could be admitted but weight thereof as evidence might well be affected.

Fawcett v Tebb [1984] Crim LR 175; (1984) 148 JP 303; [1983] TLR 719 HC QBD Not failure to provide breath specimen where specimen provided was never tested to see if adequate to give reading.

Finnigan v Sandiford; Clowser v Chaplin [1981] 2 All ER 267; (1981) 73 Cr App R 153; [1981] Crim LR 643; (1981) 145 JP 440; (1981) 131 NLJ 553; [1981] RTR 317; (1981) 125 SJ 378; [1981] 1 WLR 837 HL Power of arrest on requesting breath test does not give police officer right to enter property without invitation.

Fox v Chief Constable of Gwent [1985] 1 All ER 230; [1984] Crim LR 567; [1984] RTR 402; (1985) 129 SJ 49; [1984] TLR 406; [1985] 1 WLR 33 HC QBD Breath specimen admissible though obtained on foot of unlawful arrest.

Fox v Chief Constable of Gwent [1986] Crim LR 59; (1985) 129 SJ 757 HL Breath specimen admissible though obtained on foot of unlawful arrest.

Gage v Jones [1983] RTR 508 HC QBD Reasonable to infer from circumstances that breath test taken by police officer in uniform.

Gaimster v Marlow [1985] 1 All ER 82; (1984) 78 CrAppR 156; [1984] Crim LR 176; (1984) 148 JP 624; [1984] QB 218; [1984] RTR 49; (1983) 127 SJ 842; [1983] TLR 736; [1984] 2 WLR 16 HC QBD Computer printout from breathalyser certificate, and police operator's statement stating results related to accused's specimen were admissible in evidence.

Garner v Director of Public Prosecutions (1990) 90 Cr App R 178; [1989] Crim LR 583; (1990) 154 JP 277; [1990] RTR 208 HC QBD Intoximeter 3000 printout admissible in evidence absent police certificate.

Gill v Forster [1972] Crim LR 45; [1970] RTR 372 HC QBD On effect on validity of breath test of non-compliance with manufacturer's instructions regarding proper storage of breathalyser.

Gordon v Thorpe [1986] Crim LR 61; [1986] RTR 358 HC QBD Information ought not to have been dismissed simply on basis of discrepancy in breathalyser readings where both well above prescribed limit.

Greenaway v Director of Public Prosecutions (1994) 158 JP 27; [1994] RTR 17 HC QBD Pre- and post-test calibration to be given in evidence (may be given orally).

Haghigat-Khou v Chambers [1987] Crim LR 340; [1988] RTR 95 HC QBD On determining availability/reliability of approved breathalyser.

Hague v Director of Public Prosecutions [1995] TLR 589; [1997] RTR 146 HC QBD Breath specimen reading validly obtained (and so admissible in evidence) even though police officer had reasonably considered it to be defective and so requested blood specimen (also validly obtained).

Harris v Croson [1973] Crim LR 121; (1973) 123 NLJ 14t; [1973] RTR 57; (1973) 117 SJ 91 HC QBD That suspicion had consumed alcohol arose in course of random check did not per se render all that followed invalid; person just stopped was still driving when breath specimen requested.

Hawes v Director of Public Prosecutions [1993] RTR 116; [1992] TLR 217 HC QBD Breath test requirement of person arrested on private property was valid.

Hay v Shepherd [1974] Crim LR 114; (1973) 123 NLJ 1137t; [1974] RTR 64 HC QBD Person stopped without reasonable suspicion on part of officer that have been drinking can validly be breathalysed if officer then immediately forms reasonable suspicion.

Hayward v Eames; Kirkpatrick v Harrigan [1984] Crim LR 760; [1985] RTR 12; [1984] TLR 553 HC QBD Lion Intoximeter 3000 an approved breathalyser.

Hobbs v Clark [1988] RTR 36 HC QBD On need to specifically inform defendant that can substitute blood/urine specimen for that of breath specimen which proves to have less than 50mg of alcohol to 100ml of breath.

5

Hollingsworth v Howard [1974] Crim LR 113; [1974] RTR 58 HC QBD Police officer could request breath test of person being conveyed in ambulance; said person not a person at hospital as patient.

Hope v Director of Public Prosecutions [1992] RTR 305 HC QBD Person who provided breath specimens as required and who agreed to provide blood could then decide not to provide blood.

Horton v Twells [1983] TLR 738 HC QBD Justices should have considered whether police officer's non-compliance with instructions of breathalyser manufacturer was bona fide — if was, then ought to have continued with drunk driving trial.

Howard v Hallett [1984] Crim LR 565; [1984] RTR 353; [1984] TLR 302 HC QBD Police officer ought not to have requested more than two breath specimens at station; first specimen (had lowest reading) ought to have been considered by justices, not second (lower of second and third readings).

Hoyle v Walsh [1969] 1 All ER 38; [1969] Crim LR 91t; (1969) 133 JP 91; [1969] 2 QB 13; (1968) 112 SJ 883; [1969] 2 WLR 34 HC QBD Improper arrest for failure to provide specimen where failure arose from defective breathalyser.

Hoyle v Walsh (1969) 53 Cr App R 61 CA Not guilty of failing to provide breath specimen where failure arose through defect in breathalyser — arrest without warrant for same therefore invalid.

Hudson v Director of Public Prosecutions (1992) 156 JP 168; [1992] RTR 27; [1991] TLR 261 HC QBD Police and Criminal Evidence Act 1984, s 78(1) (discretion to exclude unduly prejudicial evidence) could apply to breath test evidence.

Hughes v McConnell [1986] 1 All ER 268; [1985] RTR 244 HC QBD Challenge to validity of breathometer reading must establish device in itself defective.

Jarvis v Director of Public Prosecutions [1996] RTR 192 HC QBD Not allowing doctor to take blood after breathalyser failed was failure to provide specimen; no judicial notice as to number of breathalysers in particular police station.

Johnson v Whitehouse [1984] RTR 38; [1983] TLR 222 HC QBD Must have reasonable basis for believing — not merely suspecting — that accused drove car involved in accident before can request breath specimen of same.

Kaye v Tyrrell [1984] TLR 436 HC QBD Breath test request deemed valid where made of passenger who had been driver when car went through red lights some minutes before request made.

Lafferty v Director of Public Prosecutions [1995] Crim LR 429 HC QBD Roadside breath test evidence validly admitted (though two police station breath samples unchallenged) as was pertinent to issue raised by driver of how much he had drunk.

Lambert v Roberts [1981] 2 All ER 15; (1981) 72 Cr App R 223; [1981] Crim LR 256; (1981) 145 JP 256; (1981) 131 NLJ 448; [1981] RTR 113 HC QBD Where police officer's licence to enter withdrawn has no legal entitlement to administer breath test.

Lloyd v Morris [1985] Crim LR 742; [1986] RTR 299 HC QBD Unusually wide discrepancy between two breathalyser readings did not per se merit decision by justices that breathalyser defective.

Lodge v Chief Constable of Greater Manchester [1988] Crim LR 533 HC QBD Valid finding by court that despite disparate readings from breathalyser it was functioning properly (question of fact for court).

Lucking v Forbes [1985] Crim LR 793; (1987) 151 JP 479; [1986] RTR 97 HC QBD Magistrates Decision on basis of expert evidence that breathalyser evidence defective not unreasonable and so valid; evidence as to excess alcohol in blood irrelevant where charged with driving with excess alcohol in breath.

Maharaj v Chand [1986] 3 All ER 107; [1987] RTR 295 HC QBD Large discrepancy in breathalyser readings did not render same inadmissible where justices satisfied device was calibrated properly when specimens taken.

Matto v Wolverhampton Crown Court [1987] RTR 337 HC QBD Bad faith, oppressive behaviour of police meant discretion ought to have been exercised and breath test evidence excluded.

Mayon v Director of Public Prosecutions [1988] RTR 281 HC QBD Conviction quashed where prosecution failed to established breathalyser was correctly calibrated.

McGrath v Field [1987] Crim LR 274; [1987] RTR 349 HC QBD Justices' decision not to exercise discretion and exclude breath test evidence obtained after procedural errors by police was valid.

McKenna v Smith [1976] Crim LR 256 HC QBD That police officer did not give detailed explanation of why was seeking breath sample did not render seeking of same invalid.

McKeown v Director of Public Prosecutions; Jones v Director of Public Prosecutions [1995] Crim LR 69 HC QBD Malfunction of clock on otherwise properly functioning breathalyser rendered print-out consequent upon breath test inadmissible.

Miller v Howe [1969] 3 All ER 451; [1969] Crim LR 491t; (1969) 133 JP 665; (1969) 119 NLJ 649t; (1969) 113 SJ 706; [1969] 1 WLR 1510 HC QBD Constable may identify breathalyser from memory — need not be produced.

Morgan v Lee [1985] Crim LR 515; (1985) 149 JP 583; [1985] RTR 409 HC QBD Police officer could not request blood specimen where had breath-tested person and breathalyser was functioning but did not give print-out (not per se evidence of unreliability).

Morris v Beardmore [1980] 2 All ER 753; [1981] AC 446; (1980) 71 Cr App R 256 (also HC QBD); [1979] Crim LR 394; (1980) 144 JP 331; (1980) 130 NLJ 707; [1980] RTR 321; (1980) 124 SJ 512; [1980] 3 WLR 283 HL Request to take breath test invalid where police officer trespassing on property of person requested at time of request.

Morris v Beardmore [1979] 3 All ER 290; (1980) 71 Cr App R 256 (also HL); (1980) 144 JP 30; [1980] QB 105; [1979] RTR 393; (1979) 123 SJ 300; [1979] 3 WLR 93 HC QBD Request to take breath test valid though police officer trespassing on property of person requested at time of request.

Moss v Jenkins [1974] Crim LR 715; [1975] RTR 25 HC QBD Absent explanation as to how station sergeant had reasonable suspicion that certain driver had been drinking, breath test request by him invalid.

Mulcaster v Wheatstone [1979] Crim LR 728; [1980] RTR 190 HC QBD Valid breath test requirement where driving did not prompt policeman's first suspicion of excess alcohol but immediately formed that opinion on approaching driver.

Neal v Evans [1976] Crim LR 384; [1976] RTR 333; (1976) 120 SJ 354 HC QBD Person who drank more alcohol pending known arrival of police and so frustrated breath test was guilty of obstructing police officer in execution of duty.

Newsome v Hayton [1974] Crim LR 112; (1973) 123 NLJ 1137t; [1974] RTR 9 HC QBD Person not driving/attempting to drive where stopped by constable over licence and minute passed before constable suspected driver of driving with excess alcohol.

Newton v Woods (1987) 151 JP 436; [1987] RTR 41 HC QBD Difference in breathalyser readings peculiar but not such as to require decision that machine unreliable (and so evidence inadmissible).

Oldfield v Anderton [1986] Crim LR 189; (1986) 150 JP 40; [1986] RTR 314 HC QBD First breath sample could be admitted in evidence where failed to give second sample and could convict of drunk driving on basis of that sample.

Owen v Chesters [1985] Crim LR 156; (1985) 149 JP 295; [1985] RTR 191; (1984) 128 SJ 856; [1984] TLR 663 HC QBD On nature of evidence as to breathalyser results ncessary to satisfy court (oral evidence of testing officer not hearsay but may not meet standard of proof).

Owen v Morgan [1986] RTR 151 HC QBD Could repeat request to person (who had refused to supply breath specimen) to supply breath specimen.

Parker v Director of Public Prosecutions (1993) 157 JP 218; [1993] RTR 283; [1992] TLR 163 HC QBD Breathalyser reading valid where reading set to Greenwich Mean Time, not British Summer Time as was appropriate.

Parsley v Beard [1978] RTR 263 HC QBD On mechanics of breath test.

Patterson v Charlton [1985] Crim LR 449; (1986) 150 JP 29; [1986] RTR 18 HC QBD Admission that were driving is enough proof of same; burden is on person over the blood-alcohol limit to prove had not driven while so.

Price v Davies [1979] RTR 204 HC QBD Breath test valid though police officer adjusted wrongly attached mouthpiece and then returned same to motorist after her first failed effort to inflate breathalyser.

Price v Nicholls [1985] Crim LR 744; [1986] RTR 155 HC QBD Breathalyser evidence as to breath-alcohol levels is definitive absent evidence that breathalyser not working properly.

Prince v Director of Public Prosecutions [1996] Crim LR 343 HC QBD Unnecessary to prove calibration of breathalyser correct where prosecution only seek to rely on blood specimen evidence.

Pritchard v Jones [1985] Crim LR 52 CA Need not be afforded second opportunity to give breath specimen where first opportunity was given in police station.

R v Aspden [1975] RTR 456 CA Bona fide actions of policeman that may/may not have breached manufacturer instructions on delay between drinking and testing breath did not render breathalyser evidence inadmissible; one year disqualification for cooperative (repeat) offender who required car for business.

R v Auker-Howlett [1974] Crim LR 52t; [1974] RTR 109 CA Whether had been refusal to provide breath specimen usually a question of fact for jury.

R v Birdwhistle [1980] Crim LR 381 CA Breath specimen evidence admitted though may have been obtained on foot of unlawful arrest.

R v Bove [1970] 2 All ER 20; [1970] Crim LR 353; (1970) 134 JP 418 HC QBD Person not driving when refused breath test was unlawfully arrested.

R v Bove (1970) 54 Cr App R 316; [1970] Crim LR 471; [1970] RTR 261; (1970) 114 SJ 418; [1970] 1 WLR 949 CA Person not driving when refused breath test was unlawfully arrested.

R v Brentford Magistrates' Court, ex parte (Clarke) Robert Anthony; Clarke v Hegarty (1986) 150 JP 495; [1987] RTR 205 HC QBD Where two breath specimen readings were the same they could both be admitted in evidence.

R v Britton [1973] Crim LR 375; [1973] RTR 502 CA Valid conviction for attempting to defeat the course of justice by drinking alcohol so as to render the laboratory specimen provided unreliable.

R v Brown (Michael) [1977] Crim LR 291t; (1977) 127 NLJ 114t; [1977] RTR 160 CA Was valid for policeman having formed opinion that defendant had excess alcohol taken to require breath test of him after detaining him but before arresting him.

R v Burdekin [1975] Crim LR 348; [1976] RTR 27 CA Constable could stop person for road traffic offence, then form opinion that driver has alcohol in body and seek to administer breath test; cannot race ahead of police to private land and then validly refuse to submit to breath test; that accused guided needle for doctor taking blood specimen did not mean specimen not taken by doctor.

R v Callum [1976] Crim LR 257; [1975] RTR 415 CA Person arrested after provided breath test specimen did not have to be told why as was apparent from circumstances; breach of manufacturer's instructions (subject not to smoke immediately before breath test) rendered breath test evidence unsafe.

R v Clarke [1969] 2 All ER 1008; (1969) 53 Cr App R 438; [1969] Crim LR 441; (1969) 113 SJ 109; (1969) 133 JP 546; (1969) 113 SJ 428; [1969] 1 WLR 1109 CA Any clearly worded request suffices in asking to take breath test; any form of declining is a refusal; any reasonable excuse a defence.

R v Clarke [1969] 1 All ER 924; [1969] 2 QB 91; (1969) 53 Cr App R 251; [1969] Crim LR 203; (1969) 133 JP 282; (1969) 119 NLJ 129t; [1969] 2 WLR 505 CA Approval of breathalyser device established by production of copy of Ministerial order in court.

R v Coates [1971] Crim LR 370; [1971] RTR 74 CA Conviction quashed where validity of arrest/proper functioning of breathalyser not left to jury.

R v Coleman [1974] RTR 359 CA Seven/eight minute wait (while not under arrest) for arrival of breathalyser from police station did not invalidate test.

R v Crowley (1977) 64 Cr App R 225; [1977] Crim LR 426; [1977] RTR 153 CA Cannot request specimen at roadside and obtain same without further ado when person removed to hospital;on requirement that medical officer in charge of suspect (normally duty casualty officer) be notified that specimen sought.

R v Dilley [1975] Crim LR 393 CrCt Arrest following breath test to be by police officer who conducted test.

R v Downey [1970] RTR 257; [1970] Crim LR 287 CA May refuse breath test where believe no circumstances arose meriting test but do so at own risk.

R v Evans (Terence) [1974] Crim LR 315; [1974] RTR 232 CA Person properly stopped and breathalysed where seen by foot constable who radioed message to police control who relayed same to patrol car which caught up with and stopped accused.

R v Fardy [1973] Crim LR 316; [1973] RTR 268 CA On reaonableness of constable's suspicion that driver had alcohol in his body.

R v Fox [1986] AC 281; (1986) 82 Cr App R 105; (1986) 150 JP 97; (1985) 135 NLJ 1058; [1985] RTR 337; [1985] 1 WLR 1126 HL Breath specimen admissible though obtained on foot of unlawful arrest.

R v Furness [1973] Crim LR 759 CA Did not invalidate breath testing request that police officer suspected appellant had breath taken before saw appellant driving.

R v Gaughan [1974] Crim LR 480; [1974] RTR 195 CA Breath test valid where constable had reasonable cause to administer same.

R v Gready [1974] Crim LR 314; [1974] RTR 16 CA Person properly convicted of drunk driving where stopped by constable who then almost immediately formed opinion that driver had been drinking.

R v Haslam [1972] RTR 297 CA Person charged with failure to provide breath/laboratory specimen could be acquitted of former charge but convicted of latter charge.

R v Holt [1968] 3 All ER 802; [1969] Crim LR 27; (1969) 133 JP 49; (1968) 112 SJ 928; [1968] 1 WLR 1942 CA Written statement of Home Office assistant-secretary that breathalyser approved by Home Secretary is admissible evidence thereof.

R v Jones (Reginald) [1969] 3 All ER 1559; (1970) 54 Cr App R 63; (1970) 134 JP 124; [1970] RTR 35; [1970] 1 WLR 16 CA Alcotest (r) 80 an approved breathalyser.

R v Kaplan [1977] Crim LR 564; [1978] RTR 119 CA That breathalyser bag burst did not affect validity of breath test.

R v Kelly (HF) [1972] Crim LR 643; [1972] RTR 447 CA On what constitutes 'failure' to provide/'opportunity' to give breath specimen; on physical disability as reasonable excuse for non-provision of breath specimen.

R v Maidment [1976] RTR 294 CA Breath test valid though passenger may (through restraint) have ceased driving at time of test as police formed opinion driver had been drinking when latter still driving.

R v Mayer [1975] RTR 411 CA Person arrested after provided breath test specimen did not have to be told why as was apparent from circumstances.

R v McGall [1974] Crim LR 482; [1974] RTR 216 CA Slow driving on empty road and mistaken use of indicators could have prompted reasonable suspicion that driver had excess alcohol.

R v Mitcham [1974] Crim LR 483; [1974] RTR 205 CA Breath test valid though issue had not been left to jury whether had been break in sequence of events from constable's suspecting driver had alcohol in body to his breathalysing latter.

R v Moore [1970] Crim LR 650; [1970] RTR 486 CA Constable to have reasonable suspicion regarding alcohol levels before requests breath specimen/may request blood or urine (neither preferable) for laboratory test; on appropriate warning to be given as to consequences of refusing blood and urine specimens.

R v Moore (George) [1979] RTR 98 CA Breath test valid where constable acted bona fide/ reasonably in administering same though told by motorist had just been drinking.

R v Needham [1974] RTR 201; [1973] Crim LR 640 CA Breath test valid where after stopping person for dangerous driving formed opinion that had alcohol in body and breathalysed him.

R v O'Boyle [1973] RTR 445 CA On what it means to 'require' a breath test of a driver.

R v Parsons (Leslie Arthur) (1972) 56 Cr App R 741; [1972] Crim LR 565t; (1972) 122 NLJ 633t; [1972] RTR 425; (1972) 116 SJ 567 CA Absent bad faith by police officer use of approved breathalyser in inferior condition was valid.

R v Pearson (Donald) [1974] Crim LR 315; [1974] RTR 92 CA Whether breath test request made as soon as reasonably practicable after road traffic offence a question of fact for jury.

R v Pico [1972] Crim LR 599; [1971] RTR 500 CA On what constituted 'accident' justifying police officer in requesting breath test of driver.

R v Price [1968] 3 All ER 814; (1969) 133 JP 47; [1968] 1 WLR 1853 CA Lawful breath test request if at time any driving offence being committed; 'any person driving' can include person in stopped car/who has alighted from car.

R v Sakhuja [1971] Crim LR 289t; [1971] RTR 261 CA Person racing ahead of police to own house could not then plead that police not entitled to demand breath test.

R v Sittingbourne Justices, ex parte Parsley [1978] RTR 153 HC QBD On degree of staining of crystals necessary in initial breath test.

R v Skinner [1968] 3 All ER 124; (1968) 52 Cr App R 599; [1968] Crim LR 451t; (1968) 132 JP 484; (1968) 118 NLJ 589t; [1968] 2 QB 700; (1968) 112 SJ 565; [1968] 3 WLR 408 CA Personal approval by Minister of breath test unnecessary.

R v Smith (Benjamin Walker) [1978] Crim LR 296 CrCt Breath test evidence excluded where subject of test wrongly arrested.

R v Storer [1969] Crim LR 204 Sessions On correct procedure as regards proving that breathalyser used in case was an approved device.

R v Thorpe (Thomas) (1974) 59 Cr App R 295; [1974] RTR 465 CA Is objective test whether have failed to provide breath specimen.

R v Tulsiani [1973] Crim LR 186 CrCt Police officer who requests breath specimen on suspicion that driver has alcohol taken must be officer who arrests driver (Road Safety Act 1967, s 2(5) applied).

R v Vardy [1978] RTR 202 CA Conviction quashed as not apparent to jury that must have been accident where constable requests breath test in reasonable belief that person driving car had been involved in accident.

R v Veevers [1971] Crim LR 174; [1971] RTR 47; (1971) 115 SJ 62 CA Drunk driving conviction quashed where issue whether arrest valid (on which validity of breath test in part depends) not left to jury.

R v Wagner [1970] Crim LR 535; [1970] RTR 422; (1970) 114 SJ 669 CA Was refusal to provide breath specimen where requesting constable did not have necessary equipment and accused refused to wait.

R v Wall (Harry) [1969] 1 All ER 968; (1969) 53 Cr App R 283; [1969] Crim LR 271; (1969) 133 JP 310; (1969) 119 NLJ 176t; (1969) 113 SJ 168; [1969] 1 WLR 400 CA Arresting officer's responsibilities upon arresting suspected drunk driver.

R v Ward and Hollister [1995] Crim LR 398 CA Successful appeal against conviction of police officers for perverting course of justice by failing to administer breath test to fellow officer involved in off-duty driving accident.

R v Wedlake [1978] RTR 529 CA On relevant time of driving for purposes of justifying breath test.

R v Weir [1972] 3 All ER 906; (1973) 137 JP 11 CrCt Not required to take breath test if arrested for non-traffic offence and not then told being detained for traffic offence.

R v Withecombe (David Ernest) (1969) 53 Cr App R 22; (1969) 133 JP 123; (1968) 112 SJ 949; [1969] 1 WLR 84 CA Blood/breath specimen not valid unless relevant procedural requirements of Road Safety Act 1967 complied with.

Ratledge v Oliver [1974] Crim LR 432; [1974] RTR 394 HC QBD Police officer could at same time notify doctor of intent to administer breath test/seek blood or urine sample of hospital patient.

Rawlins v Brown [1987] RTR 238 HC QBD Breath evidence admissible where accused (upon whom burden rested) failed to satisfy doctor that wanted to give blood sample.

Rayner v Hampshire Chief Constable [1970] Crim LR 703; (1970) 120 NLJ 993t; [1971] RTR 15; (1970) 114 SJ 913 HC QBD Excess alcohol conviction quashed where had been hole in bag of breathalyser.

Redman v Taylor [1975] Crim LR 348 HC QBD On when road traffic incident may be said to arise from 'the presence of a motor vehicle on the road' (Road Traffic Act 1972, s 8(2)).

Reference by the Attorney-General under Section 36 of the Criminal Justice Act 1972 (No 1 of 1978) (1978) 67 Cr App R 387; (1978) 128 NLJ 912t; [1978] RTR 377; (1978) 122 SJ 489 CA Breath test valid though completed in two breaths not one contrary to manufacturer's instructions.

Reference by the Attorney-General under Section 36 of the Criminal Justice Act 1972 (No 2 of 1974) (1975) 60 Cr App R 244; [1975] Crim LR 165t; (1975) 125 NLJ 44t; (1975) 119 SJ 153; [1975] 1 WLR 328 CA Breath test invalid where police officer acting in good faith breaches manufacturer's guidelines and gives test to person who has been smoking short time previously.

Regan v Anderton [1980] Crim LR 245; (1980) 144 JP 82; [1980] RTR 126 HC QBD Person could still be deemed to be driving car (for purposes of taking breath test) though car not actually moving.

Rendell v Hooper [1970] 2 All ER 72; [1970] Crim LR 285t; (1970) 134 JP 441; [1970] RTR 252; (1970) 114 SJ 248; [1970] 1 WLR 747 HC QBD Reasonable bona fide departure from manufacturer's instructions in giving breath test does not invalidate it.

Revel v Jordan; Hillis v Nicholson (1983) 147 JP 111; [1983] RTR 497; [1982] TLR 504 HC QBD Police can request second breath test.

Richards v West [1980] RTR 215 HC QBD Valid conclusion from circumstances (absent direct evidence) that officer requesting breath specimen had been uniformed.

Rickwood v Cochrane [1978] Crim LR 97; [1978] RTR 218 HC QBD Could justify breath test on one ground though originally sought to justify it on another; breath test for moving traffic offence did not have to be required while individual still driving.

Roberts v Jones [1969] Crim LR 90; (1968) 112 SJ 884 HC QBD Arrest of person in charge of motor car justified subsequent request that arrestee take breath test.

Rooney v Haughton [1970] 1 All ER 1001; [1970] Crim LR 236t; (1970) 134 JP 344; [1970] RTR 119; (1970) 114 SJ 93; [1970] 1 WLR 550 HC QBD Person arrested after first breath test cannot require second test at same station.

Sayer v Johnson [1970] RTR 286; [1970] Crim LR 589 HC QBD Unless doubt arises prosecutor is not required to prove breathalyser was stored in compliance with manufacturer's instructions.

Sharpe (George Hugo) v Director of Public Prosecutions (1994) 158 JP 595 HC QBD Magistrates could/should exclude breathalyser evidence obtained on foot of unlawful arrest/have heard evidence of two witnesses as to what transpired between police and accused.

Sharpe v Perry [1979] RTR 235 HC QBD Need not be roadside breath test when arresting person for being in charge of motor vehicle on road while unfit to drink (Road Traffic ct 1972, s 5(5)).

Shepherd v Kavulok [1978] Crim LR 170; [1978] RTR 85 HC QBD Laboratory specimen properly required where failed to fill breathalyser though constable had failed to examine crystals.

Sheridan v Webster [1980] RTR 349 HC QBD Breath test/arrest valid though had been good faith departure from manufacturer's instructions in administering test.

Shersby v Klippel [1979] Crim LR 186; [1979] RTR 116 HC QBD Constable stopping car without suspecting road traffic offence and then forming suspicion that driver had been drinking could validly require breath test.

Simpson v Spalding [1987] RTR 221 HC QBD Conviction for failure to provide blood specimen quashed where not warned that failure to provide same could result in prosecution.

Slender v Boothby (1985) 149 JP 405 HC QBD Breathalyser incapable of printing correct date on printout not a reliable device.

Smith v Director of Public Prosecutions [1989] RTR 159 HC QBD Breath specimen evidence admissible where person had first refused alternative specimen, then provided blood specimen.

Snelson v Thompson [1985] RTR 220 HC QBD Could rely on Lion Intoximeter 3000 readings though considerably different from Lion Alcolmeter SL-2 readings as former/latter designed to give exact/approximate readings.

Sparrow v Bradley [1985] RTR 122 HC QBD Evidence obtained via second breath test at police station using different breathalyser was admissible.

Steel v Goacher [1982] Crim LR 689; (1983) 147 JP 83; [1983] RTR 98; [1982] TLR 369 HC QBD Routine police stop was valid and so breath test obtained thereafter in accordance with legislation was also valid.

Stepniewski v Commissioner of Police of the Metropolis [1985] Crim LR 675; [1985] RTR 330 HC QBD Was failure to give (breath) specimen where gave one specimen and then refused to provide second.

Stoddart v Balls [1977] Crim LR 171; [1977] RTR 113 HC QBD Arrest valid though prior thereto police officer did not inspect crystals in breathalyser into which arrestee had failed to breathe.

Stokes v Sayers [1988] RTR 89 HC QBD On what prosecution must (and here failed to) prove for seeking of blood specimen in light of defectively operating breathalyser to be valid.

Stubbs v Chalmers [1974] Crim LR 257 HC QBD Person still driving when stopped on edge of road after signalled to do so by police who only then formed reasonable suspicion he had been drinking and validly required breath test.

Such v Ball [1981] Crim LR 411; [1982] RTR 140 HC QBD Police stopping motorist without reasonable belief as to moving traffic offence could validly request breath test where reasonably suspected driver to have excess alcohol taken as was not random breath test.

Sykes v Director of Public Prosecutions [1988] RTR 129 HC QBD Person could be convicted on basis of breath specimen evidence despite police officer's later invalid request for blood specimen.

Taylor v Baldwin [1976] Crim LR 137; [1976] RTR 265 HC QBD Police officer need not be in uniform when forms suspicion that driver was drinking but must be (and here was so) when administers breath test.

Teape v Godfrey [1986] RTR 213 HC QBD Failed 'reasonable excuse' plea where accused had not tried his hardest to give breath specimen/were no facts on which to ground plea.

Thom v Director of Public Prosecutions (1994) 158 JP 414; [1994] RTR 11 HC QBD Oral evidence as to level of alcohol in breath/proper functioning of breathalyser may be given by police officer.

Thomas v Director of Public Prosecutions [1991] RTR 292 HC QBD On effect of unlawful arrest on admissibility of evidence; on whether unlawfulness of arrest could be reasonable excuse for non-provision of specimen.

Thompson v Thynne [1986] Crim LR 629; [1986] RTR 293 HC QBD Policeman's reasonable decision that reliable breathalyser unavailable to be taken as correct.

Timmins v Perry [1970] Crim LR 649; [1970] RTR 477 HC QBD Police officer seeing person driving 'L'-plate vehicle on own had reasonable suspicion road traffic offence being committed to justify asking driver, whom upon stopping smelt of alcohol, to take breath test.

Trigg v Griffin [1970] Crim LR 44t; (1969) 119 NLJ 1020t; (1969) 113 SJ 962; [1970] RTR 53 HC QBD Person in car that was not moving and was on private land could not be subject of breath test request.

Waite v Smith [1986] Crim LR 405 CA On determining reliability of breathalyser.

Wakeley v Hyams [1987] Crim LR 342; [1987] RTR 49 HC QBD Blood and breath specimens rightly deemed inadmissible evidence where statutory procedure as regarded obtaining same not complied with.

Wall v Williams; Wallwork v Giles [1966] Crim LR 50t; [1970] Crim LR 109t; (1969) 119 NLJ 1142t; [1970] RTR 117; (1970) 114 SJ 36 HC QBD Police officer wearing all of uniform save for headgear is a constable in uniform and can request person to take breath test.

Walton v Rimmer [1986] RTR 31 HC QBD On what constitutes handing record of breath test results to defendant.

Watkinson v Barley [1975] 1 All ER 316; (1975) 60 Cr App R 120; [1975] Crim LR 45; (1975) 139 JP 203; [1975] RTR 136; (1975) 119 SJ 50; [1975] 1 WLR 70 HC QBD Non-compliance with breathalyser manufacturer's instructions will only invalidate test if test adversely affected.

Webber v Carey [1969] 3 All ER 406; [1969] Crim LR 549t; (1969) 133 JP 633; (1969) 119 NLJ 649t; (1969) 113 SJ 706; [1969] 1 WLR 1351 HC QBD Breath test only valid if police comply fully with instructions.

Whelehan v Director of Public Prosecutions [1995] RTR 177 HC QBD Confession by motorist that had driven to certain place admissible in evidence though no caution as caution only necessary after breath test administered.

Williams v Jones [1972] Crim LR 50; [1972] RTR 4 HC QBD Conviction quashed where constable did not have reasonable cause to suspect motorist had alcohol in his body.

Wilson v Cummings [1983] RTR 347 HC QBD Constable not required to inspect breathalyser crystals where despite offender's placing device to lips evidence showed was no air in bag.

Winter v Barlow [1980] Crim LR 51; (1980) 144 JP 77; [1980] RTR 209 HC QBD That constable did not have valid reason for stopping appellant did not mean magistrates could not hear drunk driving/failure to provide breath specimen actions.

Witts v Williams [1974] Crim LR 259 HC QBD Not having proper lights a continuing offence: person who drove to house and parked car validly required to take breath test by policeman who had been following/on appproaching driver reasonably suspected him to have been drinking.

Woon v Maskell [1985] RTR 289 HC QBD Once positive reading showed police officer no longer required to keep 'READ' button depressed.

Wright v Taplin [1986] RTR 388 HC QBD Evidence of defective breathalyser could be relied upon.

Yhnell (Paul Robert) v Director of Public Prosecutions [1989] Crim LR 384; (1989) 153 JP 364; [1989] RTR 250 HC QBD Admission of breath specimen evidence valid where did not rely on same in reaching decision in action where conviction rested on blood-alcohol level.

Young v Flint [1987] RTR 300 HC QBD Defence ought to have been allowed cross-examine prosecution witness called to testify to alterations to breathalyser (which may have rendered it an unapproved device).

BREATH TEST (hospital)

Attorney-General's Reference (No 1 of 1976) [1977] 3 All ER 557; (1977) 64 Cr App R 222; [1977] Crim LR 564; (1977) 141 JP 716; (1977) 127 NLJ 539t; [1977] RTR 284; (1977) 121 SJ 321; [1977] 1 WLR 646 CA Person in hospital precinct to be treated as patient: cannot be breathalysed.

BREATH TEST (printout)

Beck v Scammell [1985] Crim LR 794; [1986] RTR 162 HC QBD Constable's written amendments to breathalyser printout/printout copy (that should read BST not GMT) did not render printout inadmissible.

Hasler v Director of Public Prosecutions [1989] Crim LR 76; [1989] RTR 148 HC QBD Justices should not have accepted printout submitted by prosecution after 'no case to answer' plea failed.

CARELESS DRIVING (aiding and abetting)

Thornton v Mitchell [1940] 1 All ER 339; (1940) 104 JP 108; (1940) 162 LTR 296; (1940) 84 SJ 257; (1939-40) LVI TLR 296; (1940) WN (I) 52 HC KBD Only driver can be guilty of driving without due care and attention; no conviction for aiding and abetting where principal acquitted of offence.

CARELESS DRIVING (general)

Another v Probert [1968] Crim LR 564 HC QBD Giving misleading indications when driving was careless driving.

Bensley v Smith [1972] Crim LR 239; [1972] RTR 221 HC QBD Justices ought not to have considered sudden sickness/mechanical failure defence when was not preferred by defendant.

Bentley v Dickinson [1983] Crim LR 403; (1983) 147 JP 526 HC QBD Properly acquitted of driving without due care and attention where had not known of accident and was not served with notice of intended prosecution.

Broome v Perkins (1987) 85 Cr App R 321; [1987] Crim LR 271; [1987] RTR 321 HC QBD On what constitutes automatism; could not have been automatism where drove car along road for five miles.

Butty v Davey [1972] Crim LR 48; [1972] RTR 75 HC QBD Failed prosecution for driving without due care and attention where lorry driven at proper speed given road/weather conditions skidded on slippery wet road.

Coles v Underwood (1984) 148 JP 178; [1983] TLR 644 HC QBD On admissibility/usefulness of evidence of driving short way away from scene of collision following which were charged with driving without due care and attention.

Director of Public Prosecutions v Harris [1995] 1 Cr App R 170; [1995] Crim LR 73; (1994) 158 JP 896; [1995] RTR 100; [1994] TLR 151 HC QBD Emergency vehicle driver could not plead necessity to charge of driving without due care and attention.

Director of Public Prosecutions v Parker [1989] RTR 413 HC QBD Person crashing into stationary car could be found not to be guilty of driving without due care and attention.

Ex parte Newsham [1964] Crim LR 57t HC QBD Leave to seek certiorari order granted to appellant convicted of careless driving where charged with dangerous driving.

Griffin v Williams [1964] Crim LR 60g HC QBD Person whose vehicle (found by expert not to be defective) drifted across highway without explanation was guilty of driving without due care/attention.

Gubby v Littman [1976] Crim LR 386t; (1976) 126 NLJ 567t; [1976] RTR 470 HC QBD That drove with new tyres which had not 'broken in' not evidence of driving without due care and attention.

Haynes v Swain [1974] Crim LR 483; [1975] RTR 40 HC QBD Person guilty of driving without due care and attention could not plead that servicing of car obviated liability for accident arising from defect of which was/ought to have been cognisant.

Henderson v Jones [1955] Crim LR 318t; (1955) 119 JP 304 HC QBD Not defence to careless driving that were sleeping at the wheel at relevant time.

Hougham v Martin [1964] Crim LR 414; (1964) 108 SJ 138 HC QBD Direction to convict for careless driving where had pleaded that latent defect in mass-produced motor vehicle led to driving for which charged.

Hume v Ingleby [1975] Crim LR 396; [1975] RTR 502 HC QBD Was not necessarily careless driving that person reversing collided with vehicle to rear of him.

Jarvis v Fuller [1974] Crim LR 116; [1974] RTR 160 HC QBD Not driving without due care and attention where on dark, drizzly day stuck cyclist wearing dark clothes whom could only see from distance of 6-8 feet.

Jarvis v Williams [1979] RTR 497 HC QBD Absent other information could infer from car overturning after sharp bend that driver was driving without due care and attention.

Jones v Carter [1956] Crim LR 275 HC QBD Conviction quashed (without costs) where in road traffic case prosecution failed to identify driver of vehicle.

Kellett v Daisy [1977] Crim LR 566; (1977) 127 NLJ 791t; [1977] RTR 396 HC QBD Person using Crown road could be liable under Road Traffic Act 1972, s 3, for careless driving thereon.

Liddon v Stringer [1967] Crim LR 371t HC QBD Bus driver guilty of driving without due care and attention where acting on signals from conductor reversed into woman and killed her.

Lodwick v Jones [1983] RTR 273 HC QBD That had been frost on car window did not mean had to be ice on road and that another person drove safely on same road did not mean defendant (who skidded) necessarily guilty of careless driving.

McCrone v Riding [1938] 1 All ER 157; (1934-39) XXX Cox CC 670; (1938) 102 JP 109; (1938) 85 LJ 107; (1938) 158 LTR 253; (1938) 82 SJ 175; (1937-38) LIV TLR 328; (1938) WN (I) 60 HC KBD Same standard of 'due care and attention' expected of all drivers irrespective of experience.

Moses v Winder [1980] Crim LR 232; [1981] RTR 37 HC QBD Failed attempt by diabetic to establish automatism defence to driving without due care and attention charge.

Neal v Reynolds [1966] Crim LR 393; (1966) 110 SJ 353 HC QBD Justices ought to be slow to accept submissions of no case to answer.

O'Connell v Fraser [1963] Crim LR 289; (1963) 107 SJ 95 HC QBD Valid conviction of person who edged out slowly from behind parked cars but collided with motor cyclist.

Oakes v Foster [1961] Crim LR 628t HC QBD That person might have suffered dizzy spell just before involved in road traffic incident not a reasonable doubt meriting dismissal of careless driving information.

Police v Beaumont [1958] Crim LR 620 Magistrates Successful plea of automatism (consequent upon pneumonia) in careless driving prosecution.

Pratt v Bloom [1958] Crim LR 817t HC QBD Taxi-driver guilty of careless driving where indicated right, then at last moment turned left without checking and collided with another car.

R v Bristol Crown Court, ex parte Jones; Jones v Chief Constable of Avon and Somerset Constabulary (1986) 83 Cr App R 109; (1986) 150 JP 93; [1986] RTR 259 HC QBD Was not careless driving where facts giving rise to alleged offence had been immediate reaction to sudden emergency.

R v Coventry Justices, ex parte Sayers (1978) 122 SJ 264 HC QBD Justices could prefer (and convict on) charge of careless driving in respect of person brought before them charged with dangerous driving.

R v Fairbanks (John) (1986) 83 Cr App R 251; [1986] RTR 309; (1970) 130 SJ 750; [1986] 1 WLR 1202 CA Was material irregularity not to have left alternative verdict of driving without due care and attention where person charged with causing death by reckless driving.

R v Hammett [1993] RTR 275 CA Judge's introducing possibility of careless driving conviction in direction at trial for causing death by reckless driving a material irregularity.

R v Jeavons (Phillip) (1990) 91 Cr App R 307; [1990] RTR 263 CA Circumstances did not merit jury being given opportunity of convicting accused of careless driving where was charged with reckless driving.

R v Moxon-Tritsch (Leona) [1988] Crim LR 46 CrCt Ought not on basis of same facts to be prosecuted for causing death by reckless driving where had already pleaded guilty to careless driving/driving with excess alcohol.

R v Preston Justices, ex parte Lyons [1982] Crim LR 451; [1982] RTR 173; [1982] TLR 14 HC QBD Learner driver was driving without due care and attention where performed emergency stop without checking road.

R v Scammell [1967] 3 All ER 97; (1967) 51 Cr App R 398; [1967] Crim LR 594; (1967) 131 JP 462; (1967) 117 NLJ 913; (1967) 111 SJ 620; [1967] 1 WLR 1167 CA On careless driving as defence to dangerous driving.

R v Southampton Justices, ex parte Tweedie (1932) 96 JP 391; (1932) 74 LJ 168; (1933) 102 LJCL 11; (1932) 147 LTR 530; (1932) 76 SJ 545; (1931-32) XLVIII TLR 636 HC KBD Conviction quashed as charge of dangerous driving reduced to careless driving without adequate evidence of consent to reduction of charge.

R v Surrey Justices, ex parte Witherick [1931] All ER 807; (1931-34) XXIX Cox CC 414; (1931) 95 JP 219; [1932] 1 KB 450; (1932) 101 LJCL 203; (1932) 146 LTR 164; (1931) 75 SJ 853; (1931-32) XLVIII TLR 67; (1931) WN (I) 259 HC KBD Cannot charge careless driving and driving without due care in the alternative: are two separate offences.

Rabjohns v Burgar [1972] Crim LR 46; [1971] RTR 234 HC QBD Although res ipsa loquitur inapplicable to criminal action facts per se could merit inference of careless driving (without, say, prosecution having to show steering was not defective where this claim not raised).

Scott v Warren [1974] Crim LR 117; (1973) 123 NLJ 1137t; [1974] RTR 104; (1973) 117 SJ 916 HC QBD On duty on driver following another vehicle in line of traffic in motion.

Scruby v Beskeen [1980] RTR 420 HC QBD Could validly convict person of careless driving where admitted owned type of car involved in accident/had been in right place at right time.

Simpson v Peat [1952] 1 All ER 447; (1952) 116 JP 151; (1952) 102 LJ 106; [1952] 2 QB 24; (1952) 96 SJ 132; [1952] 1 TLR 469; (1952) WN (I) 97 HC QBD If not exercising ordinary care are guilty of careless driving despite cause being error of judgment.

Spencer v Silvester [1964] Crim LR 146; (1963) 107 SJ 1024 HC QBD Reversal of perverse finding by justices that was not careless driving to halt at stop sign.

St Oswald v Ball [1981] RTR 211 HC QBD Conviction quashed as CrCt put too much weight on inference drawn from nature of damage in collision and rejected evidence purely on basis of inference.

Taylor v Rogers [1960] Crim LR 271t; (1960) 124 JP 217 HC QBD Is objective test whether person was driving with due care and attention.

Venn v Morgan [1949] 2 All ER 562; (1949) 113 JP 504; (1949) LXV TLR 571; (1949) WN (I) 353 HC KBD On format of charge of careless driving.

Walker v Tolhurst [1976] Crim LR 261; [1976] RTR 513 HC QBD Essence of careless driving is whether acted without due care and here justices made reasonable decision on that basis.

Watts v Carter [1971] RTR 232 HC QBD Was case to answer where person charged with driving without due care and attention claimed steering was defective but steering upon examination was functioning properly.

Webster v Wall [1980] Crim LR 186; [1980] RTR 284 HC QBD Could reasonably be found not guilty of careless driving where struck stationary car while motor-bicycling below restricted speed limit on dark, wet evening.

Wood v Richards (1977) 65 Cr App R 300; [1977] Crim LR 295; (1977) 127 NLJ 467t; [1977] RTR 201 HC QBD Motor police officer called to emergency had to exercise same care as anyone else: absent evidence as to nature of emergency defence of necessity unavailable.

Wright v Wenlock [1972] Crim LR 49; [1971] RTR 228 HC QBD Could infer absent alternative reason being proferred that driving was careless but res ipsa loquitur doctrine per se inapplicable to criminal action.

CARRIAGE OF GOODS FOR HIRE/REWARD (general)

Steetway Sanitary Cleansers, Ltd v Bradley (1961) 105 SJ 444 HC QBD Vehicle being used to transport effluent was a goods vehicle in which goods being transported for hire/reward.

Wurzal v Houghton Main Home Coal Delivery Service, Ltd; Wurzal v Alfred Atkinson, Walter Norfolk and James Ward (Trading as The Dearne Valley Home Coal Delivery Service) [1936] 3 All ER 311; (1936) 100 JP 503; [1937] 1 KB 380; (1936) 82 LJ 315; [1937] 106 LJCL 197; (1936) 80 SJ 895; (1936-37) LIII TLR 81; (1936) WN (I) 321 HC KBD Was/was not contract for hire/reward where coal transported for limited/unincorporated mutual benefit societies.

CAUSING BODILY HARM WHEN IN CHARGE OF VEHICLE (general)

R v Cooke (Philip) [1971] Crim LR 44 Sessions On elements of offence of causing bodily harm by misconduct when in charge of a vehicle.

CAUSING DEATH BY DRIVING (general)

R v Atkinson (John Percy) (1971) 55 Cr App R 1; [1970] Crim LR 405; [1970] RTR 265 CA Latent defect suddenly becoming manifest may be good defence to dangerous driving charge.

R v Ball; R v Loughlin (1966) 50 Cr App R 266; [1966] Crim LR 451; (1965-66) 116 NLJ 978t; (1966) 110 SJ 510 CCA Offence of causing death by dangerous driving an absolute offence so driver liable though acting under instructions of another.

R v Beckford [1996] 1 Cr App R 94; [1995] RTR 251; [1995] TLR 26 CA That car unavailable did not preclude prosecution continuing where charged with causing death by careless driving after drinking.

R v Burt [1977] RTR 340 CA Direction in prosecution for causing death by dangerous driving not invalid for failure to mention the manner of driving.

R v Chapman [1978] Crim LR 172 CA Nine months' imprisonment plus two years' disqualification for causing death by dangerous driving while driving with excess alcohol (for which six months' concurrent imprisonment imposed).

R v Clancy [1979] RTR 312 CA On what is meant by 'reckless' driving.

R v Collins [1997] Crim LR 578 CA Objective test whether (as here) driving (whereby caused death) was dangerous.

R v Curphey (1957) 41 Cr App R 78 Assizes On what constitutes 'causing' death by dangerous driving.

R v Davis (William) [1979] RTR 316 CA On what is meant by 'reckless' driving.

R v Downes [1991] Crim LR 715; [1991] RTR 395 CA On admissibility of expert evidence on excess alcohol charge.

R v Evans [1962] 3 All ER 1086; (1963) 47 Cr App R 62; [1963] Crim LR 112; (1963) 127 JP 49; (1963) 113 LJ 9; [1963] 1 QB 412; (1962) 106 SJ 1013; [1962] 3 WLR 1457 CCA Degree of recklessness/carelessness not relevant in death by dangerous driving charge.

R v Gould [1963] 2 All ER 847; (1963) 47 Cr App R 241; [1963] Crim LR 645; (1963) 127 JP 414; [1964] 1 WLR 145 Assizes On death by dangerous driving.

R v Hammett [1993] RTR 275 CA Judge's introducing possibility of careless driving conviction in direction at trial for causing death by reckless driving a material irregularity.

R v Hands [1962] Crim LR 255 CCA On dangerous driving vis-a-vis causing death by dangerous driving.

R v Hart, Millar and Robert Millar (Contractors), Ltd [1969] 3 All ER 247; [1969] Crim LR 667; (1969) 133 JP 554; (1969) 119 NLJ 820t; [1970] RTR 155 CrCt Scottish employers of lorry driver causing death by dangerous driving in England with unsafe lorry liable for trial in England.

R v Hives [1991] RTR 27 CA Was reckless driving where struck woman crossing road (who was only a footstep away from the pavement) and thought had hit something like a lamp post.

R v Kimsey [1996] Crim LR 35 CA On degree of causation necessary to be guilty of 'causing' death by dangerous driving.

R v Loukes [1996] 1 Cr App R 444; [1996] Crim LR 341; [1996] RTR 164; [1995] TLR 706 CA Could not be convicted of procuring the causing of death by dangerous driving where person charged with causing same was acquitted.

R v Marison [1996] Crim LR 909; (1996) TLR 16/7/96 CA Diabetic who suffered hypoglycaemic attack while driving was legitimately found guilty of causing death by dangerous driving of other driver with whom collided following attack.

R v McBride [1961] 3 All ER 6; (1961) 45 Cr App R 262; [1961] Crim LR 625; (1961) 125 JP 544; (1961) 111 LJ 518; [1962] 2 QB 167; (1961) 105 SJ 572; [1961] 3 WLR 625 CCA Evidence of drinking admissible in death by dangerous driving prosecution; death by dangerous driving may be coupled with reckless/dangerous driving but not drunk/drugged driving charge.

R v McCabe [1995] RTR 197 CA Three years' imprisonment/five year disqualification merited for person of generally good character with learned referees where alcohol levels over prescribed limit and had caused three deaths.

R v Millington [1995] Crim LR 824; [1996] RTR 80; [1995] TLR 274 CA 18 months' imprisonment/five years' disqualification appropriate for person guilty of causing death by dangerous driving after drinking.

R v O'Neale [1988] Crim LR 122; [1988] RTR 124 CA On relevancy of driver being in violation of provisional licence-holder legislation where charged with causing death by reckless driving.

R v Parker [1957] Crim LR 468; (1958) 122 JP 17 CCA Single, brief act of negligence/non-observance of safety rules can be dangerous driving.

R v Peters [1993] Crim LR 519; [1993] RTR 133 CA That had drunk too much alcohol was admissible to explain nature of driving.

R v Reid [1991] Crim LR 269 CA On test for recklessness (Lawrence formula applied).

R v Robert Millar (Contractors) Ltd and Robert Millar [1970] 1 All ER 577; (1970) 54 Cr App R 158; (1970) 134 JP 240; (1969) 119 NLJ 1164t; [1970] 2 QB 54; [1970] RTR 147; (1970) 114 SJ 16; [1970] 2 WLR 541 CA Scottish employers of lorry driver causing death by dangerous driving in England with unsafe lorry liable for trial in England.

R v Scates [1957] Crim LR 406 CCC Whether person driving dangerously a question of fact not one of intent.

R v Sibley [1962] Crim LR 397 CCA Conviction for causing death by dangerous driving allowed stand despite witness' clearly intimating defendant had been drinking (contrary to judge's ruling that should be no such evidence admitted).

R v Skelton [1995] Crim LR 635 CA On elements of offence of causing death by dangerous driving.

R v Taziker (Attorney General's Reference (No 36 of 1994)) [1995] RTR 413 CA Generally custodial sentences merited in death by careless driving cases: here two years' imprisonment imposed.

R v Wallington [1995] RTR 112 CA Conviction of police constable for causing death by dangerous driving quashed in light of harsh criticisms of particular police behaviour by trial judge when directing jury.

R v Welburn (Derek) (1992) 94 Cr App R 297; [1992] Crim LR 203; [1992] RTR 391 CA Issue of drink only relevant to second part of R v Lawrence direction.

CLAMPING (general)

Arthur and another v Anker and another [1996] 3 All ER 783; (1996) 146 NLJ 86; [1997] QB 564; [1996] RTR 308; [1995] TLR 632; [1996] 2 WLR 602 CA Unauthorised parking on land (consequences of which were posted) meant took risk of clamping/having to pay for removal of clamp.

CONSTRUCTION AND USE (agricultural machine)

Bullen v Picking and others [1973] Crim LR 765; (1973) 123 NLJ 1043t; (1973) 117 SJ 895; [1974] RTR 46 HC QBD Agricultural machine could be used on public road for carrying non-agricultural goods for use on/off farm without having to pay more in excise.

CONSTRUCTION AND USE (articulated vehicle)

A Stevens and Co (Haulage) Ltd v Brown; Brown (EJ) v Brown [1971] Crim LR 103; (1970) 120 NLJ 1136t; [1971] RTR 43; (1971) 115 SJ 62 HC QBD Length of trailer and load relevant length when determining whether vehicle ought to hve been notified to police as in excess of 60 feet (Motor Vehicles (Construction and Use) Regulations 1969).

Bason v Vipond; Same v Robson [1962] 1 All ER 520; [1962] Crim LR 320; (1962) 126 JP 178; (1962) 112 LJ 241; (1962) 106 SJ 221 HC QBD Test whether vehicle exceeds length restrictions is in its ordinary position as constructed.

British Road Services Ltd and another v Owen [1971] 2 All ER 999; [1971] Crim LR 290; (1971) 135 JP 399; [1971] RTR 372; (1971) 115 SJ 267 HC QBD Regard to entire route to see if unsuitable at any point.

Cook and another v Briddon [1975] RTR 505; (1975) 119 SJ 462 HC QBD Was improper use of articulated vehicle of exceptional length to carry load that could be carried on standard or shorter than standard articulated vehicle.

Hunter v Towers and another [1951] 1 All ER 349; (1951) 115 JP 117; (1951) 101 LJ 77; (1951) 95 SJ 62; [1951] 1 TLR 313; (1951) WN (I) 95 HC KBD What constitutes 'articulated vehicle'.

Kingdom v Williams [1975] Crim LR 466; [1975] RTR 33 HC QBD Application of proviso as regards articulated vehicles normally used for indivisible loads of exceptional length (Motor Vehicles (Contruction and Use) Regulations 1973, reg 9(1)).

Peak Trailer and Chassis, Ltd v Jackson [1967] 1 All ER 172; (1967) 131 JP 155; (1966) 110 SJ 927; [1967] 1 WLR 155 HC QBD Forty-six from 177 short-load trips in twelve months showed lorry not normally used for loads of indivisble length.

CONSTRUCTION AND USE (axle)

Director of Public Prosecutions v Marshall and Bell (1990) 154 JP 508; [1990] RTR 384 HC QBD Construction of Road Vehicles (Construction and Use) Regulations 1986, regs 80(1) and 80(2).

CONSTRUCTION AND USE (brakes)

Bailey v Rolfe [1976] Crim LR 77 HC QBD That problem with brakes probably pre-dated collision insufficient to support conviction for driving with defective brakes.

Bowen v Wilson (1926-30) XXVIII Cox CC 298; (1927) 91 JP 3; [1927] 1 KB 507; (1926) 62 LJ 376; [1927] 96 LJCL 183; (1927) 136 LTR 310; (1925-26) 70 SJ 1161; (1926-27) XLIII TLR 77; (1926) WN (I) 308 HC KBD Flaw in brake drum not flaw in brake/s.

Cannon v Jefford (1915) 79 JP 478; [1915] 3 KB 477; [1915] 84 LJKB 1897; (1915-16) 113 LTR 701 HC KBD Conviction for improper brakes on motor car.

Cole v Young [1938] 4 All ER 39; (1938) 86 LJ 293 HC KBD That brakes fail once not prove that breaking system ineffective and hence unlawful.

Coombes v Cardiff County Council [1976] Crim LR 75; [1975] RTR 491 HC QBD Decelerometer evidence indicated that handbraking system on omnibus not an efficient braking system.

Dingwall v Walter Alexander and Sons (Midland) Ltd (1982) 132 NLJ 704 HL Failed action by widow in which had sought damages from employers of dead husband (bus driver) for his death in road accident she claimed was attributable to defective brakes on bus.

Director of Public Prosecutions v British Telecommunications plc [1991] Crim LR 532; (1991) 155 JP 869; [1990] TLR 742 HC QBD Prosecution evidence as to defectiveness of trailer brakes admissible even though nature of its examiner's scrutiny of same had rendered proper examination by defence impossible.

Director of Public Prosecutions v Young (1991) 155 JP 15; [1991] RTR 56; [1990] TLR 503 HC QBD Where brakes are attached to vehicle they must be properly maintained even if particular vehicle not required to have brakes.

Hart v Bex [1957] Crim LR 622 HC QBD Absolute discharge for driver who as driver was technically guilty for lorry having defective brakes (though maintenance actually the responsibility of another).

Hawkins v Holmes [1974] Crim LR 370; [1974] RTR 436 HC QBD That took real care to ensure brakes were in working order not defence to their being defective.

Hutchings v Giles [1955] Crim LR 784 HC QBD On what constitutes 'permitting' use of vehicle with defective brakes (here ought not to have been convicted of same).

Kennett v British Airports Authority [1975] Crim LR 106; [1975] RTR 164; (1975) 119 SJ 137 HC QBD On application of regulations pertaining to car brakes (Motor Vehicles (Construction and Use) Regulations 1973).

Kenyon v Thorley [1973] Crim LR 119; [1973] RTR 60 HC QBD Failure to apply brake on unattended vehicle did not require mandatory disqualification/endorsement.

Langton and another v Johnson [1956] 3 All ER 474; (1956) 120 JP 561; (1956) 106 LJ 698; (1956) 100 SJ 802; [1956] 1 WLR 1322 HC QBD Handbrake operating on less than half of wheels of vehicle an offence.

R v London Boroughs Transport Committee, ex parte Freight Transport Association Ltd and others [1991] RTR 13; [1990] TLR 637 CA London lorry braking requirements not compatible with EEC law.

R v The London Boroughs Transport Committee, ex parte Freight Transport Association Ltd (FTA), Road Haulage Association Ltd (RHA), Reed Transport Ltd, Wincanton Distribution Services Ltd, Conoco Ltd, Cox Plant Hire London Ltd, Mayhew Ltd (1990) 154 JP 773; [1990] RTR 109 HC QBD London road traffic order requiring quieter air brakes on lorries as was ultra vires in that went demanded more than existing construction and use regulations.

19

Stoneley v Richardson [1973] Crim LR 310; (1973) 123 NLJ 299t; [1973] RTR 229 HC QBD Irrelevant to prosecution that improperly maintained brakes discovered by constable (unauthorised examiner) in impromptu test.

Wilmott v Southwell (1908) 72 JP 491; (1908-09) XCIX LTR 839; (1908-09) XXV TLR 22 HC KBD One brake plus being able to use engine to lock wheels did not fulfil requirement of having two independent brakes (Motor Car (Use and Construction) Order 1904, Article II).

CONSTRUCTION AND USE (defective tyres)

Eden v Mitchell [1975] Crim LR 467; [1975] RTR 425; (1975) 119 SJ 645 HC QBD That did not intend to drive car with defective tyres did not mean were not guilty of use of car with defective tyres.

CONSTRUCTION AND USE (general)

Badham v Lamb's, Ltd [1945] 2 All ER 295; [1946] KB 45; [1946] 115 LJ 180; (1945) 173 LTR 139; (1944-45) LXI TLR 569 HC KBD No action for damages for sale of car with defective brakes: is punishable as offence.

Baldwin v Worsman [1963] 2 All ER 8; [1963] Crim LR 364; (1963) 127 JP 287; (1963) 113 LJ 349; (1963) 107 SJ 215; [1963] 1 WLR 326 HC QBD Vehicle capable of reversing does not cease to be such because means of reverse shut off.

Balfour Beatty and Co Ltd v Grindey [1974] Crim LR 120; [1975] RTR 156 HC QBD Was wrong to charge company with 'use' of vehicle with defective indicator and not 'causing or permitting' same to be on road.

Childs v Coghlan [1968] Crim LR 225t; (1968) 118 NLJ 182t; (1968) 112 SJ 175 HC QBD Earth excavator was a motor vehicle for purposes of Road Traffic Act 1960.

Claude Hughes and Co (Carlisle), Ltd v Hyde [1963] 1 All ER 598; [1963] Crim LR 287; (1963) 127 JP 226; (1963) 113 LJ 267; [1963] 2 QB 757; (1963) 107 SJ 115; [1963] 2 WLR 381 HC QBD Motor vehicle exceeding length restrictions when detachable container attached does breach regulations.

Cording v Halse [1954] 3 All ER 287; [1955] Crim LR 43; (1954) 118 JP 558; (1954) 104 LJ 761; [1955] 1 QB 63; (1954) 98 SJ 769; [1954] 3 WLR 625 HC QBD Simply adding something to motor vehicle does not mean it is alternative body (so not to be included in weight).

Corp v Toleman International Ltd and another; Toleman Delivery Services Ltd and another v Pattison [1981] RTR 385 HC QBD Were guilty of violating length restrictions where trailer plus plate joining trailer to tow vehicle together exceeded permitted maximum length.

Creek v Fossett, Eccles and Supertents Ltd [1986] Crim LR 256 HC QBD Application of operator licence, test certificate and tachograph legislation in circus vehicles context.

Daley and others v Hargreaves [1961] 1 All ER 552; [1961] Crim LR 488; (1961) 125 JP 193; (1961) 111 LJ 222; (1961) 105 SJ 111; [1961] 1 WLR 487 HC QBD Dumpers not intended/not adapted for road use.

Drysdale and others v Harrison [1972] Crim LR 573; [1973] RTR 45 HC QBD Employer continued to be user of vehicle driven by employee though latter inter alia stopped by police/required by Department of Environment inspectors to be taken to weighbridge.

Elieson v Parker (1917) 81 JP 265; (1917-18) 117 LTR 276; (1916-17) 61 SJ 559; (1916-17) XXXIII TLR 380 HC KBD Electrically propelled one-quarter horse-power bath chair capable of travel at about 2mph is a motor car under Motor Car Act 1903.

Gaunt v Nelson and another [1987] RTR 1 HC QBD On what constitutes a 'specialised vehicle for door to door selling' and so is immune from tachograph requirements.

Hollands v Willliamson [1920] 89 LJCL 298; (1919) WN (I) 271 HC KBD Automated tricycle a carriage/motor car and so regulated by motor vehicle legislation.

Keeble v Miller [1950] 1 All ER 261; (1950) 114 JP 143; [1950] 1 KB 601; (1950) 94 SJ 163; [1950] 66 (1) TLR 429; (1950) WN (I) 59 HC KBD 'Construction' of car determined on date of charge: here heavy motor car had become light locomotive.

Kennet v Holding and Barnes Ltd and another; TL Harvey Ltd v Hall and another [1986] RTR 334 HC QBD Break-down/recovery vehicles did not breach goods vehicle testing and plating/ excise licensing provisons where each bore two disabled vehicles (one carried, one towed).

Keyte v Dew [1970] RTR 481; (1970) 114 SJ 621 HC QBD Rusty body and absent bolts not construction defects.

Lowe and another v Stone [1948] 2 All ER 1076; (1949) 113 JP 59; [1949] 118 LJR 797; (1948) WN (I) 487 HC KBD Detachable boards fitted to side of lorry meant was being used in altered condition.

Markham v Stacey [1968] 3 All ER 758; [1969] Crim LR 35t; (1969) 133 JP 63; (1968) 118 NLJ 1006t; (1968) 112 SJ 866; [1968] 1 WLR 1881 HC QBD Meaning of 'land implement' in Motor Vehicles (Construction and Use) Regulations 1966, reg 3.

McDermott Movements Ltd v Horsfield [1982] Crim LR 693; [1983] RTR 42 HC QBD Owners of insecurely loaded lorry which did not spill any goods wrongfully prosecuted under provision of motor construction regulations concerning spilling of loads.

Mickleborough v BRS (Contracts) Ltd [1977] Crim LR 568; [1977] RTR 389 HC QBD Use of vehicle on road that does not comply with regulations made by Secretary of State for Transport and absolute offence; company hiring out vehicle and driver liable where vehicle did not meet regulatory requirements.

NFC Forwarding Ltd v Director of Public Prosecutions [1989] RTR 239 HC QBD On offence of using defective vehicle on road contrary to Road Traffic Act 1972, s 40(5).

O'Brien v Anderton [1979] RTR 388 HC QBD 'Italjet' a motor vehicle (had two wheels/22cc engine/seat/handle bars).

O'Connell v Murphy [1981] Crim LR 256; [1981] RTR 163 HC QBD Breach of motor vehicle construction and use law where transported exhaust pipe part of which sticking from window exceeded permissible lateral projection limits.

O'Neill v Brown [1961] 1 All ER 571; [1961] Crim LR 317; (1961) 125 JP 225; (1961) 111 LJ 206 [1961] 1 QB 420; (1961) 105 SJ 208; [1961] 2 WLR 224 HC QBD Guilty of endangering with trailer where improperly joined — though car and trailer independently in good condition.

Patterson v Redpath Brothers Ltd [1979] Crim LR 187; (1979) 129 (1) NLJ 193; [1979] RTR 431; (1979) 123 SJ 165; [1979] 1 WLR 553 HC QBD On what constituted 'indivisible load' for purposes of Motor Vehicles (Construction and Use) Reegulations 1973, regs 3(1) and 9(1).

Percy and another v Smith [1986] RTR 252 HC QBD Fork lift truck was intended for use on public roads and not a 'works truck'.

R v Stally [1959] 3 All ER 814; (1960) 44 Cr App R 5; [1960] Crim LR 199; (1960) 124 JP 65; (1959) 109 LJ 719; (1960) 104 SJ 108; [1960] 1 WLR 79 CCA Not guilty of taking and driving away if no part in taking but later joined for drive knowing vehicle to have been taken.

Reeve v Webb [1973] Crim LR 120; [1973] RTR 130; (1973) 117 SJ 127 HC QBD Motor Vehicles (Construction and Use) Regulations 1969, reg 76(1) requiring vehicle to be in safe condition not concerned solely with maintenance.

Scott v Gutteridge Plant Hire Ltd [1974] Crim LR 125; [1974] RTR 292 HC QBD Low loader used to transport disabled vehicle was not a 'recovery vehicle'.

Siddle C Cook, Ltd and another v Holden [1962] 3 All ER 984; [1963] Crim LR 53; (1963) 127 JP 55; (1963) 113 LJ 9; [1963] 1 QB 248; (1962) 106 SJ 920; [1962] 3 WLR 1448 HC QBD Violation of special types vehicles legislation.

Simpson v Vant [1986] Crim LR 473; [1986] RTR 247 HC QBD Sheepdog sitting on owner-driver's lap a badly distributed load which made driving a cause of danger.

Smart v Allan and another [1962] 3 All ER 893; [1963] Crim LR 54t; (1963) 127 JP 35; (1962) 112 LJ 801; [1963] 1 QB 291; (1962) 106 SJ 881; [1962] 3 WLR 1325 HC QBD Not 'mechanically propelled vehicle' if so transformed will never again be mobile.

St Albans Sand and Gravel Co Ltd v Minnis [1981] RTR 231 HC QBD On what constitutes unlawful loading.

CONSTRUCTION AND USE (land implement)

Thomas v Hooper [1986] Crim LR 191; [1986] RTR 1 HC QBD Inoperable car whose wheels did not turn as was towed along road was not in use on road.

Thurrock Borough Council v William Blythe and Co Ltd [1977] Crim LR 297; [1977] RTR 301 HC QBD Information properly dismissed where method of weighing articulated vehicle unsatisfactory.

Travel-Gas (Midlands) Ltd v Reynolds; Walton v Reynolds; JH Myers Ltd and others v Licensing Area for the North Eastern Traffic Area [1989] RTR 75 HC QBD Could be prosecuted for excess gross weight and excess axle weight offences arising from same facts.

Wilkinson v Barrett [1958] Crim LR 478t; (1958) 122 JP 349 HC QBD Mobile car jack was a trailer (and in this case contravened Motor Vehicles (Construction and Use) Regulations 1955/482).

CONSTRUCTION AND USE (land implement)

Hockin v Reed and Co (Torquay) Ltd [1962] Crim LR 400; (1962) 112 LJ 337; (1962) 106 SJ 198 HC QBD On what constituted a land implement for purposes of Motor Vehicles (Construction and Use) Regulations 1955.

CONSTRUCTION AND USE (latent defect)

Phillips v Britannia Hygienic Laundry Company, Limited [1923] 1 KB 539; (1923) CC Rep XII 28; [1923] 92 LJCL 389; (1923) 128 LTR 690; (1922-23) 67 SJ 365; (1922-23) XXXIX TLR 207; (1923) WN (I) 47 HC KBD Apportionment of liability for motor car with hidden defect.

Phillips v Britannia Hygienic Laundry Company, Limited [1924] 93 LJCL 5; (1923) 129 LTR 777; (1922-23) XXXIX TLR 530 CA Regulations concerning use and construction of motor cars did not create personal remedy against repairers who returned car with latent defect which meant car did not comply with regulations.

CONSTRUCTION AND USE (mechanically propelled vehicle)

Carrimore Six Wheelers, Ltd v Arnold [1949] 2 All ER 416; (1949) 113 JP 456; (1949) LXV TLR 506; (1949) WN (I) 349 HC KBD Tractor covered by limited trade licence carrying goods in attached trailer not so licenced an offence: trailer not part of tractor.

CONSTRUCTION AND USE (mirrors)

Mawdsley v Walter Cox (Transport), Ltd; Same v Allen [1965] 3 All ER 728; [1966] Crim LR 110; (1966) 130 JP 62; (1965-66) 116 NLJ 189; (1966) 110 SJ 36; [1966] 1 WLR 63 HC QBD Vehicles of which mirrors violations alleged to be considered in unloaded/normally loaded state.

CONSTRUCTION AND USE (moped)

G (a minor) v Jarrett [1980] Crim LR 652; [1981] RTR 186 HC QBD Missing pedal rest did not mean moped ceased to be moped.

McEachran v Hurst [1978] Crim LR 499; [1978] RTR 462 HC QBD Immobile moped still mechanically propelled vehicle where being brought to repairers to be fixed.

R v Tahsin [1970] Crim LR 160; [1970] RTR 88; (1970) 114 SJ 56 CA Moped a motor vehicle though may have suffered non-permanent cut-off of engine power.

CONSTRUCTION AND USE (motor vehicle)

Chief Constable of Avon and Somerset v F (1987) 84 Cr App R 345 HC QBD Scrambler motor bicycle not proved to be motor vehicle.

Floyd v Bush [1953] 1 All ER 265; [1953] 1 WLR 242 HC QBD Motorised pedal cycle a 'motor vehicle' requiring licence.

Lawrence v Howlett [1952] 2 All ER 74 HC QBD Pedal bicycle with motor not a motor vehicle when used as pedal cycle.

Newberry v Simmonds [1961] 2 All ER 318; [1961] Crim LR 556; (1961) 125 JP 409; (1961) 111 LJ 306; [1961] 2 QB 345; (1961) 105 SJ 324; [1961] 2 WLR 675 HC QBD Motor car from which engine removed may still be 'mechanically propelled vehicle' if engine soon to be replaced.

R v Challinor (Robert) [1985] Crim LR 53; (1985) 149 JP 358; [1985] RTR 373 CA Towed car may have been 'motor vehicle' for purposes of Road Traffic Act 1972, s 190(1).

CONSTRUCTION AND USE (motorcycle)

Brown v Anderson [1965] 2 All ER 1; [1965] Crim LR 313t; (1965) 129 JP 298; [1965] 1 WLR 528 HC QBD Bubble car a 'motor cycle' under Road Traffic Act 1960 but not a motor bicycle under Motor Vehicles (Driving Licences) Regulations 1963.

CONSTRUCTION AND USE (special vehicle)

Dixon v BRS (Pickfords) Ltd [1959] 1 All ER 449; [1959] Crim LR 369; (1959) 123 JP 207; (1959) 103 SJ 241 HC QBD 'Vehicle' in Motor Vehicles (Authorisation of Special Types) General Order 1955, art 18(1) does not include joined vehicles.

Hollis Brothers Ltd v Bailey; Buttwell v Bailey [1968] Crim LR 393; (1968) 112 SJ 381; [1968] 1 WLR 663 HC QBD On meaning of when use of vehicle 'is so unsuitable as to cause or be likely to cause danger' (Motor Vehicles (Construction and Use) Regulations 1966, reg 75(3)).

CONSTRUCTION AND USE (steering)

Sandford Motor Sales v Habgood [1962] Crim LR 487t HC QBD Motor car auctioneers found liable for defective steering on motor vehicle they sold.

CONSTRUCTION AND USE (trailer)

Amalgamated Roadstone Corporation, Ltd v Bond [1963] 1 All ER 682; [1963] Crim LR 290; (1963) 127 JP 254; (1963) 113 LJ 234; (1963) 107 SJ 316; [1963] 1 WLR 618 HC QBD Spreader used in conjunction with tractor a 'land implement'.

Baker v Esau [1972] Crim LR 559; [1973] RTR 49 HC QBD Ambulance trailer and car together constituted single four-wheeled trailer.

British Transport Commission and others v McKelvie and Co, Ltd (1960) 110 LJ 733 TrTb On use of trailers by A-licence holders.

Cripps v Cooper [1936] 2 All ER 48 HC KBD Carriage of indivisible load contravened road traffic regulations as was danger to persons on trailer/highway.

Garner v Burr and others [1950] 2 All ER 683; (1950) 114 JP 484; [1951] 1 KB 31; (1950) 94 SJ 597; [1950] 66 (2) TLR 768; (1950) WN (I) 445 HC KBD Poultry shed with iron wheels a 'trailer'.

Robinsons Limited v Richards (1926-30) XXVIII Cox CC 498; (1928) 92 JP 73; [1928] 2 KB 234; (1928) 65 LJ 401; (1928) 97 LJCL 483; (1928) 139 LTR 164; (1928) WN (I) 110 HC KBD Breach of Motor Cars (Use and Construction) Order 1904 where had to take both hands off wheel to apply tractor and trailer brakes.

Union Cartage Co, Ltd v Heamon; Eggleton v Heamon [1937] 1 All ER 538 HC KBD On requirement of person to watch over trailer pulled by motor tractor.

CONSTRUCTION AND USE (tyres)

Carmichael and Sons Ltd v Cottle [1971] Crim LR 45t; (1970) 120 NLJ 1040t; [1971] RTR 11; (1970) 114 SJ 867 HC QBD Persons hiring car to another could not be said to be using it/were not accessories before the fact to the car being used with threadbare tyres.

Connor v Graham and another [1981] RTR 291 HC QBD Tyres (though must have capacity to bear maximum axle weight of vehicle) need only be inflated to extent required for actual use to which being put.

Coote v Parkin [1977] Crim LR 172; [1977] RTR 61 HC QBD Tread pattern requirements related to that area of wheel normally touching road when car moving.

CONSTRUCTION AND USE (unroadworthy vehicle)

Goosey v Adams [1972] Crim LR 49; [1971] RTR 365 HC QBD Tyres on double wheel may be treated as separate tyres.

Phillips v Thomas [1974] RTR 28 HC QBD Issue before justices in defective tyre case was whether tyres defective not whether police officer/authorised examiner discovered defect/s.

Renouf v Franklin [1972] Crim LR 115; [1971] RTR 469 HC QBD Ply/cord structure of tyre not exposed where had to lift tear on tyre to view same.

Sandford and another v Butcher [1977] Crim LR 567; (1977) 127 NLJ 913t; [1978] RTR 132 HC QBD On what constitutes tyre 'tread'/'tread pattern'.

CONSTRUCTION AND USE (unroadworthy vehicle)

British Car Auctions Ltd v Wright [1972] 3 All ER 462; [1972] Crim LR 562t; (1972) 122 NLJ 680t; [1972] RTR 540; [1972] 1 WLR 1519 HC QBD Car auctioneer seeking bids not offering to sell car — conviction for selling unroadworthy vehicle quashed.

CONSTRUCTION AND USE (use)

Brown v Roberts and another [1963] 2 All ER 263; [1963] Crim LR 435; (1963) 107 SJ 666; [1965] 1 QB 1; [1963] 3 WLR 75 HC QBD On what constitutes use of motor vehicle on road (Road Traffic Act 1930, s 36(1)).

Burns v Currell [1963] 2 All ER 297; [1963] Crim LR 436; (1963) 127 JP 397; (1963) 113 LJ 350; [1963] 2 QB 433; (1963) 107 SJ 272; [1963] 2 WLR 1106 HC QBD Test whether vehicle is for road use is what reasonable person looking on would say.

Cobb v Williams [1973] Crim LR 243; [1973] RTR 113 HC QBD Owner of car was 'using' it when travelled as passenger in own car for own aims.

Garrett v Hooper [1973] Crim LR 61; [1973] RTR 1 HC QBD That partner drove motor vehicle in course of partnership work did not mean other party guilty of 'use' of vehicle.

Gosling v Howard [1975] RTR 429 HC QBD Having 'use' of vehicle means using or keeping/ causing or permitting to be used for purposes of vehicle excise/road traffic legislation.

Hill and Sons (Botley and Denmead) Ltd v Hampshire Chief Constable [1972] Crim LR 538; [1972] RTR 29; (1971) 115 SJ 675 HC QBD Company's practice of examining vehicle every four weeks not reckless.

Leathley v Tatton [1980] RTR 21 HC QBD Person not driving/controlling car could still be 'using' it.

Pumbien v Vines [1996] Crim LR 124; [1996] RTR 37; [1995] TLR 337 HC QBD Vehicle on road that was irremovable unless repaired was being 'used' on road.

West Yorkshire Trading Standards Service v Lex Vehicle Leasing Ltd [1996] RTR 70 HC QBD On what constitutes 'use' of vehicle on road.

Windle v Dunning and Son Ltd [1968] 2 All ER 46; [1968] Crim LR 337; (1968) 132 JP 284; (1968) 118 NLJ 397; (1968) 112 SJ 196; [1968] 1 WLR 552 HC QBD Persons renting lorry from haulage contractors not using lorry but causing it to be used.

Young and another v Director of Public Prosecutions [1992] RTR 194; [1991] TLR 217 HC QBD On determining suitability of trailer for purpose for which being used.

CONSTRUCTION AND USE (vehicle)

R v Evans (Stanley) [1964] 3 All ER 666; (1965) 49 Cr App R 10; (1965) 129 JP 29; (1964) 108 SJ 880; [1964] 1 WLR 1388 CCA Condition of bicycle though examined on earlier date is taken to apply on later date (of certificate) — if does not are issuing false certificate.

CONSTRUCTION AND USE (vehicle testing)

Artingstoll v Hewen's Garages Ltd [1973] RTR 197 HC QBD On contractual duty of care of examiner authorised to carry out MOT tests.

Dial Contracts Ltd v Vickers [1972] Crim LR 27; [1971] RTR 386 HC QBD Owner of vehicles not liable as aiders and abettors of breaches of plating and testing regulations by hirers of vehicles.

Essendon Engineering Co Ltd v Maile [1982] Crim LR 510; [1982] RTR 260 HC QBD Company not liable for employee's false issuing of test certificate as not proven that had fully delegated relevant responsibilities to employee.

Hewer v Cutler [1973] Crim LR 762; [1974] RTR 155 HC QBD Car which could not be moved but left on road not in use so did not require valid test certificate.

Murphy v Griffiths [1967] 1 All ER 424; [1967] Crim LR 181t; (1967) 131 JP 204; (1965-66) 116 NLJ 1600t; (1967) 111 SJ 76; [1967] 1 WLR 333 HC QBD Back-dating of test certificate makes it false in material particular.

R v Minister of Transport, ex parte Males [1970] 3 All ER 434; (1970) 134 JP 657; (1970) 120 NLJ 662t; [1970] RTR 436 HC QBD Procedure when testing motor vehicles.

R v Pilditch [1981] Crim LR 184; [1981] RTR 303 CA Was fraudulent use of test certificate to complete stolen blank certificate and submit as valid — blank test form was 'test certificate'.

Rowley v Chatham [1970] RTR 462 HC QBD That vehicle certified as roadworthy by authorised examiner was found to be defective three weeks later (defect resulting in accident) did not per se render examiner liable in negligence/for breach of statutory duty.

Wurzal v Reader Brothers Ltd and another [1973] Crim LR 640; [1974] RTR 383 HC QBD Traffic examiner not required to produce authorisation when acting pursuant to Motor Vehicles (Construction and Use) Regulations 1969, reg 121.

CONSTRUCTION AND USE (weight)

Dent v Coleman and another [1977] Crim LR 753; [1978] RTR 1 HC QBD Were guilty of carrying improperly secured load where straps securing load defective though did not know were so.

Director of Public Prosecutions v Marshall and Bell (1990) 154 JP 508; [1990] RTR 384 HC QBD Construction of Road Vehicles (Construction and Use) Regulations 1986, regs 80(1) and 80(2).

Hudson and another v Bushrod [1982] RTR 87 HC QBD Defence to excess load charge that were bringing lorry to weighbridge did not apply to grossly overweighted vehicle.

Lloyd v Ross (1913-14) XXIII Cox CC 460; (1913) 48 LJ 229; [1913] 82 LJKB 578; (1913-14) 109 LTR 71; (1912-13) XXIX TLR 400; (1913) WN (I) 108 HC KBD Proper conviction for driving overweight car over bridge.

Lovett v Payne [1979] Crim LR 729; (1979) 143 JP 756; [1980] RTR 103 HC QBD On what was ment by 'nearest' weighbridge in Road Traffic Act 1972, s 40(6).

Prosser v Richings and another [1936] 2 All ER 1627; (1934-39) XXX Cox CC 457; (1936) 100 JP 390; (1936) 82 LJ 175; (1936) 155 LTR 284; (1935-36) LII TLR 677; (1936) 80 SJ 794; (1936) WN (I) 260 HC KBD On proper testing of weight of heavy motor car.

Thurrock District Council v LA and A Pinch Ltd [1974] Crim LR 425; [1974] RTR 269 HC QBD On pleading defence that overweight vehicle had been en route to weighbridge for weighing.

CONSTRUCTION AND USE (works truck)

Hayes v Kingsworthy Foundry Co Ltd and another [1971] Crim LR 239; [1971] RTR 286 HC QBD Dumper driving three-fifths of a mile from foundry to tip not a 'works truck'.

CROSSING (general)

Bailey v Geddes [1938] 1 KB 156; (1937) 84 LJ 97; [1938] 107 LJCL 38; (1937) 157 LTR 364; (1937) 81 SJ 684; (1936-37) LIII TLR 975; (1937) WN (I) 317 CA Motorist fully liable to pedestrian whom struck when failed to observe pedestrian crossing regulations.

Buckoke and others v Greater London Council (1970) 134 JP 465; (1970) 120 NLJ 337t; [1970] RTR 327; (1970) 114 SJ 269; [1970] 1 WLR 1092 HC ChD London Fire Brigade Order 144/8 not unlawful as did not require though did in essence permit violation of red traffic lights.

Burns v Bidder [1966] 3 All ER 29; [1966] Crim LR 395t; (1966) 130 JP 342; (1965-66) 116 NLJ 1173; [1967] 2 QB 227; (1966) 110 SJ 430; [1966] 3 WLR 99 HC QBD Not offence if failure to yield to pedestrian at crossing result of latent vehicle defect.

Cantwell v Revill [1940] LJNCCR 240 CyCt Driver who drove onto crossing while lights just went in his favour not liable in damages to pedestrian who stepped onto crossing when lights just went against her.

Chisholm v London Passenger Transport Board [1938] 2 All ER 579; (1938) 85 LJ 345; (1938) 82 SJ 396; (1937-38) LIV TLR 773 HC KBD Driver negligent where struck pedestrian already on crossing.

Chisholm v London Passenger Transport Board [1938] 4 All ER 850; [1939] 1 KB 426; (1939) 87 LJ 27; [1939] 108 LJCL 239; (1939) 160 LTR 79; (1938) 82 SJ 1050; (1938-39) LV TLR 284; (1939) WN (I) 15 CA Driver not negligent in striking pedestrian already on crossing: case decided by reference to common law principles.

Clifford v Drymond [1976] RTR 134; (1976) 120 SJ 149 CA Pedestrian on zebra crossing contributorily negligent (20%) in failing to keep an eye on oncoming traffic whilst crossing.

Connor v Paterson [1977] 3 All ER 516; [1977] Crim LR 428; (1978) 142 JP 20; (1977) 127 NLJ 639t; [1977] RTR 379; (1977) 121 SJ 392; [1977] 1 WLR 1450 HC QBD Was unlawful under 'Zebra' Pedestrian Crossings Regulations 1971/1524 to overtake car which had stopped at crossing to let people cross even though all people crossing had crossed at moment of overtaking.

Crank v Brooks [1980] RTR 441 HC QBD Person on foot pushing pedal bicycle across 'Zebra' crossing was 'foot passenger'.

Franklin v Langdown [1971] 3 All ER 662; [1972] Crim LR 543; (1971) 135 JP 615; (1971) 121 NLJ 690t; [1971] RTR 471; (1971) 115 SJ 688 HC QBD Must stop until lollipop lady no longer showing stop sign.

Gibbons v Kahl [1955] 3 All ER 345; (1956) 120 JP 1; [1956] 1 QB 59; [1955] Crim LR 788; (1955) 99 SJ 782; [1955] 3 WLR 596 HC QBD Offence where drive towards pedestrian crossing in such a way that cannot stop if person thereon.

Gullen v Ford; Prowse v Clarke [1975] 2 All ER 24; [1975] Crim LR 172; (1975) 139 JP 405; (1975) 125 NLJ 111t; [1975] RTR 302; (1975) 119 SJ 153; [1975] 1 WLR 335 HC QBD Overtaking of person stopped because sees person about to cross crossing is offence.

Hoy v Smith [1964] 3 All ER 670; [1965] Crim LR 49; (1965) 129 JP 33; (1964) 114 LJ 790; (1964) 108 SJ 841; [1964] 1 WLR 1377 HC QBD Must be able to see 'Stop' on school crossing sign for school crossing requirements to apply.

Hughes v Hall [1960] 2 All ER 504; [1960] Crim LR 710; (1960) 124 JP 411; (1960) 110 LJ 462; (1960) 104 SJ 566; [1960] 1 WLR 733 HC QBD Guilty of absolute offence of not stopping at crossing even though car already over studs when pedestrian stepped onto road.

Issatt v Greenwood [1971] RTR 476 HC QBD Not required to stop at school crossing once children are no longer on crossing.

Kayser v London Passenger Transport Board [1950] 1 All ER 231; (1950) 114 JP 122; [1950] 1 All ER 231 HC KBD Driver stopped for pedestrians may proceed once considers it safe but must be able to stop again should pedestrians act dangerously/negligently.

Knight v Sampson (1938) 82 SJ 524; (1937-38) LIV TLR 974 HC KBD Person at pedestrian crossing who stepped onto road at last moment when car approaching could not recover personal injury damages notwithstanding pedestrian crossing regulations.

Kozimor v Adey and another [1962] Crim LR 564; (1962) 106 SJ 431 HC QBD On duty of driver on approaching pedestrian crossing.

Leicester v Pearson [1952] 2 All ER 71; (1952) 116 JP 407; (1952) 102 LJ 331; [1952] 2 QB 668; (1952) 96 SJ 397; [1952] 1 TLR 1537; (1952) WN (I) 317 HC QBD Driver not under absolute duty to give precedence to pedestrian at unmarked crossing.

Lockie v Lawton [1959] Crim LR 856; (1960) 124 JP 24; (1959) 103 SJ 874 HC QBD On responsibilities of motorist at pedestrian crossing.

London Passenger Transport Board v Upson and another [1949] 1 All ER 60; [1949] AC 155; (1948) 98 LJ 701; [1949] 118 LJR 238; (1949) LXV TLR 9; (1948) WN (I) 492 HL Driver's duties at crossing extant even though light green.

Maynard v Rogers (1970) 114 SJ 320 HC QBD Pedestrian (two thirds)/driver (one third) responsible for collision at pedestrian crossing; on driver's duty on approaching pedestrian crossing.

Moulder v Neville [1974] Crim LR 126; [1974] RTR 53; (1974) 118 SJ 185 HC QBD Driver must yield to person stepping onto stripes at zebra crossing before driver enters onto that area.

Neal v Bedford [1965] 3 All ER 250; [1965] Crim LR 614; (1965) 129 JP 534; (1965) 115 LJ 677; [1966] 1 QB 505; (1965) 109 SJ 477; [1965] 3 WLR 1008 HC QBD Must yield to pedestrians at pedestrian crossing.

Oakley-Moore v Robinson [1982] RTR 74 HC QBD Erroneous belief that did not have petrol not a defence to parking inside prohibited area of Pelican crossing.

Parkinson and another v Parkinson [1973] RTR 193 CA Not contributory negligence for pedestrians to walk on left hand side of road.

R v Greenwood [1962] Crim LR 639t HC QBD No duty to stop for lollipop lady unless children present who are crossing/seeking to cross.

Sparks v Edward Ash, Limited [1943] KB 223; [1943] 112 LJ 289; (1943) 168 LTR 118; (1943) WN (I) 21 CA Despite black-out were still required to approach pedestrian crossing at speed that could safely stop if pedestrian on it.

Sparks v Edward Ash, Ltd [1942] 2 All ER 214; (1942) 106 JP 239; [1942] 111 LJ 587; (1942) 167 LTR 64; (1941-42) LVIII TLR 324; (1942) WN (I) 152 HC KBD Despite black-out were still required to approach pedestrian crossing at speed that could safely stop if pedestrian on it but pedestrian could be contributorily negligent.

Stimson v Pitt [1947] KB 668; [1948] 117 LJR 351; (1947) 177 LTR 187; (1947) 91 SJ 309; (1947) LXIII TLR 293; (1947) WN (I) 157 HC KBD Each of four cross-roads with traffic-lighted pedestrian crossings is 'controlled' by lights (Pedestrian Crossing Places (Traffic) Regulation 1941, r 5).

Sulston v Hammond [1970] 2 All ER 830; [1970] Crim LR 473t; (1970) 134 JP 601; [1970] RTR 361; (1970) 114 SJ 533; [1970] 1 WLR 1164 HC QBD Notice of intended prosecution not needed for pelican crossing violations.

Toole v Sherbourne Pouffes Ltd and another; Newport Corporation (Third Party) (1971) 121 NLJ 710t; [1971] RTR 479 CA Corporation vicariously liable for negligence of lollipop man.

Upson v London Passenger Transport Board [1947] KB 930; (1947) 97 LJ 431; [1947] 116 LJR 1382; (1947) 177 LTR 475; (1947) LXIII TLR 452; (1947) WN (I) 252 CA Driver's duties at crossing extant even though light green.

Upson v London Passenger Transport Board (1947) 176 LTR 356 HC KBD Driver's duties at crossing extant even though light green.

Wall v Walwyn [1973] Crim LR 376; [1974] RTR 24 HC QBD Must stop upon approaching school crossing at which 'Stop' sign being displayed and must continue to stop as long as sign displayed.

Ward v Miller [1940] LJNCCR 174 CyCt Driver who drove onto crossing while lights against him liable in damages to pedestrian who stepped onto crossing without checking for traffic when lights were in her favour.

Wilkinson v Chetham-Strode [1940] 2 All ER 643; (1940) 104 JP 283; (1940) 89 LJ 260; [1940] 109 LJCL 823; (1940) 163 LTR 26; (1940) 84 SJ 573; (1939-40) LVI TLR 767; (1940) WN (I) 193 CA Refuge not part of pedestrian crossing.

Wilkinson v Chetham-Strode [1940] 1 All ER 67; [1940] 1 KB 309; (1940) 104 JP 81; [1940] 109 LJCL 193; (1940) 162 LTR 157; (1940) 84 SJ 188; (1939-40) LVI TLR 228; (1939) WN (I) 417 HC KBD Refuge not part of pedestrian crossing so stepping from same to crossing could be (and was) contributorily negligent to collision.

Wright v Hunt [1984] TLR 309 HC QBD On offence of overtaking at zebra crossing.

DANGEROUS DRIVING (general)

Archer v Woodard [1959] Crim LR 461 HC QBD That drank whisky on empty stomach not special reason justifying reduced disqualification period.

Baker v Williams [1956] Crim LR 204t HC QBD Justices found to have misdirected themselves on facts (which showed defendant must have been driving dangerously).

Beresford and another v Richardson (1918-21) XXVI Cox CC 673; (1921) 85 JP 60; [1921] 1 KB 243; [1921] LJCL 313; (1921) 124 LTR 274; (1920-21) XXXVII TLR 53 HC KBD Speed/circumstantial evidence held to prove driving was dangerous.

Bracegirdle v Oxley (1947) 111 JP 131 HC KBD Speed on occasion can of itself be indicative of dangerous driving.

Cook v Atchison [1968] Crim LR 266; (1968) 112 SJ 234 HC QBD Once credible defence raised in road traffic prosecution is for prosecutor to disprove the defence.

Director of Public Prosecutions v Khan [1997] RTR 82 HC QBD Acquittal of inconsiderate driving not autrefois acquit to charge of dangerous driving.

Ex parte Newsham [1964] Crim LR 57t HC QBD Leave to seek certiorari order granted to appellant convicted of careless driving where charged with dangerous driving.

Hallett v Warren (1929) 93 JP 225 HC KBD Not the case that evidence before/a while away from collision inadmissible because of distance.

Hill v Baxter [1958] 1 All ER 193; (1958) 42 Cr App R 51; [1958] Crim LR 192; (1958) 122 JP 134; (1958) 108 LJ 138; [1958] 1 QB 277; (1958) 102 SJ 53; [1958] 2 WLR 76 HC QBD Dangerous driving/non-conformity with road sign are absolute offences; automatism defence rejected.

Johnstone v Hawkins [1959] Crim LR 459t HC QBD Driver on main road who passed junction at speed of 85-90 mph was prima facie guilty of dangerous driving.

Johnstone v Hawkins [1959] Crim LR 854t HC QBD Failed appeal against acquittal of dangerous driving of driver who drove past junction on main road at speed of 85-90 mph.

Kay v Butterworth (1945) 173 LTR 191; (1945) 89 SJ 381; (1944-45) LXI TLR 452 HC KBD That driver of car was asleep and so unaware of offences not defence to charges of careless and dangerous driving.

Kay v Butterworth (1946) 110 JP 75 CA Person driving car while asleep was guilty of careless and dangerous driving.

Kingman v Seager (1934-39) XXX Cox CC 639; (1937) 101 JP 543; [1938] 1 KB 397; [1938] 107 LJCL 97; (1937) 157 LTR 535; (1937) 81 SJ 903; (1937-38) LIV TLR 50; (1937) WN (I) 369 HC KBD In dangerous driving charge justices to consider real/likely danger caused by accused's fast driving.

Maguire v Crouch (1939-40) XXXI Cox CC 441; (1940) 104 JP 445; [1941] 1 KB 108; [1941] 110 LJ 71; (1941) 164 LTR 171; (1940) 84 SJ 608; (1940-41) LVII TLR 75; (1940) WN (I) 366 HC KBD Valid conviction under Road Traffic Act 1930, s 50, for causing vehicle to remain at rest in manner likely to endanger another.

Marson v Thompson [1955] Crim LR 319t HC QBD HC QBD review of dangerous driving acquittal justified as decision of justices that defendant had not been driving dangerously was unsupportable.

Middleton v Rowlett (1954) 98 SJ 373 HC QBD Refusal to re-open case of dangerous driving originally dismissed where prosecution failed to prove accused was driver of vehicle at issue.

Parson v Tomlin [1956] Crim LR 192; (1956) 120 JP 129 HC QBD Wife not competent (prosecution) witness against husband; evidence of police officer as to general level of traffic on particular road was admissible in dangerous driving prosecution.

R v Clow [1963] 2 All ER 216; [1963] Crim LR 514; (1963) 107 SJ 537; [1963] 3 WLR 84 CA Different offences arising from same occurrence can be charged conjunctively.

R v Gosney [1971] 3 All ER 220; (1971) 55 Cr App R 502; [1972] Crim LR 534; (1971) 135 JP 529; (1971) 121 NLJ 617t; [1971] 2 QB 674; [1971] RTR 321; (1971) 115 SJ 608; [1971] 3 WLR 343 CA Dangerous driving not absolute offence.

R v Hands [1962] Crim LR 255 CCA On dangerous driving vis-a-vis causing death by dangerous driving.

R v Hennigan [1971] 3 All ER 133; (1971) 55 Cr App R 262; [1971] Crim LR 285; (1971) 135 JP 504; (1971) 121 NLJ 178t; [1971] RTR 305; (1971) 115 SJ 268 CA Prosecution to show that dangerous driving more than minimal cause of accident.

R v Howell (1939-40) XXXI Cox CC 190; (1939) 103 JP 9; (1939) 160 LTR 16 CCA Jury finding of guilty of dangerous driving through error of judgment an acquittal of dangerous driving.

R v Jennings [1956] 3 All ER 429; (1956) 40 Cr App R 147; [1956] Crim LR 698; (1956) 100 SJ 861; [1956] 1 WLR 1497 CMAC Could convict of dangerous driving though not notified of prosecution fourteen days beforehand.

R v Johnson [1960] Crim LR 430 CCC On appropriate test when determining whether driving was dangerous.

R v Leicester Justices, ex parte Walker; Myers v Walker (1936) 80 SJ 54 HC KBD Failure to inform defendant of right to jury trial rendered trial a nullity; failure to order endorsement of licence upon conviction for dangerous driving was improper under the Road Traffic Act 1930, s 11(1).

R v Mills [1977] RTR 188 CA Disqualification lifted as offender guilty of dangerous driving had very ill child who might require rapid removal to hospital and offender was only driver available.

R v Norrington [1960] Crim LR 432 Assizes Exclusion (in trial for dangerous driving) of evidence that defendant had been drinking alcohol.

R v Parker (Capel Francis) (1957) 41 Cr App R 134 CCA Momentary negligence when driving could amount to dangerous driving.

R v Scammell [1967] 3 All ER 97; (1967) 51 Cr App R 398; [1967] Crim LR 594; (1967) 131 JP 462; (1967) 117 NLJ 913; (1967) 111 SJ 620; [1967] 1 WLR 1167 CA On careless driving as defence to dangerous driving.

R v Segal [1976] RTR 319 CA Person properly convicted of driving at dangerous speed but acquitted of driving in dangerous manner.

R v Shipton, ex parte Director of Public Prosecutions [1957] 1 All ER 206; (1957) 101 SJ 148 HC QBD Whether quarter sessions could try dangerous driving when indictment for manslaughter/accused acquitted of manslaughter.

R v Southampton Justices, ex parte Tweedie (1932) 96 JP 391; (1932) 74 LJ 168; (1933) 102 LJCL 11; (1932) 147 LTR 530; (1932) 76 SJ 545; (1931-32) XLVIII TLR 636 HC KBD Conviction quashed as charge of dangerous driving reduced to careless driving without adequate evidence of consent to reduction of charge.

R v Spurge [1961] 2 All ER 688; (1961) 45 Cr App R 191; [1961] Crim LR 627; (1961) 125 JP 502; (1961) 111 LJ 438; [1961] 2 QB 205; (1961) 105 SJ 469; [1961] 3 WLR 23 CCA Accused may advance (and prosecution to disprove) that sudden manifestation of latent mechanical defect caused 'dangerous' driving.

R v Stringer (1931-34) XXIX Cox CC 605; (1933) 97 JP 99; [1933] 1 KB 704; (1933) 75 LJ 96; (1933) 102 LJCL 206; (1933) 148 LTR 503; (1933) 77 SJ 65; (1932-33) XLIX TLR 189; (1933) WN (I) 28 CCA Could be tried together for manslaughter/dangerous driving and convicted of latter.

R v Strong [1995] Crim LR 428 CA Failed conviction for dangerous driving where manner of driving arose from vehicle defect that would not have been obvious to careful/competent driver.

R v Telford [1954] Crim LR 137 Sessions On what is meant by 'wanton driving'/'wilful neglect'.

R v Thorpe [1972] 1 All ER 929; (1972) 56 Cr App R 293; [1972] Crim LR 240t; (1972) 136 JP 301; (1972) 122 NLJ 57t; [1972] RTR 118; (1972) 116 SJ 80; [1972] 1 WLR 342 CA Evidence of drinking admissible at death by dangerous driving trial.

Richards v Gardner [1974] Crim LR 119; [1976] RTR 476 HC QBD That were driving vehicle dangerously inferred (absent other evidence) from peculiar behaviour of vehicle when brakes applied.

Squire v Metropolitan Police Commissioner [1957] Crim LR 817t HC QBD Was prima facie case of dangerous driving against person simultaneously overtaking bus and two preceding cars while was oncoming traffic.

DANGEROUS VEHICLE (general)

Trentham v Rowlands [1974] Crim LR 118; (1974) 124 NLJ 10t; [1974] RTR 164 HC QBD Driver harassing driver who remained in overtaking lane and who overtook latter on near side was guilty of dangerous driving.

Walker v Dowswell [1977] RTR 215 HC QBD Was dangerous driving where failed to comply with traffic sign.

Watmore v Jenkins [1962] 2 All ER 868; [1962] Crim LR 562t; (1962) 126 JP 432; (1962) 112 LJ 569; [1962] 2 QB 572; (1962) 106 SJ 492 HC QBD Diabetic driver's behaviour upon reaction to insulin not automatism; diabetic having injected insulin and driven dangerously acquitted of driving under influence of drug.

DANGEROUS VEHICLE (general)

Shave v Rosner [1954] 2 All ER 280; [1954] Crim LR 552; (1954) 118 JP 364; (1954) 104 LJ 346; [1954] 2 QB 113; (1954) 98 SJ 355; [1954] 2 WLR 1057 HC QBD Mechanic who returned car to owner's control could not later be said to have caused vehicle to be used while dangerous.

DEFENCES (alcoholism)

R v Chichester Justices, ex parte Crouch (1982) 146 JP 26 HC KBD Secretary of State ought not to have been notified accused suffered disease where counsel claimed at accused's first drunk driving prosecution that accused had a drink problem.

DEFENCES (automatism)

Attorney-General's Reference No 2 of 1992 [1993] 4 All ER 683; (1993) 97 Cr App R 429; [1994] Crim LR 692; (1994) 158 JP 741; (1993) 143 NLJ 919; [1994] QB 91; [1993] RTR 337; (1993) 137 SJ LB 152; [1993] TLR 303; [1993] 3 WLR 982 CA 'Driving without awareness' not automatism as partial not total loss of control.

Broome v Perkins (1987) 85 Cr App R 321; [1987] Crim LR 271; [1987] RTR 321 HC QBD On what constitutes automatism; could not have been automatism where drove car along road for five miles.

Moses v Winder [1980] Crim LR 232; [1981] RTR 37 HC QBD Failed attempt by diabetic to establish automatism defence to driving without due care and attention charge.

R v Burgess [1991] 2 All ER 769; [1991] Crim LR 548; (1991) 141 NLJ 527; [1991] 2 QB 92; [1991] TLR 161; [1991] 2 WLR 1206 CA Passing disorder prompting violent behaviour and caused by internal factor comes within M'Naghten Rules and is not non-insane automatism.

R v Isitt (Douglas Lance) (1978) 67 Cr App R 44; [1978] Crim LR 159; [1978] RTR 211 CA That mind not functioning properly was neither defence nor automatism and so not defence to offence charged.

R v Quick [1973] Crim LR 434t CA On what constitutes a 'disease of the mind' (in context of diabetic suffering hypoglycaemia after taking medically prescribed insulin).

R v Stripp (David Peter) (1979) 69 Cr App R 318 CA Once ground for automatism established for jury prosecution must prove accused acted wilfully.

DEFENCES (duress)

Director of Public Prosecutions v Bell [1992] Crim LR 176; [1992] RTR 335 HC QBD Person with excess alcohol fleeing from dangerous scene in terror/not proven to have driven longer than necessary/ who did not cease being terrified while driving could rely on defence of duress.

Director of Public Prosecutions v Davis; Director of Public Prosecutions v Pittaway [1994] Crim LR 600 HC QBD On test when determining whether duress present (here in context of offence of driving with excess alcohol).

DEFENCES (necessity)

Director of Public Prosecutions v Harris [1995] 1 Cr App R 170; [1995] Crim LR 73; (1994) 158 JP 896; [1995] RTR 100; [1994] TLR 151 HC QBD Emergency vehicle driver could not plead necessity to charge of driving without due care and atttention.

Director of Public Prosecutions v Jones [1990] RTR 33 HC QBD Person using car to escape attack by another could not succeed on defence of necessity insofar as entire journey was concerned.

R v Conway [1988] 3 All ER 1025; (1989) 88 Cr App R159; [1989] Crim LR 74; (1988) 152 JP 649; [1989] QB 290; [1989] RTR 35; (1988) 132 SJ 1244; [1988] 3 WLR 1238 CA Necessity a defence to reckless driving if duress of circumstances present.

R v Denton (Stanley Arthur) (1987) 85 Cr App R 246; [1987] RTR 129; (1987) 131 SJ 476 CA Defence of necessity unavailable where person charged with reckless driving claimed had driven carefully at relevant time.

R v Martin [1989] 1 All ER 652; (1989) 88 Cr App R 343; [1989] Crim LR 284; (1989) 153 JP 231; [1989] RTR 63; [1988] 1 WLR 655 CA Defence of necessity arises if objectively viewed person acted reasonably/proportionately to avoid death/serious injury threat.

DEFENCES (sleep)

Higgins v Bernard [1972] 1 All ER 1037; [1972] Crim LR 242t; (1972) 136 JP 314; (1972) 122 NLJ 153t; [1972] RTR 304; (1972) 116 SJ 179; [1972] 1 WLR 455 HC QBD Drowsiness not an emergency justifying stopping on motorway.

Kay v Butterworth (1945) 173 LTR 191; (1945) 89 SJ 381; (1944-45) LXI TLR 452 HC KBD That driver of car was asleep and so unaware of offences not defence to charges of careless and dangerous driving.

Kay v Butterworth (1946) 110 JP 75 CA Person driving car while asleep was guilty of careless and dangerous driving.

DOCUMENTS (general)

B and S Displays Ltd and others v Inland Revenue Commissioners (1978) 128 NLJ 389t HC ChD On validity of notices (requiring production of documents) issued under the Taxes Management Act 1970, s 20.

Boyce v Absalom [1974] Crim LR 192; [1974] RTR 248 HC QBD Constable cannot under Road Traffic Acts demand driving licence/other driving documents of person not driving.

Davey v Towle [1973] Crim LR 360; [1973] RTR 328 HC QBD Conviction for failure to produce test certificate/driving car without necessary insurance.

Holloway v Brown [1979] Crim LR 58; [1978] RTR 537 HC QBD Successful appeal against conviction for use of forged international road haulage permit.

Hough v Liverpool City Council [1980] Crim LR 443; [1981] RTR 67 HC QBD On what constituted 'reasonable cause' to fail to give authorised officer information pursuant to Liverpool Corporation Act 1972, s 36(1).

R v Aworinde [1995] Crim LR 825; (1995) 159 JP 68; [1996] RTR 66; [1995] TLR 239 CA Blank false insurance forms could form basis on which to ground conviction for possession of documents calculated to deceive.

R v Pilditch [1981] Crim LR 184; [1981] RTR 303 CA Was fraudulent use of test certificate to complete stolen blank certificate and submit as valid — blank test form was 'test certificate'.

Tremelling v Martin [1972] Crim LR 596; [1971] RTR 196 HC QBD Was not production of licence at police station to produce same but withdraw it before constable could examine it.

DRIVER'S HOURS (general)

Alcock v GC Griston Ltd [1980] Crim LR 653; (1980) 130 NLJ 1211; [1981] RTR 34 HC QBD For purposes of Transport Act 1968, s 103, firm using driver provided by agency was employer of driver.

Appleby (RW) Ltd and another v Vehicle Inspectorate [1994] RTR 380 HC QBD Non-liability of parent company for non-provision of tachograph records by subsidiary to employee; tachograph necessary when driving from depot to docks to collect passengers (no notion of 'positioning' journey under the tachograph legislation).

Beer v Clench (WH) (1930) Limited (1934-39) XXX Cox CC 364; (1936) 80 SJ 266 HC KBD Employer's working hour records admissible in prosecution for permitting worker-driver to drive above permitted work limit.

Beer v TM Fairclough and Sons Limited (1934-39) XXX Cox CC 551; (1937) 101 JP 157; (1937) 156 LTR 238; (1937) 81 SJ 180; (1936-37) LIII TLR 345 HC KBD Employer to ensure worker-driver has rest periods, not to police how he spends them.

Birkett and another v Wing; Fisher v Dukes Transport (Craigavon) Ltd [1997] TLR 303 HC QBD Is an offence under the Transport Act 1968, s 97, not to keep record sheets that were used in tachograph.

Blakey Transport Ltd v Casebourne [1975] Crim LR 169; (1975) 125 NLJ 134t; [1975] RTR 221; (1975) 119 SJ 274 HC QBD Is not preserving driver's record book intact for employers to whom it is returned re-issue same to another within six months of receipt.

Bowra v Dann Catering Ltd [1982] RTR 120 HC QBD Catering firm providing portable equipment (including toilets) at entertainment venues were not travelling showmen; portable toilets here not meant for circus/fun-fair so firm/its drivers not exempt from driver-records legislation.

British Gypsum Ltd v Corner [1982] RTR 308 HC QBD Bowser which carried goods subject to driver's records requirements and required operator's licence.

Brown (TH) Ltd v Baggott (No 2) [1970] Crim LR 643; [1970] RTR 323 HC QBD Consignment notes admissible to prove that driver's records not entirely correct.

Browne v Anelay [1997] TLR 302 HC QBD On when driver deemed to have taken over vehicle for purposes of tachograph legislation.

Carter v Walton [1985] RTR 378; [1984] TLR 356 HC QBD Whether and when fact that lorry driver not driving means is off-duty a subjective matter to be decided from individual circumstances.

Cassady v Reg Morris (Transport) Ltd [1975] Crim LR 398; [1975] RTR 470 HC QBD Employer's not punishing non-delivery/encouraging delivery by driver of completed record sheets within seven days did not in instant circumstances render employer guilty as aider and abettor to employee's offence.

Cassady v Ward and Smith Ltd [1975] Crim LR 399; [1975] RTR 353 HC QBD Requirement that new driver be provided with new driver's book not met by providing worker-driver with driver's book used by several previous temporary worker-drivers.

Cook v Alfred Plumpton, Ltd and another [1935] All ER 806; (1935) 99 JP 308; (1935) 80 LJ 27; (1935) 153 LTR 462; (1935) 79 SJ 504; (1934-35) LI TLR 513; (1935) WN (I) 131 HC KBD Sum of driving periods in one day contravened maximum driving times allowed under Road Traffic Act 1930, s 19(1).

Cook v Henderson (1934-39) XXX Cox CC 270; (1935) 99 JP 308 HC KBD Driver calculated to have breached permitted working hours by adding of separate periods punctuated by breaks.

Creek v Fossett, Eccles and Supertents Ltd [1986] Crim LR 256 HC QBD Application of operator licence, test certificate and tachograph legislation in circus vehicles context.

Director of Public Prosecutions v Aston [1989] RTR 198 HC QBD Grocer carrying stock from market to shop for retail purposes not carrying 'material' for own use in course of work so not exempt from tachograph legislation.

Director of Public Prosecutions v Cargo Handling Ltd (1992) 156 JP 486; [1992] RTR 318; [1991] TLR 575 HC QBD On what constituted public road so that EEC-inspired tachograph legislation applied.

Director of Public Prosecutions v Digby (1992) 156 JP 420; [1992] RTR 204; [1992] TLR 61 HC QBD Lorry delivering coal to miners/retired miners for free not engaged in door-to-door selling and so ought to have had tachograph.

Director of Public Prosecutions v Guy [1997] TLR 354 HC QBD Person guilty of tachograph offence though had finished work and was driving home at time when challenged.

Director of Public Prosecutions v Ryan and another (1991) 155 JP 456; [1992] RTR 13; [1991] TLR 88 HC QBD On what constitutes a goods vehicle (and so is subject to tachograph legislation).

Gaunt v Nelson and another [1987] RTR 1 HC QBD On what constitutes a 'specialised vehicle for door to door selling' and so is immune from tachograph requirements.

Geldart v Brown and others [1986] RTR 106 HC QBD Absent emergency driver's driving excess hours out of consideration towards passengers was unlawful.

Grays Haulage Co, Ltd v Arnold [1966] 1 All ER 896; [1966] Crim LR 224; (1966) 130 JP 196; (1965-66) 116 NLJ 445t; (1966) 110 SJ 112; [1966] 1 WLR 534 HC QBD Must be actual knowledge on employer's part to sustain charge of permitting driver to drive for longer than law allows.

Green v Harrison [1979] Crim LR 395; (1979) 129 (2) NLJ 734; [1979] RTR 483 HC QBD Puncture did not cause unavoidable delay in completion of journey so were guilty of driving for excess hours.

Kelly v Shulman [1988] Crim LR 755; [1989] RTR 84; (1988) 132 SJ 1036; [1988] 1 WLR 1134 HC QBD On what constitutes a day for purposes of EEC driver-worktime legislation (Council Regulation 3820/85/EEC): successful prosecution for breach of UK driver-worktime legislation (Transport Act 1968, s 96(11A)).

Knowles Transport Ltd v Russell [1974] Crim LR 717; [1975] RTR 87 HC QBD No grounds on which to infer that company knew/could have known of excess hours worked by lorry driver.

Lackenby v Browns of Wem Ltd [1980] RTR 363 HC QBD Employer not required to issue unused driver's record book to employee drivers.

Lawson v Fox and others [1974] 1 All ER 783; [1974] AC 803; (1974) 138 JP 368; [1974] RTR 353; [1974] 2 WLR 247 HL Acts outside England relevant in calculation lorry driver's breach of working time regulations.

Light v Director of Public Prosecutions [1994] RTR 396 HC QBD Transport manager found to have permitted breaches of tachograph legislation where he had not done everything possible to end same.

Oxford v Spencer [1983] RTR 63 HC QBD Driver's non-entry of name/address/date of birth in record book not a criminal act.

Oxford v Thomas Scott and Sons Bakers Ltd and another [1983] Crim LR 481; [1983] RTR 369 HC QBD On what constitute 'specialised vehicles' engaged in 'door-to-door selling' and so are exempt from EEC tachograph legislation (as implemented).

Paterson v Richardson and another [1982] RTR 49 HC QBD On what constitutes a 'national'/ 'international' journey for purposes of drivers' hours legislation.

Pearson v Rutterford and another [1982] RTR 54 HC QBD Period when driver acting as instructor not a 'daily rest period' but another period of work.

Potter v Gorbould and another [1969] 3 All ER 828; [1970] Crim LR 46; (1969) 133 JP 717; [1970] 1 QB 238; (1969) 113 SJ 673; [1969] 3 WLR 810 HC QBD Non-driving overtime work not part of rest period.

Prime v Hosking (1995) 159 JP 755; [1995] RTR 189; [1994] TLR 686 HC QBD Driver working overtime in non-driving capacity ought to have entered same on tachograph record sheet.

R v British Fuel Company Limited [1983] Crim LR 747 CrCt Conviction for not requiring driver-employee to keep record book of driving times.

Redhead Freight Ltd v Shulman [1988] Crim LR 696; [1989] RTR 1 HC QBD Mere acquiesence to driver's breach of tachograph legislation was not causing of same.

Swain v McCaul and another [1997] RTR 102; (1996) 140 SJ LB 142; (1996) TLR 10/7/96 HC QBD Commercial skip firm was required to comply with tachograph legislation requirements.

Wells and Son, Ltd v Sidery [1939] 4 All ER 54; (1939) 103 JP 375; [1940] 1 KB 85; (1939) 88 LJ 265; [1939] 108 LJCL 871; (1939) 161 LTR 352; (1939) 83 SJ 891; (1939-40) LVI TLR 25; (1939) WN (I) 348 HC KBD Driver 'standing by' to drive is not driving.

Whitby v Stead [1975] Crim LR 240; [1975] RTR 169 HC QBD Failed attempt to rely on defence of unavoidable delay in prosecution of driver for working excessive hours.

Wing v Nuttall (1997) 141 SJ LB 98; [1997] TLR 225 HC QBD Employers are required to check tachograph charts: failure to do so is reckless blindness to workers' tachograph offences, so are guilty of permitting same.

DRIVING CAR ON PAVEMENT (general)

Curtis v Geeves (1931-34) XXIX Cox CC 126 HC KBD Conviction for driving car on footway.

DRIVING INSTRUCTION (general)

Toms v Hurst; Toms v Langton [1996] RTR 226 HC QBD Admission inadmissible where obtained in breach of Code of Practice; succesful prosecution for using driving instruction register certificate with intent to deceive.

DRIVING TEST (general)

Geraghty v Morris [1939] 2 All ER 269; (1939-40) XXXI Cox CC 249; (1939) 103 JP 175; (1939) 87 LJ 288; (1939) 160 LTR 397; (1939) 83 SJ 359; (1938-39) LV TLR 599; (1939) WN (I) 153 HC KBD No review of driving test results.

DRIVING WHILE DISQUALIFIED (aiding and abetting)

R v Gommo [1964] Crim LR 469 CCA That passenger had been sitting next to driver who was disqualified was not per se evidence that knew of disqualification (and so was guilty of aiding and abetting offence of driving while disqualified).

DRIVING WHILE DISQUALIFIED (general)

Aichroth v Cottee [1954] 2 All ER 856; [1954] Crim LR 795; (1954) 118 JP 499; (1954) 104 LJ 538; [1954] 1 WLR 1124 HC QBD Sudden hazard may constitute special factor justifying substitution of fine for imprisonment.

R v Bowsher [1973] Crim LR 373; [1973] RTR 202 CA Driving while disqualified an absolute offence; disqualifications ordered under Road Traffic Act 1962, s 5(3) to be consecutive.

R v Derwentside Justices, ex parte Heaviside [1996] RTR 384 HC QBD On proper identification of accused as person previously convicted (comparing personal details insufficient).

R v MacDonagh [1974] 2 All ER 257; (1974) 59 Cr App R 55; [1974] Crim LR 317; (1974) 138 JP 488; (1974) 124 NLJ 222t; [1974] QB 448; [1974] RTR 372; (1974) 118 SJ 222; [1974] 2 WLR 529 CA 'Driving' must be driving within common understanding.

R v Miller [1975] 2 All ER 974; (1975) 61 Cr App R 182; [1975] Crim LR 723; (1975) 139 JP 613; [1975] RTR 479; (1975) 119 SJ 562; [1975] 1 WLR 1222 CA Not defence to driving on road while disqualified that believed place driving on not road.

R v Munning [1961] Crim LR 555 Magistrates Pushing scooter not driving of same so not guilty of driving while disqualified.

R v Saddleworth Justices, ex parte Staples [1968] 1 All ER 1189; (1968) 132 JP 275; (1968) 118 NLJ 373; (1968) 112 SJ 336; [1968] 1 WLR 556 HC QBD Prosecution could choose between under-age driving charge or driving while disqualified charge.

R v Spindley [1961] Crim LR 486 Sessions Steering car being pushed from behind constituted 'driving' while disqualified (though fine, not imprisonment imposed).

R v Thames Magistrates' Court, ex parte Levy [1997] TLR 394 HC QBD Where are subject of disqualification order but offence in respect of which disqualification imposed has been quashed one can nonetheless (as here) be guilty of driving while disqualified.

Reader v Bunyard (1987) 85 Cr App R 185; [1987] Crim LR 274; [1987] RTR 406 HC QBD Not clear whether steering towed vehicle that might not be used as mechanical motor vehicle in the future could justify driving while disqualified conviction.

Scott v Jelf [1974] Crim LR 191; (1974) 124 NLJ 148t; [1974] RTR 256 HC QBD Person breaching condition of provisional licence (validly obtained) while subject to disqualification order (in force until passed new driving test) was guilty of driving while disqualified.

Taylor v Kenyon [1952] 2 All ER 726; (1952) 116 JP 599; (1952) 96 SJ 749; (1952) WN (I) 478 HC QBD Driving while disqualified an absolute offence.

DRIVING WITH EXCESS ALCOHOL (general)

Director of Public Prosecutions v H [1997] TLR 238 HC QBD Insanity not a defence to strict liability offence (here driving with excess alcohol).

Hoyle v Walsh [1969] 1 All ER 38; [1969] Crim LR 91t; (1969) 133 JP 91; [1969] 2 QB 13; (1968) 112 SJ 883; [1969] 2 WLR 34 HC QBD Improper arrest for failure to provide specimen where failure arose from defective breathalyser.

Jowett-Shooter v Franklin [1949] 2 All ER 730; (1949) 113 JP 525 HC KBD That did not/was not going to drive special reasons justifying non-disqualification upon drunk driving conviction.

Monaghan v Corbett (1983) 147 JP 545; [1983] TLR 438 HC QBD Cannot base reasonable suspicion of drunken driving on facts unrelated to relevant incident of driving.

R v Bolton Justices, ex parte Scally; R v Bolton Justices, ex parte Greenfield; R v Eccles Justices, ex parte Meredith; R v Trafford Justices, ex parte Durran-Jorda [1991] Crim LR 550; (1991) 155 JP 501; [1991] 1 QB 537; [1991] RTR 84; (1990) 134 SJ 1403; [1990] TLR 639; [1991] 2 WLR 239 HC QBD Excessive blood-alcohol concentration convictions quashed where very likely that blood samples taken had been contaminated.

R v Chapman [1969] 2 All ER 321; (1969) 53 Cr App R 336; [1969] Crim LR 269; (1969) 113 SJ 229; (1969) 133 JP 405; (1969) 119 NLJt 224; [1969] 2 QB 436; [1969] 2 WLR 1004 CA Police may give evidence that medic did not object to taking of breath test.

R v Newcastle-upon-Tyne Justices, ex parte Hindle; Hindle v Thynne [1984] 1 All ER 770; [1984] RTR 231 HC QBD Chance accused drank after stopped driving but before test requires acquittal unless prosecution disprove possibility; prohibition order where was confusing summons which obstructed/embarrassed accused and failed to detail one charge.

Woodage v Jones (No 2) (1975) 60 Cr App R 260; [1975] Crim LR 169; [1975] RTR 119; (1975) 119 SJ 304 HC QBD Offender properly convicted of being drunk and in charge of motor vehicle though was half-mile from vehicle when arrested.

DRIVING WITHOUT AUTHORITY (general)

Boldizsar v Knight [1980] Crim LR 653; [1981] RTR 136 HC QBD Person accepting lift in uninsured car which discovered was being driven without owner's consent guilty/not guilty of allowing self to be carried in same/using same without insurance.

DRIVING/ATTEMPTING TO DRIVE (general)

Adams v Valentine [1975] Crim LR 238; [1975] RTR 563 HC QBD Constable having validly stopped car for other reason could then form opinion that driver had been drinking and seek to administer breath test (driver still deemed to be driving).

Anthony v Jenkins [1972] Crim LR 596; [1971] RTR 19 HC QBD Person who had stopped car and run away from it was not driving when constable formed opinion that had been drinking alcohol.

Burgoyne v Phillips [1983] Crim LR 265; (1983) 147 JP 375; [1983] RTR 49; [1982] TLR 559 HC QBD On what constitutes 'driving' a car.

Caise v Wright; Fox v Wright [1981] RTR 49 HC QBD Person steering towed vehicle was 'driving' so could be convicted of driving while disqualified.

Campbell v Tormey [1969] 1 All ER 961; (1969) 53 Cr App R 99; [1969] Crim LR 150t; (1969) 133 JP 267; (1968) 118 NLJ 1196t; (1968) 112 SJ 1023; [1969] 1 WLR 189 HC QBD Cannot ask person who has fully completed drive to take breath test; must arrest person before can request breath test/laboratory specimen.

Gifford v Whittaker [1942] 1 All ER 604; (1942) 106 JP 128; [1942] 1 KB 501; [1942] 111 LJ 461; (1942) 166 LTR 324; (1942) 86 SJ 154; (1941-42) LVIII TLR 195; (1942) WN (I) 85 HC KBD Driver was 'using' lorry even though did so under instructions of foreman.

Gunnell v Director of Public Prosecutions [1993] Crim LR 619; [1994] RTR 151 HC QBD Person fully controlling motion/direction of moped (engine of which not running) was driving same.

Haines v Roberts [1953] 1 All ER 344; (1953) 117 JP 123; (1953) 97 SJ 117; [1953] 1 WLR 309 HC QBD Person in charge of vehicle until places in charge of another.

Harman v Wardrop (1971) 55 Cr App R 211; [1971] Crim LR 242t; (1971) 121 NLJ 130t CA Could not be convicted of drunk driving where had been impeded from so driving.

Hay v Shepherd [1974] Crim LR 114; (1973) 123 NLJ 1137t; [1974] RTR 64 HC QBD Person stopped without reasonable suspicion on part of officer that have been drinking can validly be breathalysed if officer then immediately forms reasonable suspicion.

Jones v Brooks and Brooks (1968) 52 Cr App R 614; [1968] Crim LR 498 HC QBD On elements necessary to establish attempt.

Jones v Pratt [1983] RTR 54 HC QBD Passenger briefly seizing control of wheel could be (but here was not) 'driving'.

Jones v Prothero [1952] 1 All ER 434; (1952) 116 JP 141; (1952) 102 LJ 121 HC QBD Driver is driver from taking of vehicle on road to completion of trip.

Kelly v Hogan [1982] Crim LR 507; [1982] RTR 352 HC QBD Person unfit to drive sitting into car and placing wrong key in ignition switch was attempting to drive.

Leach and another v Director of Public Prosecutions [1993] RTR 161 HC QBD Person sitting up in driver's seat with hands on wheel and engine on but car not moving was not driving; police officer requesting licence in those circumstances was acting in course of duty as reasonably believed driver to have committed offence.

McKoen v Ellis [1987] Crim LR 54; (1987) 151 JP 60; [1987] RTR 26 HC QBD On what constitutes 'driving' (here on motorcycle).

McQuaid v Anderton [1980] 3 All ER 540; (1980) 144 JP 456; [1980] RTR 371; (1981) 125 SJ 101; [1981] 1 WLR 154 HC QBD Steering/braking a car being towed is 'driving'.

Mendham v Lawrence [1972] Crim LR 113; [1972] RTR 153; (1972) 116 SJ 80 HC QBD Person in driver seat of car just stopped was a 'person driving'.

Newsome v Hayton [1974] Crim LR 112; (1973) 123 NLJ 1137t; [1974] RTR 9 HC QBD Person not driving/attempting to drive where stopped by constable over licence and minute passed before constable suspected driver of driving with excess alcohol.

Northfield v Pinder [1968] 3 All ER 854; (1969) 53 Cr App R 72; [1969] Crim LR 96t; (1969) 133 JP 107; (1968) 118 NLJ 1053t; [1969] 2 QB 7; (1968) 112 SJ 884; [1969] 2 WLR 50 HC QBD Accused in charge of motor vehicle while drunk must show no chance of driving car while over blood-alcohol limit.

Purvis v Hogg [1969] Crim LR 379; (1969) 113 SJ 388 HC QBD Person slumped over car wheel had not ceased to be driver and become instead the person in charge of the car.

R v Arnold [1964] Crim LR 664 Sessions Was not 'driving' to be steering one motor vehicle being pushed along by another.

R v Bates [1973] 2 All ER 509; (1973) 57 Cr App R 757; [1973] Crim LR 449t; (1973) 137 JP 547; (1973) 123 NLJ 493t; [1973] RTR 264; (1973) 117 SJ 395; [1973] 1 WLR 718 CA Whether driving/attempting to drive ultimately a matter for jury.

R v Bove [1970] 2 All ER 20; [1970] Crim LR 353; (1970) 134 JP 418 HC QBD Person not driving when refused breath test was unlawfully arrested.

R v Bove (1970) 54 Cr App R 316; [1970] Crim LR 471; [1970] RTR 261; (1970) 114 SJ 418; [1970] 1 WLR 949 CA Person not driving when refused breath test was unlawfully arrested.

R v Brown (Michael) (1977) 64 Cr App R 231; [1977] Crim LR 291t; (1977) 127 NLJ 114t; [1977] RTR 160 CA Was valid for policeman having formed opinion that defendant had excess alcohol taken to require breath test of him after detaining him but before arresting him.

R v Cooper [1974] RTR 489 CA On whether/when handing over of car-keys to constable constituted ceasing driving.

R v Farrance (Gordon Arthur) (1978) 67 Cr App R 136; [1978] Crim LR 496; [1978] RTR 225 CA Was attempt to drive where person in driving seat sought to mechanically propel car (though car did not move).

R v Garforth [1970] Crim LR 704t; (1970) 120 NLJ 945t; (1970) 114 SJ 770 CA On what constitutes 'driving' a vehicle (locking it up to make it safe from theft did not).

R v Gready [1974] Crim LR 314; [1974] RTR 16 CA Person properly convicted of drunk driving where stopped by constable who then almost immediately formed opinion that driver had been drinking.

R v Guttridge [1973] Crim LR 314; [1973] RTR 135 CA Issue whether when breath test administered person was driving (if controverted) to be left to jury.

R v Herd (1973) 57 Cr App R 560; [1973] Crim LR 315t; (1973) 123 NLJ 321t; [1973] RTR 165 CA Still driving where between car being stopped/breath test requested opened car boot to show constable were not conveying stolen sheep.

R v Ingram (1975) 125 NLJ 461t; [1977] RTR 420 CA Direction that person driving as matter of law if police evidence accepted improper (as was not matter of law) but allowed stand as no unfairness resulted.

R v Kelly [1970] 2 All ER 198; (1970) 54 Cr App R 334; [1970] Crim LR 352t; (1970) 134 JP 482; (1970) 120 NLJ 384t; [1970] RTR 301; (1970) 114 SJ 357; [1970] 1 WLR 1050 CA Person not driving when voluntarily stopped car, got out to make telephone call and was questioned on different matter.

R v Kitson (Herbert) (1955) 39 Cr App R 66; [1955] Crim LR 436t CCA Was driving under influence under drink where accused who had been drinking awoke in passenger seat to find car moving of own accord and sought to control it.

R v MacDonagh [1974] 2 All ER 257; [1974] Crim LR 317; (1974) 138 JP 488; (1974) 124 NLJ 222t; [1974] QB 448; [1974] RTR 372; (1974) 118 SJ 222; [1974] 2 WLR 529 CA 'Driving' must be driving within common understanding.

R v Mallender [1975] Crim LR 725; [1975] RTR 246 CA Order to re-take driving test quashed (but three year disqualification continued) for driver guilty of driving with excess alcohol but whose competency as a driver was not in issue.

R v Masters [1972] Crim LR 644t; [1972] RTR 492; (1972) 116 SJ 667 CA Person stopped by plain clothes officer (who took ignition keys) on suspicion that had no authority to be driving car not driving/attempting to drive when uniformed police constable arrived some minutes later to administer breath test.

R v Mitcham [1974] Crim LR 483; [1974] RTR 205 CA Breath test valid though issue had not been left to jury whether had been break in sequence of events from constable's suspecting driver had alcohol in body to his breathalysing latter.

R v Neilson [1978] Crim LR 693; [1978] RTR 232 CA Conviction quashed where inadequate direction of jury as to whether car could be driven (and so whether accused could be guilty of driving same).

R v Price [1968] 3 All ER 814; (1969) 53 Cr App R 25; (1969) 133 JP 47; [1968] 1 WLR 1853 CA Lawful breath test request if at time any driving offence being committed; 'any person driving' can include person in stopped car/who has alighted from car.

R v Rees (Anthony George) (1977) 64 Cr App R 155; [1977] RTR 181 CA On stopping as an incident of driving.

R v Richardson (John) [1975] 1 All ER 905; (1975) 60 Cr App R 136; [1975] Crim LR 163t; (1975) 139 JP 362; (1975) 125 NLJ 17t; [1975] RTR 173; (1975) 119 SJ 152; [1975] 1 WLR 321 CA Crown must show accused drove/attempted to drive prior to specimen being requested to sustain conviction for refusing latter.

R v Wedlake [1978] RTR 529 CA On relevant time of driving for purposes of justifying breath test.

Reader v Bunyard (1987) 85 Cr App R 185; [1987] Crim LR 274; [1987] RTR 406 HC QBD Not clear whether steering towed vehicle that might not be used as mechanical motor vehicle in the future could justify driving while disqualified conviction.

Regan v Anderton [1980] Crim LR 245; (1980) 144 JP 82; [1980] RTR 126 HC QBD Person could still be deemed to be driving car (for purposes of taking breath test) though car not actually moving.

Richmond London Borough Council v Pinn and Wheeler Ltd [1989] Crim LR 510; [1989] RTR 354; (1989) 133 SJ 389 HC QBD Company could not be convicted of unlawful driving of goods vehicle.

Saycell v Bool [1948] 2 All ER 83; (1948) 112 JP 341; (1948) 98 LJ 315; (1948) LXIV TLR 421; (1948) WN (I) 232 HC KBD Freewheeling is 'driving'.

Thoms v Cooper (1986) 150 JP 53 HC QBD Not guilty of driving uninsured motor vehicle where vehicle at isuue was completely incapable of being directed by person ostensibly controlling it.

Tyler v Whatmore [1976] Crim LR 315; [1976] RTR 83 HC QBD Where one person steering and other person maintaining fuelled movement of vehicle both could be said to be driving.

Wright v Brobyn [1971] Crim LR 241; [1971] RTR 204; (1971) 115 SJ 310 HC QBD Delay in taking second breath specimen did not mean driver lost quality of driving.

DRUGGED/DRUNK DRIVING (general)

Armstrong v Clark [1957] 1 All ER 433; (1957) 41 Cr App R 56; [1957] Crim LR 256; (1957) 121 JP 193; (1957) 107 LJ 138; [1957] 2 QB 391; (1957) 101 SJ 208; [1957] 2 WLR 400 HC QBD Diabetic having injected insulin can be charged of driving under influence of drug.

Blakeley and another v Director of Public Prosecutions [1991] Crim LR 763; (1991) 141 NLJ 860t; [1991] RTR 405; [1991] TLR 288 HC QBD Recklessness an inappropriate consideration when deciding whether person had necessary mens rea to be guilty of procuring principal to commit offence.

Bradford v Wilson (1984) 78 Cr App R 77; [1983] Crim LR 482; (1983) 147 JP 573; [1984] RTR 116 HC QBD Driving after inhalation of toluene (element of glue) was driving under influence of drug.

Butcher v Catterall (1975) 61 Cr App R 221; (1975) 125 NLJ 621t; (1975) 119 SJ 508 CA Non-compliance with manufacturer's instructions did not render breath test useless.

Carter v Richardson [1974] Crim LR 190; [1974] RTR 314 HC QBD Supervising driver validly convicted of aiding and abetting learner driver in offence of driving with excess alcohol.

Chatters v Burke [1986] 3 All ER 168; (1986) 150 JP 581; (1986) 136 NLJ 777; [1986] RTR 396; (1970) 130 SJ 666; [1986] 1 WLR 1321 HC QBD Special reasons that might be raised in defence to charge of drunk driving.

Clark v Price (1984) 148 JP 55; [1983] TLR 419 HC QBD Proper that police man should have reasonable suspicion that defendant guilty of drunk driving where abandoned car in middle of night in remote area after followed by police.

Collins v Lucking [1983] Crim LR 264; (1983) 147 JP 307; [1983] RTR 312; [1982] TLR 545 HC QBD Drunk driving conviction quashed where rested on accused's being diabetic and was no evidence either way as to whether was/was not diabetic.

Corkery v Carpenter [1950] 2 All ER 745; (1950) 114 JP 481; [1951] 1 KB 102; (1950) 94 SJ 488; [1950] 66 (2) TLR 333; (1950) WN (I) 442 HC KBD Being drunk in charge of 'carriage' under Licensing Act 1872, s 12 includes bicycles.

Cronkshaw v Rydeheard [1969] Crim LR 492t; (1969) 113 SJ 673 HC QBD Successful appeal against drunk driving conviction where specimen handed to appellant capable of analysis only by way of gas chromotography.

Dawson v Lunn (1985) 149 JP 491; [1986] RTR 234; [1984] TLR 720 HC QBD Justices ought not to have dismissed drunk driving charge purely on basis of medical journal report produced by defence.

Dhillon (Surinder Singh) v Director of Public Prosecutions (1993) 157 JP 420 HC QBD Person not properly given option to provide blood/urine specimen where police interpreted vague answer as refusal of same.

Director of Public Prosecutions v Anderson (1991) 155 JP 157; [1990] RTR 269 HC QBD On what constitutes aiding and abetting another in committing offence of driving with excess alcohol.

Director of Public Prosecutions v Beech [1992] Crim LR 64; (1992) 156 JP 31; (1991) 141 NLJ 1004t; [1992] RTR 239; [1991] TLR 342 HC QBD Not reasonable excuse for non-provision of breath specimen that had made self so drunk could not understand what were instructed to do.

Director of Public Prosecutions v Davis; Director of Public Prosecutions v Pittaway [1994] Crim LR 600 HC QBD On test when determining whether duress present (here in context of offence of driving with excess alcohol).

Director of Public Prosecutions v Frost [1989] Crim LR 154; (1989) 153 JP 405; [1989] RTR 11 HC QBD On nature of offences of being in charge of vehicle while unfit to drive (matter which ordinary person could decide)/in charge with excess alcohol (provable by reference to expert evidence).

Director of Public Prosecutions v Johnson [1995] 4 All ER 53; [1994] Crim LR 601; (1994) 158 JP 891; [1994] TLR 144; [1995] RTR 9; [1995] 1 WLR 728 HC QBD Prescribed injection of painkiller containing alcohol was consumption of alcohol for purposes of Road Traffic Act 1988.

Director of Public Prosecutions v Lowden [1993] RTR 349; [1992] TLR 185 HC QBD Successful defence to excess alcohol charge whereby established it was extra alcohol consumed after finished driving which pushed driver over prescribed limit.

Director of Public Prosecutions v O'Connor; Director of Public Prosecutions v Allatt; Director of Public Prosecutions v Connor; Director of Public Prosecutions v Chapman; R v Crown Court at Chichester, ex parte Moss; Director of Public Prosecutions v Allen (1992) 95 Cr App R 135; (1991) 141 NLJ 1004t; [1992] RTR 66; [1991] TLR 335 HC QBD Guidelines on non-disqualification from driving of those convicted of drunk driving.

Director of Public Prosecutions v Parkin [1989] Crim LR 379 CA Once (could be oral) evidence as to calibration given machine deemed in proper functioning order unless is contrary evidence.

Director of Public Prosecutions v Singh [1988] RTR 209 HC QBD Could convict person for driving with excess alcohol who claimed (but did not adduce expert evidence to effect) that was post- not pre-driving drinking which led to blood alcohol exceeding prescribed limit.

Director of Public Prosecutions v Warren (Frank John) [1992] Crim LR 200; (1992) 156 JP 753; [1992] RTR 129 HC QBD Failed prosecution for driving with excess alcohol as accused not given adequate opportunity to comment on what type of specimen would prefer to give (where breathalyser not functioning properly).

Director of Public Prosecutions v Watkins [1989] 1 All ER 1126; (1989) 89 Cr App R 112; [1989] Crim LR 453; (1990) 154 JP 370; (1989) 139 NLJ 365; [1989] QB 821; [1989] RTR 324; (1989) 133 SJ 514; [1989] 2 WLR 966 HC QBD If prima facie case of being in charge of vehicle while drunk defendant must prove no chance of him driving vehicle.

Director of Public Prosecutions v Welsh (1997) 161 JP 57; (1996) TLR 15/11/96 HC QBD Rounding down of readings/deduction of 6mg from final alcohol reading are common practices not immutable rules.

Director of Public Prosecutions v Williams [1989] Crim LR 382 HC QBD Under Road Traffic Act 1972, s 10(2) (as amended) burden of proof on accused to show that when last driving level of alcohol in body lower than that contained in specimen provided.

Director of Public Prosecutions v Wilson (Eric Leslie) [1991] Crim LR 441; (1992) 156 JP 916; [1991] RTR 284; [1991] TLR 70 HC QBD Police could act on anonymous telephone call to apprehend person guilty of driving with excess alcohol.

Drake v Director of Public Prosecutions [1994] Crim LR 855; (1994) 158 JP 828; [1994] RTR 411; [1994] TLR 192 HC QBD Not guilty of being in charge of car with excess alcohol where car had been wheel-clamped.

Dryden v Johnson [1961] Crim LR 551t HC QBD Failed drunk driving prosecution sent back to justices with direction to convict as no reasonable bench of magistrates could have failed to convict.

Duck v Peacock [1949] 1 All ER 318; (1949) 113 JP 135; (1949) 99 LJ 49; (1949) LXV TLR 87; (1949) WN (I) 36 HC KBD Stopping car as soon as feel effects of drinking not special reason justifying non-disqualification.

Edkins v Knowles [1973] 2 All ER 503; (1973) 57 Cr App R 751; [1973] Crim LR 446; (1973) 137 JP 550; [1973] RTR 257; (1973) 123 NLJ 469t; [1973] QB 748; (1973) 117 SJ 395; [1973] 2 WLR 977 HC QBD Must form opinion that person driving in excess of limit when person is driving, not after.

Elkins v Cartlidge [1947] 1 All ER 829; (1947) 177 LTR 519; (1947) 91 SJ 573 HC KBD Parking enclosure a public place.

Gilligan v Wright [1968] Crim LR 276 Sessions That were only slightly in excess of proscribed blood-alcohol concentration was not a basis for non-disqualification.

Hallett v Newton (1951) 95 SJ 712 HC KBD On serious nature of drunk driving offence.

Harman v Wardrop (1971) 55 Cr App R 211; [1971] Crim LR 242t; (1971) 121 NLJ 130t CA Could not be convicted of drunk driving where had been impeded from so driving.

Havell v Director of Public Prosecutions [1993] Crim LR 621; (1994) 158 JP 680 HC QBD Private members' club car park not a public place.

Hirst v Wilson [1969] 3 All ER 1566; [1970] Crim LR 106; [1970] RTR 67; (1969) 113 SJ 906; [1970] 1 WLR 47 HC QBD Arrest for failing to provide proper specimen — though medical reason — lawful: drunk driving conviction sustained.

Hopper v Stansfield (1950) 114 JP 368 HC KBD That car was stopped/battery flat not special reasons justifying non-disqualification following drunk driving conviction.

Hoyle v Walsh (1969) 53 Cr App R 61 CA Not guilty of failing to provide breath specimen where failure arose through defect in breathalyser — arrest without warrant for same therefore invalid.

John v Bentley [1961] Crim LR 552; (1961) 105 SJ 406 HC QBD Conviction for being drunk in charge of motor vehicle quashed where was so drunk that could not have driven same.

John v Humphreys (1955) 119 JP 309; (1955) 105 LJ 202; (1955) 99 SJ 222; [1955] 1 WLR 325 HC QBD Prosecution/defendant not required to/must prove that defendant possesses licence.

Jones v English [1951] 2 All ER 853; (1951) 115 JP 609; (1951) 101 LJ 625; (1951) 95 SJ 712; [1951] 2 TLR 973; (1951) WN (I) 552 HC KBD Must be/hear evidence to back special reasons justifying non-disqualification.

Lafferty v Director of Public Prosecutions [1995] Crim LR 429 HC QBD Roadside breath test evidence validly admitted (though two police station breath samples unchallenged) as was pertinent to issue raised by driver of how much he had drunk.

Leach v Evans [1952] 2 All ER 264 HC QBD What constitutes being 'in charge' of motor vehicle.

Lomas v Bowler [1984] Crim LR 178 HC QBD Way in which defendant drove considered in tandem with conflicting analyses of urine samples meant drunk driving conviction valid.

Marshall v Lloyd [1956] Crim LR 483 Magistrates Successful prosecution of individual for being drunk in the highway contrary to the Licensing Act 1872, s 12.

Matto (Jit Singh) v Director of Public Prosecutions [1987] Crim LR 641 HC QBD Bad faith behaviour of police at defendant's property tainted provision of breath specimen: drunk driving conviction quashed.

Morton v Confer [1963] 2 All ER 765; [1963] Crim LR 577; (1963) 127 JP 433; (1963) 113 LJ 530; (1963) 107 SJ 417; [1963] 1 WLR 763 HC QBD Accused to show (civil burden) that had no intention to drive until fit to do so; claim to be considered in light of all the facts.

Norman v Magill [1972] RTR 81 HC QBD Issue whether were driving when approached by police officer not relevant when arrested for being in charge of car while unfit to drive.

Northfield v Pinder [1968] 3 All ER 854; (1969) 53 Cr App R 72; [1969] Crim LR 96t; (1969) 133 JP 107; (1968) 118 NLJ 1053t; [1969] 2 QB 7; (1968) 112 SJ 884; [1969] 2 WLR 50 HC QBD Accused in charge of motor vehicle while drunk must show no chance of driving car while over blood-alcohol limit.

Oswald (Robert John) v Director of Public Prosecutions (1989) 153 JP 590; [1989] RTR 360 HC QBD Where prosecution/defence rounded-down figures (rounded down to allow for error differential) were 82/80mg alcohol per ml of blood could be convicted of driving with over 80mg of alcohol per ml of blood.

Patterson v Charlton [1985] Crim LR 449; (1986) 150 JP 29; [1986] RTR 18 HC QBD Admission that were driving is enough proof of same; burden is on person over the blood-alcohol limit to prove had not driven while so.

Pinner v Everett [1969] 3 All ER 257; (1977) 64 Cr App R 160; [1969] Crim LR 607; (1969) 133 JP 653; [1970] RTR 3; (1969) 113 SJ 674; [1969] 1 WLR 1266 HL On when (unlike here) person may be said to be driving car.

Pinner v Everett [1969] Crim LR 378t; (1969) 119 NLJ 438 HC QBD On when (unlike here) person may be said to be driving car.

Police v Liversidge and Featherstone [1956] Crim LR 59 Magistrates Two parties guilty of being drunk in charge of same motor vehicle.

Prince v Director of Public Prosecutions [1996] Crim LR 343 HC QBD Unnecessary to prove calibration of breathalyser correct where prosecution only seek to rely on blood specimen evidence.

Purvis v Hogg [1969] Crim LR 379; (1969) 113 SJ 388 HC QBD Person slumped over car wheel had not ceased to be driver and become instead the person in charge of the car.

R v Arnold [1964] Crim LR 664 Sessions Was not 'driving' to be steering one motor vehicle being pushed along by another.

R v Ayres [1970] Crim LR 114; (1970) 114 SJ 16 CA Drunk driving conviction quashed where issue whether defendant had been properly arrested was not left to the jury.

R v Beaumont [1964] Crim LR 665; (1964) 114 LJ 739 CCA Occupation road deemed not to road on which could be guilty of offence of drunk driving.

R v Cambridge Magistrates' Court and the Crown Prosecution Service, ex parte Wong (1992) 156 JP 377; [1992] RTR 382 HC QBD Where taking prescribed cough medicine had unaware to person pushed them slightly over alcohol limit this could be special reason justifying non-disqualification — whether should be non-disqualification a matter for discretion of justices.

R v Chichester Justices, ex parte Crouch (1982) 146 JP 26 HC KBD Secretary of State ought not to have been notified accused suffered disease where counsel claimed at accused's first drunk driving prosecution that accused had a drink problem.

R v Clwd Justices, ex parte Charles (1990) 154 JP 486 HC QBD Drunk driving conviction quashed where accused had not been given option of providing blood/urine specimens.

R v Court [1962] Crim LR 697 CCA Improper to admit medical evidence as to drunkenness where obtained by doctor under assurance to accused that would not be used at later trial.

R v Coward (Darnley Alton) (1976) 63 Cr App R 54; [1976] RTR 425 CA On adequacy of urine specimen.

R v Curran [1975] 2 All ER 1045; [1975] Crim LR 464t; (1975) 139 JP 631; (1975) 125 NLJ 554/600t; [1975] RTR 445; (1975) 119 SJ 425; [1975] 1 WLR 876 CA If lawful arrest can be convicted of failure to give specimen even if acquitted of drunk driving.

R v Davies [1962] 3 All ER 97; (1962) 46 Cr App R 292; [1962] Crim LR 547; (1962) 126 JP 455; (1962) 112 LJ 569; (1962) 106 SJ 393; [1962] 1 WLR 1111 CMAC Ordinary witness may testify in belief that accused had drunk but cannot testify as to driving capacity.

R v Dick-Cleland [1965] Crim LR 440t; (1965) 109 SJ 377 CCA Misdirection to tell jury that must (rather than could) hold against defendant fact that had refused to supply urine specimen.

R v Dooley [1964] 1 All ER 178; [1964] Crim LR 315; (1964) 128 JP 119; (1964) 114 LJ 106; (1964) 108 SJ 384; [1964] 1 WLR 648 CrCt Analyst's certificate admissible/inadmissible where sample requested by doctor/police officer and portion offered/not offered to accused.

R v Ealing Stipendiary Magistrate, ex parte Woodman [1994] Crim LR 372; (1994) 158 JP 997; [1994] RTR 189 HC QBD Non-disqualification not merited for diabetic negligent as to taking insulin/careless as to blood-sugar levels who was found guilty of driving while unfit through drugs.

R v Hamilton [1970] 3 All ER 284; [1970] Crim LR 651t; (1970) 120 NLJ 968; [1970] RTR 417 CA Analyst's blood certificate when alcohol obtained after driving is not certificate of blood whilst driving.

R v Hawkes (James Albert) (1930-31) Cr App R 172; (1931) 75 SJ 247 CCA Guilty of drunk driving under Road Traffic Act 1930, s 15(1) if were under influence of drink and were incapable of controlling vehicle.

R v Hegarty [1977] RTR 337 CA Drunk driving conviction quashed where prosecution failed to explain discrepancy between doctor's clinical examination of accused and later specimen results.

R v Hillman [1977] Crim LR 752; [1978] RTR 124 CA On refusal to provide laboratory specimens as evidence that drove while unfit to drive through drink.

R v Hunt (Reginald) [1980] RTR 29 CA Jury could infer impaired driving ability from collision with fully visible stationary vehicle/that intoxication led to impairment where blood-alcohol concentration over twice the prescribed limit.

R v Kitson (Herbert) (1955) 39 Cr App R 66; [1955] Crim LR 436t CCA Was driving under influence under drink where accused who had been drinking awoke in passenger seat to find car moving of own accord and sought to control it.

R v Kwame (Emmanuele) (1975) 60 Cr App R 65; [1974] Crim LR 676; [1975] RTR 106 CA Condition of bail that party not to drive before trial was valid and was not special reason later justifying non-disqualification.

R v Lawrence [1973] 1 All ER 364; (1973) 57 Cr App R 285; [1973] Crim LR 242; (1973) 137 JP 168; [1973] RTR 64; (1973) 117 SJ 225; [1973] 1 WLR 329 CA State of car relevant in deciding whether it c/would have been driven by accused while drunk.

R v McGall [1974] Crim LR 482; [1974] RTR 216 CA Slow driving on empty road and mistaken use of indicators could have prompted reasonable suspicion that driver had excess alcohol.

R v McKenzie [1971] 1 All ER 729; [1972] Crim LR 655; (1971) 135 JP 26 Assizes Invalid recognisance could not be subject of forgery; recognisance invalid where imposed on person who would not be detained.

R v Mills [1974] RTR 215 CA Two year disqualification for driving with excess alcohol was not excessive.

R v Mitten [1965] 2 All ER 59; (1965) 49 Cr App R 216; (1965) 129 JP 371; (1965) 115 LJ 316; [1966] 1 QB 10; [1965] 3 WLR 268 CCA Non-compliance with law in requesting specimen need not mean evidence inadmissible; police to offer to supply part of specimen when requesting it/shortly before.

R v Moore (Richard) [1975] Crim LR 722; [1975] RTR 285 CA Onus on defendant to establish defence that arrest invalid; jury could reasonably decide that was adequate connection between specimen analysed and driving behaviour three hours prior to giving specimen.

R v Morris [1972] 1 All ER 384; (1972) 56 Cr App R 175; [1972] Crim LR 116; (1972) 136 JP 194; (1971) 121 NLJ 1074t; [1972] RTR 201; (1972) 116 SJ 17; [1972] 1 WLR 228 CA 'Accident' in Road Safety Act 1967, s 2(2), to be given ordinary meaning.

R v Nowell [1948] 1 All ER 794; (1946-48) 32 Cr App R 173; (1948) 112 JP 255; (1948) 98 LJ 245; (1948) LXIV TLR 277; (1948) WN (I) 154 CCA Police doctor's persuading person to give sample as in their own interest did not make doctor's evidence inadmissible.

R v Palfrey; R v Sadler [1970] 2 All ER 12; (1970) 54 Cr App R 217; [1970] Crim LR 231t; (1970) 134 JP 397; (1970) 120 NLJ 81t; [1970] RTR 127; (1970) 114 SJ 92; [1970] 1 WLR 416 CA Practice regarding blood tests.

R v Payne [1963] 1 All ER 848; (1963) 47 Cr App R 122; [1963] Crim LR 288; (1963) 127 JP 230; (1963) 113 LJ 285; (1963) 107 SJ 97; [1963] 1 WLR 637 CCA Evidence of doctor though admissible ought not to be admitted where person submits to medical examination on basis of express assurance that results would not be admissible in evidence.

R v Presdee (Herbert) (1927-28) 20 Cr App R 95 CCA Drunkeness' an issue of fact for jury.

R v Price [1968] 3 All ER 814; (1969) 53 Cr App R 25; (1969) 133 JP 47; [1968] 1 WLR 1853 CA Lawful breath test request if at time any driving offence being committed; 'any person driving' can include person in stopped car/who has alighted from car.

R v Pursehouse [1970] 3 All ER 218; (1970) 54 Cr App R 478; [1970] Crim LR 651; (1970) 134 JP 682; (1970) 120 NLJ 920; [1970] RTR 494 CA Police need not specify statutory requirements after two refusals to give any sample.

R v Rees (Anthony George) (1977) 64 Cr App R 155; [1977] RTR 181 CA On stopping as an incident of driving.

R v Reid (Philip) [1973] 3 All ER 1020; (1973) 57 Cr App R 807; [1973] Crim LR 760; (1974) 138 JP 51; [1973] RTR 536; (1972) 116 SJ 565; (1973) 117 SJ 681; [1973] 1 WLR 1283 CA That get out of car need not mean not driving/attempting to drive; mistaken belief that not obliged to give specimen not failure for not doing so; whether was refusal a factual question for jury.

R v Richardson (John) [1975] 1 All ER 905; (1975) 60 Cr App R 136; [1975] Crim LR 163t; (1975) 139 JP 362; (1975) 125 NLJ 17t; [1975] RTR 173; (1975) 119 SJ 152; [1975] 1 WLR 321 CA Crown must show accused drove/attempted to drive prior to specimen being requested to sustain conviction for refusing latter.

R v Seward [1970] 1 All ER 329; (1970) 54 Cr App R 85; [1970] Crim LR 113t; (1970) 134 JP 195; (1969) 119 NLJ 1069t; [1970] RTR 102; (1969) 113 SJ 984; [1970] 1 WLR 323 CA Matter for jury whether presence of car on road prompted accident.

R v Sharman [1974] Crim LR 129; [1974] RTR 213 CA Two year disqualification for driving with excess alcohol was not excessive.

R v Shaw (Kenneth) [1975] RTR 160 CA Drunk driving conviction of person steering towed vehicle quashed.

R v Stimpson [1977] Crim LR 114 CA Twelve months' imprisonment, order to resit driving test plus three (not five) years' disqualification for driving with excess alcohol.

R v Vaughan [1976] RTR 184 CA Analyst's certificate admissible though only gave urine-alcohol concentration of specimen and not also blood-alcohol concentration.

R v Veevers [1971] Crim LR 174; [1971] RTR 47; (1971) 115 SJ 62 CA Drunk driving conviction quashed where issue whether arrest valid (on which validity of breath test in part depends) not left to jury.

R v Way [1970] Crim LR 469; (1970) 120 NLJ 481; [1970] RTR 348; (1970) 114 SJ 418 CMAC That person's breath smelt of alcohol was not per se evidence that driving ability was impaired.

R v Weir [1972] 3 All ER 906; (1973) 137 JP 11 CrCt Not required to take breath test if arrested for non-traffic offence and not then told being detained for traffic offence.

R v Whitlow [1965] Crim LR 170 CrCt Manual moving of motor cycle from stationary position was 'driving' of same.

R v Wilson (Richard) [1975] RTR 485 CA Breathalyser evidence obtained in relation to accused's alleged drunk driving of one car but pertinent to his drunk driving of a second car was admissible in evidence on second charge.

R v Woodward [1995] 3 All ER 79; [1995] 2 Cr App R 388; [1995] Crim LR 487; (1995) 159 JP 349; [1995] RTR 130; (1995) 139 SJ LB 18; [1994] TLR 632; [1995] 1 WLR 375 CA That a driver was adversely affected by alcohol is relevant to death by dangerous driving charge; that he merely consumed alcohol is not; failure to warn jury of this latter point a misdirection.

Rowlands v Hamilton [1971] 1 All ER 1089; (1971) 55 Cr App R 347; [1971] Crim LR 366t; (1971) 135 JP 241; [1971] RTR 153; (1971) 115 SJ 268; [1971] 1 WLR 647 HL Specimen test inadmissible if drank after stopped driving nor evidence to show would have been over limit anyway.

Rynsard v Spalding [1985] Crim LR 795; [1986] RTR 303 HC QBD Failed prosecution of person charged with driving after drinking where only rose above prescribed limit after ceased driving.

Sandy v Martin [1974] Crim LR 258; (1975) 139 JP 241; [1974] RTR 263 HC QBD Inn car park a public place during hours that persons invited to use it but absent contrary evidence not thereafter (so were not drunk and in charge of motor vehicle in public place).

Sharpe v Perry [1979] RTR 235 HC QBD Need not be roadside breath test when arresting person for being in charge of motor vehicle on road while unfit to drink (Road Traffic Act 1972, s 5(5)).

Sheldon v Jones [1970] Crim LR 38t; (1969) 119 NLJ 997t; [1970] RTR 38; (1969) 113 SJ 943 HC QBD Man (full licence-holder) with excess alcohol supervising wife (provisional licence-holder) could prove was no chance of his driving.

Smith v Mellors and Soar (1987) 84 Cr App R 279; [1987] Crim LR 421; [1987] RTR 210 HC QBD Could convict both driver and passenger of driving with excess alcohol where one aided and abetted another without having to prove who drove/was passenger.

Stevens v Thornborrow [1969] 3 All ER 1487; [1970] Crim LR 41t; (1970) 134 JP 99; [1970] RTR 31; (1969) 113 SJ 961; [1970] 1 WLR 23 HC QBD Person is driving after vehicle stops if what is doing is connected with driving.

Taylor v Austin [1969] Crim LR 152; (1969) 133 JP 182; (1968) 112 SJ 1024; [1969] 1 WLR 264 HC QBD That excess alcohol did not affect driving ability/were not responsible for accident/would suffer greatly financially did not justify non-disqualification.

Thomson v Knights [1947] 1 All ER 112; (1947) 111 JP 43; [1947] KB 336; [1947] 116 LJR 445; (1947) 176 LTR 367; (1947) 91 SJ 68; (1947) LXIII TLR 38; (1947) WN (I) 37 HC KBD Conviction for driving while 'under the influence of drink or a drug' not void for vagueness.

Walker v Lovell [1975] 3 All ER 107; (1978) 67 Cr App R 249; [1975] Crim LR 720; (1975) 139 JP 708; (1975) 125 NLJ 820t; [1975] RTR 377; (1975) 119 SJ 544; [1975] 1 WLR 1141 HL Where enough breath to provide specimen arrest for failure to provide specimen unlawful and this being so there can be no conviction for drunk driving.

Walker v Lovell [1975] Crim LR 102; (1975) 125 NLJ 43; [1975] RTR 61; (1975) 119 SJ 258 HC QBD Where enough breath to provide specimen arrest for failure to provide specimen unlawful and this being so there can be no conviction for drunk driving.

Wareing (Alan) v Director of Public Prosecutions (1990) 154 JP 443 [1990] TLR 41 HC QBD Dismissal of drunk driving charge on basis that defendant not given option by police of providing urine specimen was not dismissal on basis of technicality.

Wiltshire v Barrett [1965] 2 All ER 271; (1965) 129 JP 348; [1966] 1 QB 312; (1965) 109 SJ 274; [1965] 2 WLR 1195 CA Valid arrest if police officer has reasonable suspicion are committing road traffic offence/are released once discover innocent or no case.

Winter v Barlow [1980] Crim LR 51; (1980) 144 JP 77; [1980] RTR 209 HC QBD That constable did not have valid reason for stopping appellant did not mean magistrates could not hear drunk driving/failure to provide breath specimen actions.

Woodage v Jones [1975] Crim LR 47 HC QBD On when person may correctly be stated to be in charge of motor vehicle.

DUTY TO STOP/REPORT (general)

Beard v Wood [1980] Crim LR 384; [1980] RTR 454 HC QBD Road Traffic Act 1972, s 159 empowered uniformed police constable to stop traffic independent of any common law powers.

Bentley v Mullen [1986] RTR 7 HC QBD Supervisor of learner driver convicted of aiding and abetting latter in not stopping after accident occurred.

Bulman v Bennett [1974] Crim LR 121; (1973) 123 NLJ 1113t; [1974] RTR 1; (1973) 117 SJ 916 HC QBD Successful prosecution for failure to report accident as soon as was reasonably practicable.

Bulman v Lakin [1981] RTR 1 HC QBD Absent explanatory reason ten-hour delay in reporting accident was not reporting of same as soon as reasonably practicable.

Butler v Whittaker [1955] Crim LR 317 Sessions Person hearing knock as van touched one of herd of cows on road ought to have stopped to ascertain whether (as here) had caused injury.

Dawson v Winter (1931-34) XXIX Cox CC 633; (1933) 149 LTR 18; (1933) 77 SJ 29; (1932-33) XLIX TLR 128 HC KBD Not defence to refusing personal details to interested party that reported accident to police inside time limit.

Director of Public Prosecutions v Bennett [1993] Crim LR 71; (1993) 157 JP 493; [1993] RTR 175; [1992] TLR 250 HC QBD Failing to stop and failing to give address a single offence.

Director of Public Prosecutions v Drury (1989) 153 JP 417; [1989] RTR 165 HC QBD If person unaware of car accident becomes aware of same within twenty-four hours is under duty to report it to police.

Green v Dunn [1953] 1 All ER 550 HC QBD Person giving all details to other driver but not telling police within twenty-four hours not guilty of offence.

Grew v Cubitt (1951) 95 SJ 452; [1951] 2 TLR 305 HC KBD Failed prosecution of driver involved in road traffic accident who did not stop as was inadequate evidence that accused's vehicle had been involved.

Harding v Price [1948] 1 All ER 283; [1948] 1 KB 695 HC KBD Where driver does not know of accident is not guilty of failure to report it.

Johnson v Finbow (1983) 5 Cr App R (S) 95; [1983] Crim LR 480; (1983) 147 JP 563; [1983] RTR 363; (1983) 127 SJ 411; [1983] TLR 203; [1983] 1 WLR 879 HC QBD Not stopping after accident and not reporting same though at different times were on same occasion for purposes of Transport Act 1981, s 19(1): on appropriate penalty points for same.

Lee v Knapp [1966] 3 All ER 961; [1967] Crim LR 182; (1967) 131 JP 110; (1965-66) 116 NLJ 1712; [1967] 2 QB 442; (1966) 110 SJ 981; [1967] 2 WLR 6 HC QBD Duty to stop after accident involves stopping for as long as is needed for necessary information to be recorded by authorities.

Lodwick v Sanders [1985] 1 All ER 577; (1985) 80 Cr App R 304; [1985] Crim LR 210; [1985] RTR 385; (1985) 129 SJ 69; [1984] TLR 686; [1985] 1 WLR 382 HC QBD Police constable stopping vehicle suspected was stolen could stop it for as long as necessary to arrest driver and explain basis for arrest.

McDermott v Director of Public Prosecutions (1997) 161 JP 244; (1996) TLR 27/11/96 HC QBD Issue of fact whether requirement that stop after accident has been complied with.

Mutton v Bates (1983) 147 JP 459; [1984] RTR 256; [1983] TLR 490 HC QBD On duty under Road Traffic Act 1972, s 25 to stop as soon as practicable after accident and report it to the police.

North v Gerrish [1959] Crim LR 462t; (1959) 123 JP 313 HC QBD Must both stop and give name and address after motor accident if are not to contravene Road Traffic Act 1930, s 22(1).

Pagett v Mayo [1939] 2 All ER 362; (1939-40) XXXI Cox CC 251; (1939) 103 JP 177; [1939] 2 KB 95; (1939) 87 LJ 303; [1939] 108 LJCL 501; (1939) 160 LTR 398; (1939) 83 SJ 418; (1938-39) LV TLR 598; (1939) WN (I) 140 HC KBD Need not report accident where only damage done was to property.

Peek v Towle [1945] 2 All ER 611; [1945] KB 458; (1945) 109 JP 160; [1945] 114 LJ 540; (1945) 173 LTR 360; (1944-45) LXI TLR 399 HC KBD On duty to report car accident to police as soon as is reasonably possible within twenty-four hours (Road Traffic Act 1930, s 22(2)).

Quelch v Phipps [1955] 2 All ER 302; [1955] Crim LR 382; (1955) 119 JP 430; (1955) 105 LJ 298; [1955] 2 QB 107; (1955) 99 SJ 355; [1955] 2 WLR 1067 HC QBD Must report accident if direct connection between vehicle/occurrence: bus driver must report that person fell from bus.

R v Criminal Injuries Compensation Board, ex parte Carr [1980] Crim LR 643; [1981] RTR 122 HC QBD Person beginning to but not succeeding in leaving scene of accident had not contravened prohibition on leaving scene of accident.

Roper v Sullivan [1978] Crim LR 233; [1978] RTR 181 HC QBD Not giving name and address to other driver/not reporting accident to police are two separate offences; statement of facts not made by witness (being cross-examined) not to be put to same as being evidence disparate from court testimony.

Selby v Chief Constable of Avon and Somerset [1988] RTR 216 HC QBD Once proved that person was involved in accident and did not report it to police, burden falls on person to prove did not know of accident.

EMERGENCY VEHICLE (general)

Director of Public Prosecutions v Hawkins [1996] RTR 160 CA Not offence to use emergency vehicle for alternative purpose when blue light not lit.

EMISSION OF OIL FROM LORRY (general)

Tidswell v Llewellyn [1965] Crim LR 732t HC QBD ROAD Restoration of conviction for emission of oil from lorry where 'spilling' occurred on private land but dripping continued on public road.

EXPRESS CARRIAGE (general)

Birmingham and Midland Motor Omnibus Company Limited v Nelson [1932] All ER 351; (1931-34) XXIX Cox CC 529; (1932) 96 JP 385; [1933] 1 KB 188; (1933) 102 LJCL 47; (1932) 147 LTR 435; (1931-32) XLVIII TLR 620; (1932) WN (I) 200 HC KBD Conviction for unlicensed use of coaches as express carriages (not for conveying private party/for special event).

Browning v JWH Watson (Rochester), Ltd [1953] 2 All ER 775; (1953) 117 JP 479; (1953) 103 LJ 590; [1953] 1 WLR 1172 HC QBD 'Special occasion' means special to locality; no defence that did not know were not conveying private party — were under duty to so ensure.

Drew v Dingle [1933] All ER 518; (1934) 98 JP 1; [1934] 1 KB 187; (1933) 76 LJ 304; (1933) 77 SJ 799; (1933-34) L TLR 101; (1933) WN (I) 255; (1934-39) XXX Cox CC 53; (1934) 103 LJCL 97; (1934) 150 LTR 219 HC KBD Payment of fares meant vehicle unlawfully used as express carriage (even though passengers each carrying produce).

East Midland Area Traffic Commissioners v Tyler (1938) 82 SJ 416 HC KBD Person ferrying co-workers to/from work in own car in return for contributions from each to running expenses was operating vehicle for hire/reward at separate fares.

Evans v Hassan and another [1936] 2 All ER 107; (1936) 81 LJ 384; (1936) 80 SJ 409 HC KBD 'Special occasion' provisions of Road Traffic Act 1930, s 72 requires no mens rea to be contravened; to be special occasion must conform to provisions of Road Traffic Act 1934, s 25.

Lyons v Denscombe [1949] 1 All ER 977; (1949) 99 LJ 259; (1949) 113 JP 305; (1949) WN (I) 257 HC KBD Weekly lump sum payment from fellow employees for conveying home not use as express carriage.

Nelson v Blackford [1936] 2 All ER 109 HC KBD Special occasion to be same must be limited in time to a few days.

Newell v Cross; Newell v Cook; Newell v Plume; Newell v Chenery [1936] 2 All ER 203; (1934-39) XXX Cox CC 437; (1936) 100 JP 371; [1936] 2 KB 632; (1936) 81 LJ 437; [1936] 105 LJCL 742; (1936) 155 LTR 173; (1936) 80 SJ 674; (1935-36) LII TLR 489 HC KBD Passengers paying portion of hackney carriage fare a contravention of Road Traffic Act 1930, ss 61 and 72.

Reynolds v GH Austin and Sons, Ltd [1951] 1 All ER 606; (1951) 115 JP 192; [1951] 2 KB 135; (1951) 101 LJ 135; (1951) 95 SJ 173; [1951] 1 TLR 614; (1951) WN (I) 135 HC KBD Not guilty of offence by doing act not unlawful in itself made unlawful by actions of third party (non-servant/agent) unknown to self.

Wurzal v Dowker [1953] 2 All ER 88; (1953) 117 JP 336; (1953) 103 LJ 349; [1954] 1 QB 52; (1953) 97 SJ 390; [1953] 2 WLR 1196 HC QBD 'Special occasion' means occasion special to place where made not to persons on trip.

FURIOUS DRIVING (general)

Chatterton v Parker (1914-15) XXIV Cox CC 312; (1914) 78 JP 339; (1914-15) 111 LTR 380; (1914) WN (I) 206 HC KBD Proper conviction for furious driving while asleep of horse and trap.

GENERAL (miscellaneous)

Claude Hughes and Co (Carlisle), Ltd v Hyde [1963] 1 All ER 598; [1963] Crim LR 287; (1963) 127 JP 226; (1963) 113 LJ 267; [1963] 2 QB 757; (1963) 107 SJ 115; [1963] 2 WLR 381 HC QBD Motor vehicle exceeding length restrictions when detachable container attached does breach regulations.

Cording v Halse [1954] 3 All ER 287; [1955] Crim LR 43; (1954) 118 JP 558; (1954) 104 LJ 761; [1955] 1 QB 63; (1954) 98 SJ 769; [1954] 3 WLR 625 HC QBD Simply adding something to motor vehicle does not mean it is alternative body (so not to be included in weight).

R v Littell [1981] 3 All ER 1; [1981] Crim LR 642; (1981) 145 JP 451; [1981] RTR 449; (1981) 125 SJ 465; [1981] 1 WLR 1146 CA Provision of breath specimen in manner other than indicated a failure to provide breath specimen.

GOODS VEHICLE (access permit)

TNT Express (UK) Ltd v Richmond upon Thames London Borough Council [1995] TLR 370 HC QBD When deciding reasonableness of route to be taken in order to comply with goods vehicle access permit financial factors are not to be taken into account.

GOODS VEHICLE (driver)

Gross Cash Registers, Ltd v Vogt [1965] 3 All ER 832; [1966] Crim LR 109; (1966) 130 JP 113; (1965-66) 116 NLJ 245; [1967] 2 QB 77; (1966) 110 SJ 174; [1966] 2 WLR 470 HC QBD Person not obliged to but using company van in course of employment a 'part-time driver'.

GOODS VEHICLE (driver's records)

Lackenby v Browns of Wem Ltd [1980] RTR 363 HC QBD Employer not required to issue unused driver's record book to employee drivers.

GOODS VEHICLE (farmer's vehicle)

Cambrian Land Ltd v Allan [1981] RTR 109 HC QBD Use of vehicle by farmers to transport slaughtered animals from slaughterhouse to butchers not use of same as farmer's goods vehicle.

GOODS VEHICLE (general)

Alcock v GC Griston Ltd [1980] Crim LR 653; (1980) 130 NLJ 1211; [1981] RTR 34 HC QBD For purposes of Transport Act 1968, s 103, firm using driver provided by agency was employer of driver.

Bennington v Peter; R v Swaffham Justices, ex parte Peter [1984] RTR 383; [1984] TLR 55 HC QBD That were diabetic did not mean that as driver would be likely to cause danger to public; on whether licence to continue pending hearing of appeal against revocation of same.

Blakey Transport Ltd v Casebourne [1975] Crim LR 169; (1975) 125 NLJ 134t; [1975] RTR 221; (1975) 119 SJ 274 HC QBD Is not preserving driver's record book intact for employers to whom it is returned re-issue same to another within six months of receipt.

Burmingham v Russell [1936] 2 All ER 159 HC KBD Sound-recording van a goods vehicle.

Burningham v Lindsell (1936) 80 SJ 367 HC KBD Vehicle with permanent apparatus a goods vehicle.

Cassady v Reg Morris (Transport) Ltd [1975] Crim LR 398; [1975] RTR 470 HC QBD Employer's not punishing non-delivery/encouraging delivery by driver of completed record sheets within seven days did not in instant circumstances render employer guilty as aider and abettor to employee's offence.

Cassady v Ward and Smith Ltd [1975] Crim LR 399; [1975] RTR 353 HC QBD Requirement that new driver be provided with new driver's book not met by providing worker-driver with driver's book used by several previous temporary worker-drivers.

Connor v Graham and another [1981] RTR 291 HC QBD Tyres (though must have capacity to bear maximum axle weight of vehicle) need only be inflated to extent required for actual use to which being put.

Dent v Coleman and another [1977] Crim LR 753; [1978] RTR 1 HC QBD Were guilty of carrying improperly secured load where straps securing load defective though did not know were so.

Director of Public Prosecutions v Derbyshire (listed sub nom Jardine v Derbyshire) [1994] RTR 351 HC QBD Vehicle principally geared towards carrying asphalt — albeit adapted for other purposes — was not exempt from goods vehicle licensing regulations.

Director of Public Prosecutions v Howard (1991) 155 JP 198; [1991] RTR 49 HC QBD User of cement-mixing lorry did not require operator's licence (Transport Act 1968, s 60(1)).

Director of Public Prosecutions v Ryan and another (1991) 155 JP 456; [1992] RTR 13; [1991] TLR 88 HC QBD On what constitutes a goods vehicle (and so is subject to tachograph legislation).

Director of Public Prosecutions v Scott Greenham Ltd [1988] RTR 426 HC QBD On determining whether trailer vehicle carrying mobile crane and hoist was a vehicle carrying goods necessary for running of vehicle or for use with fitted machine (was not).

Flatman v Poole; Flatman v Oatey [1937] 1 All ER 495 HC KBD No need to keep records in respect of agricultural lorry transporting mixture of non-/agricultural goods from one farm to another.

George Cohen 600 Group Ltd v Hird [1970] 2 All ER 650; [1970] Crim LR 473t; (1970) 134 JP 598; (1970) 120 NLJ 550t; [1970] RTR 386; (1970) 114 SJ 552; [1970] 1 WLR 1226 HC QBD Notice to highway and bridge authority needed for use of heavy goods vehicle.

Halliday v Burl [1983] RTR 21 HC QBD Closest weighbridge was weighbridge nearest available notwithstanding difficulties associated with its use.

Hickman v Chichester District Council (1992) 156 JP 218; [1992] RTR 121; [1991] TLR 471 HC QBD Can be convicted of using vehicle in public car park in connection with trade/business where place advertising pamphlet under windscreen wiper of another's car.

Hubbard v Messenger [1937] 4 All ER 48; (1934-39) XXX Cox CC 624; (1937) 101 JP 533; [1938] 1 KB 300; (1937) 84 LJ 291; [1938] 107 LJCL 44; (1937) 157 LTR 512; (1937) 81 SJ 846; (1937-38) LIV TLR 1; (1937) WN (I) 340 HC KBD Utility car a goods vehicle.

James v Davies [1952] 2 All ER 758; (1952) 116 JP 603; (1952) 102 LJ 625; [1953] 1 QB 8; (1952) 96 SJ 729; [1952] 2 TLR 662; (1952) WN (I) 480 HC QBD Unloaded vehicle pulling loaded trailer was a goods vehicle.

Knowles (JM), Ltd v Rand [1962] 2 All ER 926; [1962] Crim LR 561t; (1962) 126 JP 442; (1962) 112 LJ 650; (1962) 106 SJ 513 HC QBD Van carrying hatching eggs for business purposes a farmer's goods/not a carriers' vehicle.

London Boroughs Transport Committee v Freight Transport Association Ltd and others (1992) 156 JP 69; [1991] RTR 337; (1991) 135 SJ LB 101; [1991] 1 WLR 828 HL London lorry braking requirements compatible with EEC law.

London County Council v Hay's Wharf Cartage Co Ld [1953] 2 All ER 34; (1953) 97 SJ 334; [1953] 1 WLR 677 HC QBD Iron block used as ballast to be included in unladen weight of haulage vehicle.

Manley v Dabson; Manley v Same [1949] 2 All ER 578; (1949) 113 JP 501; [1950] 1 KB 100; (1949) 99 LJ 455; [1949] 118 LJR 1427; (1949) LXV TLR 491; (1949) WN (I) 376 HC KBD Farmer selling goods to wholesaler/retailer within twenty-five miles of farm not required to maintain record of trip under Goods Vehicles (Keeping of Records) regulations 1935, reg 6(3).

McDermott Movements Ltd v Horsfield [1982] Crim LR 693; [1983] RTR 42 HC QBD Owners of insecurely loaded lorry which did not spill any goods wrongfully prosecuted under provision of motor construction regulations concerning spilling of loads.

McKenzie v Griffiths Contractors (Agricultural) Ltd [1976] Crim LR 69; [1976] RTR 140 HC QBD Farmers' moving manure (formerly their straw) from another's stable to third party's mushroom farm not using transporting vehicle as farmer's goods vehicle.

Metropolitan Traffic Area Licensing Authority v Blackman [1973] RTR 525 HC QBD Longtime driver who was disqualified for six months/did not drive for another eighteen days in certain year was not 'in the habit of driving' such vehicles.

Nutland v R [1991] Crim LR 630 HC QBD On what is meant by 'structural attachment' to goods vehicle that renders it a tanker (Dangerous Substances (Conveyance by Road in Road Tankers and Tank Containers) Regulations 1981, regulation 2).

Oxford v Spencer [1983] RTR 63 HC QBD Driver's non-entry of name/address/date of birth in record book not a criminal act.

Oxford v Thomas Scott and Sons Bakers Ltd and another [1983] Crim LR 481; [1983] RTR 369 HC QBD On what constitute 'specialised vehicles' engaged in 'door-to-door selling' and so are exempt from EEC tachograph legislation (as implemented).

Parkinson and another v Axon [1951] 2 All ER 647; (1951) 115 JP 528; [1951] 2 KB 678; (1951) 101 LJ 469; (1951) 95 SJ 641; (1951) WN (I) 473 HC KBD Post-driving depot work not '[t]ime spent . . . on other work in connection with a vehicle or the load carried thereby' contrary to Road Traffic Act 1930, s 19.

Plume v Suckling [1977] RTR 271 HC QBD Coach modified to carry passengers, kitchen items and stock car was a goods vehicle.

Police v Dormer [1955] Crim LR 252 Magistrates Electrician carrying testing tools and equipment in his van was not carrying 'goods'.

R v Southend-on-Sea Justices, ex parte Sharp and another [1980] RTR 25 HC QBD Person's licence wrongfully endorsed as his age/weight of vehicle brought him within exempting provisions of Road Traffic (Drivers' Ages and Hours of Work) Act 1976, Schedule 2.

Redhead Freight Ltd v Shulman [1988] Crim LR 696; [1989] RTR 1 HC QBD Mere acquiesence to driver's breach of tachograph legislation was not causing of same.

Series v Poole [1967] 3 All ER 849; [1967] Crim LR 712; (1968) 132 JP 82; (1967) 117 NLJ 1140t; [1969] 1 QB 676; (1967) 111 SJ 871; [1968] 2 WLR 261 HC QBD Employer delegating task of seeing complies with road traffic legislation remains liable for non-compliance.

Small v Director of Public Prosecutions [1995] Crim LR 165; [1995] RTR 95; [1994] TLR 208 HC QBD On determining permissible maximum weight of goods vehicle.

Sweetway Sanitary Cleansers, Ltd v Bradley [1961] 2 All ER 821; [1961] Crim LR 553; (1961) 125 JP 470; (1961) 111 LJ 582; [1962] 2 QB 108; [1961] 3 WLR 196 HC QBD Vehicle transporting effluent a goods vehicle; because bringing effluent to farm (without charge) part of removing it from tank (for which charge) was for reward.

Taylor v Thompson [1956] 1 All ER 352; [1956] Crim LR 206; (1956) 120 JP 124; (1956) 106 LJ 106; (1956) 100 SJ 133; [1956] 1 WLR 167 HC QBD Car a goods vehicle as constructed for and used as goods vehicle.

Theobald (J) (Hounslow) Ltd and another v Stacy [1979] Crim LR 595; [1979] RTR 411 HC QBD Where gross weight/two axle weights exceeded prescribed maxima were guilty of three offences.

Thurrock District Council v LA and A Pinch Ltd [1974] Crim LR 425; [1974] RTR 269 HC QBD On pleading defence that overweight vehicle had been en route to weighbridge for weighing.

Walker-Trowbridge Ltd and another v Director of Public Prosecutions [1992] RTR 182; [1992] TLR 92 HC QBD Load knocked from lorry had fallen from same for purposes of Road Vehicles (Construction and Use) Regulations 1986, reg 100(2); on determining adequacy of securing of load.

Wardhaugh (AF) Ltd v Mace (1952) 116 JP 369; (1952) 102 LJ 318; (1952) 96 SJ 396; [1952] 1 TLR 1444; (1952) WN (I) 305 HC QBD On what constitutes 'meat' for purposes of Transport Act 1947, s 125.

Witchell v Abbott and another [1966] 2 All ER 657; [1966] Crim LR 283; (1966) 130 JP 297; (1965-66) 116 NLJ 528; (1966) 110 SJ 388; [1966] 1 WLR 852 HC QBD Voluntary travel home albeit in company car was time for rest.

Woolley v Moore [1952] 2 All ER 797; (1952) 116 JP 601; (1952) 102 LJ 639 [1953] 1 QB 43; (1952) 96 SJ 749; [1952] 2 TLR 673; (1952) WN (I) 480 HC QBD Unloaded goods vehicle not subject to special conditions of carrier's licence.

Wurzal v WGA Robinson (Express Haulage), Ltd [1969] 2 All ER 1021; [1969] Crim LR 666; (1969) 113 SJ 408; [1969] 1 WLR 996 HC QBD Unless rigorously policed record-keeping system do not conform to Goods Vehicles (Keeping of Records) Regulations 1935.

Young and another v Director of Public Prosecutions [1992] RTR 194; [1991] TLR 217 HC QBD On determining suitability of trailer for purpose for which being used.

GOODS VEHICLE (licence)

British Gypsum Ltd v Corner [1982] RTR 308 HC QBD Bowser which carried goods subject to driver's records requirements and required operator's licence.

North West Traffic Area Licensing Authority v Post Office [1982] RTR 304 HC QBD Post Office guilty of unlicensed use of goods vehicle used in part to transport items incidental to but not necessary for operation of machine on vehicle.

Stirk v McKenna [1984] RTR 330 HC QBD Operator's licence not needed by garage owner for vehicle in which transported stock car to races as was not using same for hire/reward or in course of trade/business.

HEALTH AND SAFETY AT WORK (general)

Coult v Szuba [1982] RTR 376 HC QBD Person who had 'clocked on' but was two miles away from actual place of work could not be said to be acting in course of his employment and so was not liable for breach of Health and Safety at Work etc Act 1974, s 33(1)(a).

HEAVY MOTOR VEHICLE (general)

Andrews v HE Kershaw, Ltd and another [1951] 2 All ER 764; (1951) 115 JP 568; [1952] 1 KB 70; (1951) 101 LJ 581; (1951) 95 SJ 698; [1951] 2 TLR 867; (1951) WN (I) 510 HC KBD Overhang measured when tailboard up.

Churchill v Norris (1939-40) XXXI Cox CC 1; (1938) 158 LTR 255; (1938) 82 SJ 114 HC KBD Employer underestimating weight of load later found to exceed permitted weight guilty of offence.

Guest Scottish Carriers, Ltd and another v Trend [1967] 3 All ER 52; [1967] Crim LR 595; (1967) 131 JP 468; (1967) 117 NLJ 913; (1967) 111 SJ 812; [1967] 1 WLR 1371 HC QBD Vehicle length measurable with tailboard down when design of vehicle designed to be increased in latter way.

Hawkins v Harold A Russett Ltd [1983] Crim LR 116; (1983) 133 NLJ 154; [1983] RTR 406 HC QBD Detachable container attached to lorry was part of vehicle and so to be included as part of overhang.

Mackinnon v Peate and another [1956] 2 All ER 240 HC KBD Interpretation of Motor Vehicles (Construction and Use) (Amendment) Provisional Regulations 1931, reg 10(ii).

McCrory v Director of Public Prosecutions (1990) 154 JP 520; [1991] RTR 187; [1990] TLR 119 HC QBD On what consttitues 'heavy motor car' for purposes of Motorways Traffic (England and Wales) Regulations 1982.

Moscrop and Wills v Blair [1962] Crim LR 323; (1961) 105 SJ 950 HC QBD Two ton car attached to one ton trailer validly found to be one entity and so a heavy motor car.

Pilgrim and others v Simmonds (1911-13) XXII Cox CC 579; (1911) 75 JP 427; (1911-12) 105 LTR 241 HC KBD On registering of vehicle as heavy motor car.

Pritchard v Dyke (1934-39) XXX Cox CC 1; (1933) 97 JP 179; (1933) 149 LTR 493; (1932-33) XLIX TLR 473 HC KBD No evidence on speed necessary to sustain speeding conviction where weight of vehicle over twelve tons.

Staunton v Coates (1924) 88 JP 193; (1924) 59 LJ 681; [1925] 94 LJCL 95; (1925) 132 LTR 199; (1924-25) 69 SJ 126; (1924-25) XLI TLR 33 HC KBD Driver of heavy motor-car to be prosecuted for speeding need not be given warning of intention to prosecute (Locomotives on Highways Act 1896, s 6, as amended).

Sunter Brothers Ltd and another v Arlidge [1962] 1 All ER 510; [1962] Crim LR 320; (1962) 126 JP 159; (1962) 112 LJ 240; (1962) 106 SJ 154 HC QBD Load an 'abnormal indivisible load' if would be undue risk/damage in dividing.

Windle v Dunning and Son Ltd [1968] 2 All ER 46; [1968] Crim LR 337; (1968) 132 JP 284; (1968) 118 NLJ 397; (1968) 112 SJ 196; [1968] 1 WLR 552 HC QBD Persons renting lorry from haulage contractors not using lorry but causing it to be used.

IDENTITY (general)

Bingham v Bruce [1962] 1 All ER 136; [1962] Crim LR 114t; (1962) 126 JP 81; (1962) 112 LJ 106; (1961) 105 SJ 1086 HC QBD 'Any person' bound to give information means any person other than owner (including driver).

Boss v Measures [1989] Crim LR 582; [1990] RTR 26 HC QBD Local authority could validly require that information as to identity of driver be given in writing.

Cattermole v Millar [1977] Crim LR 553; [1978] RTR 258 HC QBD Dangerous driving conviction quashed where was inadequate evidence linking car that was driven and car of registration number noted in police form.

Clarke (Simon Robert) v Director of Public Prosecutions (1992) 156 JP 605 HC QBD On presumption that owner of car was driver of car: burden not upon owner to show was not driver of car but upon justices to decide whether are sure owner was driver.

Cooke v McCann [1973] Crim LR 522; [1974] RTR 131 HC QBD On determining whether person before court is person properly charged: name and address from licence matching those of person in court.

Creed v Scott [1976] Crim LR 381; [1976] RTR 485 HC QBD On identifying defendant to speeding charge as person who was in fact driving vehicle.

Director of Public Prosecutions v Mansfield [1997] RTR 96 HC QBD On proving identity of person charged with driving while disqualified.

Elliott v Loake [1983] Crim LR 36 HC QBD Proper inference that owner of car was driver of car when offence committed.

Ellis v Smith [1962] 3 All ER 954; [1963] Crim LR 128; (1963) 127 JP 51; (1963) 113 LJ 26; (1962) 106 SJ 1069; [1962] 1 WLR 1486 HC QBD Until relief (bus) driver assumes charge last (bus) driver is person in charge of vehicle.

Ex parte Beecham (1913-14) XXIII Cox CC 571; [1913] 82 LJKB 905; (1913-14) 109 LTR 442; (1912-13) XXIX TLR 586 HC KBD On prosecution of car owner who refuses to divulge identity of driver who committed offence.

Hamilton v Jones (1925-26) XLII TLR 148 HC KBD On proving that person of whom driving licence demanded was 'the driver' of the vehicle (Motor Car Act 1903, s 3(4)).

Hampson v Powell [1970] 1 All ER 929; (1970) 134 JP 321; (1970) 120 NLJ 338; [1970] RTR 293; [1970] Crim LR 351 HC QBD Circumstantial evidence suffices to identify driver of truck in accident.

Hateley v Greenough [1962] Crim LR 329t HC QBD On who is owner of car bought on hire purchase.

Hodgson v Burn [1966] Crim LR 226; (1965-66) 116 NLJ 501t; (1966) 110 SJ 151 HC QBD Obligation to name driver of car fell on owner personally.

Hunter v Mann [1974] 2 All ER 414; (1974) 59 Cr App R 37; [1974] Crim LR 260; (1974) 138 JP 473; (1974) 124 NLJ 202t; [1974] QB 767; [1974] RTR 328; [1974] 2 WLR 742 HC QBD Doctor's professional duty superseded by duty to give information identifying driver in accident.

Jacob v Garland [1974] Crim LR 194; [1974] RTR 40; (1973) 117 SJ 915 HC QBD On ingredients of offence of non-provision of information by keeper of vehicle to police as to identity of driver when properly requested to do so.

Lowe v Lester [1986] Crim LR 339; [1987] RTR 30 HC QBD Person required by local authority to give identity of car-driver must do so immediately/within reasonable period.

Neal v Fior [1968] 3 All ER 865; [1969] Crim LR 95; (1969) 133 JP 78; (1968) 118 NLJ 1150; (1968) 112 SJ 948; [1968] 1 WLR 1875 HC QBD If factual basis of information charge denied justices to satisfy themselves (criminal burden) that vehicle was owner's vehicle; if so, that it was reasonable for owner to know identity of driver.

Nelms v Roe [1969] 3 All ER 1379; (1970) 54 Cr App R 43; [1970] Crim LR 48t; (1970) 134 JP 88; (1969) 119 NLJ 997t; [1970] RTR 45; (1969) 113 SJ 942; [1970] 1 WLR 4 HC QBD Police inspector was acting under implied delegated responsibility of Police Commissioner when completed notice under Road Traffic Act 1960, s 232(2).

Osgerby v Walden [1967] Crim LR 307t; (1967) 111 SJ 259 HC QBD Successful prosecution for non-disclosure of identity of driver to police.

Pamplin v Gorman [1980] Crim LR 52; [1980] RTR 54 HC QBD On proving that document in which identity of driver sought was genuine and validly served.

Pulton v Leader [1949] 2 All ER 747; (1949) 113 JP 537; (1949) 99 LJ 612; (1950) 94 SJ 33; (1949) LXV TLR 687; (1949) WN (I) 400 HC KBD Owner must give details of car driver suspected of offence immediately upon police officer's request.

R v Collins (George) [1994] RTR 216 CA Jury to be warned that joint owner of car may have loaned car to third party where description evidence indicates that car actually driven by other owner at relevant time.

R v Derwentside Justices, ex parte Swift; R v Sunderland Justices, ex parte Bate [1997] RTR 89 HC QBD On proving identity of person charged with driving while disqualified.

R v Hankey and another (1907-09) XXI Cox CC 1; (1905) 69 JP 219; [1905] 2 KB 687; (1905-06) XCIII LTR 107; (1904-05) XXI TLR 409; (1905-06) 54 WR 80 HC KBD On conviction of car-owner for non-disclosure of identity of driver who committed offence.

R v Hewett [1977] Crim LR 554; [1978] RTR 174 CA Direction like that required where identification based on momentary glimpse unnecessary where prosecution/defence stated driver one person/another — jury to decide on criminal standard whether prosecution case proved.

Rathbone v Bundock [1962] 2 All ER 257; [1962] Crim LR 327; (1962) 126 JP 328; (1962) 112 LJ 404; [1962] 2 QB 260; (1962) 106 SJ 245 HC QBD Requirement to give information under Road Traffic Act 1960.

Record Tower Cranes Ltd v Gisbey [1969] Crim LR 94t; (1969) 133 JP 167; (1968) 118 NLJ 1174t; (1969) 113 SJ 38; [1969] 1 WLR 148 HC QBD Conviction quashed where not proved that form requesting company to identify employee driver had been properly authorised.

Scruby v Beskeen [1980] RTR 420 HC QBD Could validly convict person of careless driving where admitted owned type of car involved in accident/had been in right place at right time.

INCONSIDERATE DRIVING (general)

Pawley v Whardall (1965) 115 LJ 529 HC QBD '[O]ther persons using the Road' (in the Road Traffic Act 1960, s 3(1) includes persons in vehicle being driven.

INTERFERENCE WITH FREE PASSAGE (general)

Midlands Electricity Board v Stephenson [1973] Crim LR 441 HC QBD Interpretation of the Highways Act 1835, s 78.

Watson v Lowe (1950) 114 JP 85; (1950) 100 LJ 20; (1950) 94 SJ 15; [1950] 66 (1) TLR 169; (1950) WN (I) 26 HC KBD Person in driving seat of stopped car negligently causing injury to passing cyclist by opening door guilty of interference with free passage of person on highway (Highway Act 1835, s 78).

INTERFERENCE WITH MOTOR VEHICLE (general)

Reynolds and Warren v Metropolitan Police [1982] Crim LR 831 CrCt On elements of 'interference' with motor vehicle (Criminal Attempts Act 1981, s 9).

LICENCE (articulated vehicle)

British Transport Commission and others v McKelvie and Co, Ltd (1959) 109 LJ 394 TrTb Reduction in number of articulated vehicles licensed where did not come to Tribunal with clean hands.

LICENCE (disabled person)

Guest v Kingsnorth [1972] Crim LR 243t; (1972) 122 NLJ 153t; [1972] RTR 265 HC QBD Valid refusal of driving licence to disabled person.

LICENCE (epileptic)

Re Alborough's Application [1958] Crim LR 691 Magistrates Non-interference with local authority exercise of discretion in refusing driving licence to epileptic.

Re Leddington's Application [1958] Crim LR 550 Magistrates Epileptic whose fits were now under control through use of drugs allowed to have driving licence.

Secretary of State for Transport v Adams [1982] Crim LR 530; [1982] RTR 369 HC QBD On eligibility of epileptic for driving licence.

LICENCE (excise)

Algar v Shaw [1986] Crim LR 750; [1987] RTR 229 HC QBD Excise licence prosecution properly brought by Secretary of State for Transport within six months of knowing had right to bring same.

Binks v Department of the Environment [1975] Crim LR 244; [1975] RTR 318; (1975) 119 SJ 304 HC QBD Seriously damaged car without engine a mechanically propelled vehicle which (as was kept on public road) required licence as could (and was accused's intention that would) move again.

Brook v Friend [1954] Crim LR 942 HC QBD Improper use of tractor in manner for which should have obtained more expensive licence.

Chief Constable of Kent v Mather [1986] RTR 36 HC QBD Justices could not reduce back duty payable.

Cobb v Whorton [1971] Crim LR 372; [1971] RTR 392 HC QBD Towed van a trailer and a mechanically propelled vehicle and so subject to tax.

Cook v Lanyon [1972] Crim LR 570t; (1972) 122 NLJ 657t; [1972] RTR 496 HC QBD On fraudulent intent necessary to be guilty of fraudulent use of excise licence contrary to Vehicles (Excise) Act 1971, s 26(1).

French v Champkin (1919) WN (I) 260 HC KBD Interpretation of Customs and Inland Revenue Act 1888, s 4(3) — horse-drawn Ralli car not a 'carriage'.

Gosling v Howard [1975] RTR 429 HC QBD Having 'use' of vehicle means using or keeping/causing or permitting to be used for purposes of vehicle excise/road traffic legislation.

Heumann v Evans [1977] Crim LR 229; [1977] RTR 250 HC QBD Was fraudulent use of excise licence to remove from wrecked car and use on similar new car.

Holliday v Henry [1974] Crim LR 126; [1974] RTR 101 HC QBD Vehicle sitting on roller skates sitting on road was itself sitting on road so licence required.

Nattrass v Gibson [1969] Crim LR 96t; (1968) 112 SJ 866 HC QBD Person guilty of driving without excise licence (albeit that had mailed off for one and believed self to be acting lawfully).

Patterson v Helling [1960] Crim LR 562 HC QBD Information for use of motor vehicle absent excise licence ought not to have been dismissed in light of mitigating factors — these should have only counted towards sentence.

R v Johnson [1995] Crim LR 250; (1994) 158 JP 788; [1995] RTR 15; [1994] TLR 104 CA Conviction for fraudulent use of vehicle excise licence quashed where vehicle was on private land and was inadequate proof of intention to use vehicle on public road.

R v Reigate Justices, ex parte Holland [1956] 2 All ER 289; [1956] Crim LR 492; (1956) 120 JP 355; (1956) 100 SJ 436; [1956] 1 WLR 638 HC QBD No limit on county council's right to prosecute for excise duties offences.

R v Terry [1984] 1 All ER 65; [1984] AC 374; (1984) 78 Cr App R 101; (1984) 148 JP 613; [1984] RTR 129; (1984) 128 SJ 34; [1983] TLR 771; [1984] 2 WLR 23 HL Crown need not show intent not to pay licence fee when prosecuting for fraudulent use of licence.

R v Terry (Neil William) (1983) 77 Cr App R 173; [1983] Crim LR 557; [1983] RTR 321; [1983] TLR 309 CA Must be intent not to pay licence fee when prosecuting for fraudulent use of licence.

Sly v Randall (1916) WN (I) 64 HC KBD On time by which were expected to acquire excise licence once assumed possession of car. (Revenue Act 1869, s 18 considered).

Smith v Koumourou [1979] Crim LR 116; [1979] RTR 355 HC QBD Was obtaining pecuniary advantage by deception where driver without excise licence displayed undated police receipt for earlier expired licence to avoid non-payment of duty being discovered and so avoid paying excise duty.

Strowger v John [1974] Crim LR 123; (1974) 124 NLJ 57t; [1974] RTR 124; (1974) 118 SJ 101 HC QBD Non-display of valid vehicle excise licence an absolute offence.

LICENCE (express carriage)

Webb v Maidstone and District Motor Services, Limited (1934) 78 SJ 336 HC KBD Motor vehicle unlawfully used as express carriage without road service licence.

LICENCE (false)

R v Howe [1982] RTR 455 CA Was not 'use' of false driving licence to hand over same to police in circumstances unconnected with driving.

LICENCE (false declaration)

Bloomfield v Williams [1970] Crim LR 292t; [1970] RTR 184 HC QBD Person who did not know that what were signing was false could not be guilty of false declaration.

LICENCE (farmer's goods vehicle)

Howard v Grass Products Ltd [1972] 3 All ER 530; [1972] Crim LR 572; (1972) 136 JP 813; [1972] RTR 547; (1972) 116 SJ 786; [1972] 1 WLR 1323 HC QBD If not possessor/likely possessor of land cannot get lower rate of duty for farmer's goods vehicles.

LICENCE (forgery)

R v Macrae [1994] Crim LR 363; (1995) 159 JP 359 CA Need not intend to cause another economic loss to be guilty of forging a licence.

LICENCE (general)

A One Transport (Leeds), Ltd v British Railways Board and others and CNC Transport Ltd v the Same (1965) 115 LJ 266 TrTb On need to consider interests of public (other hauliers) when deciding whether to grant 'A' licences.

Abercromby v Morris [1932] All ER 676; (1931-34) XXIX Cox CC 553; (1932) 96 JP 392; (1932) 147 LTR 529; (1932) 76 SJ 560; (1931-32) XLVIII TLR 635; (1932) WN (I) 201 HC KBD Owner not liable for act of person (not his agent or servant) in use of car.

Alderton v Richard Burgon Associates Ltd [1974] Crim LR 318; [1974] RTR 422 HC QBD Failure to provide operator's licence for driver rested with company using him and not employment agency from which obtained him (and who inter alia acted as agent through whom wages were paid).

Arnold Transport (Rochester), Ltd v British Transport Commission and others [1963] 1 QB 457 CA Transport Tribunal could not refuse licence on basis that was not in applicant's best interests.

Arthur Sanderson (Great Broughton), Ltd v Vickers [1964] Crim LR 474g; (1964) 108 SJ 425 HC QBD Licensing authority when issuing 'A' carrier's licence could specify which trailers might be used.

Barham v Castell (1951) 115 JP 603; (1951) 101 LJ 611; [1951] 2 TLR 923; (1951) WN (I) 526 HC KBD On effect of holding 'A' licence in tandem with Road Haulage Executive permit allowing carriage of goods over twenty-five miles from operating centre.

Barnes v Gevaux [1981] RTR 236 HC QBD Justices bald statement that were special reasons for non-endorsement procedurally improper but non-disqualfication allowed stand.

Barnett v Fieldhouse [1987] RTR 266 HC QBD Father's being away on holiday at relevant time was special reason for not endorsing licence where was convicted of permitting uninsured use of car by son.

BCF Transport Co Ltd v Townend (1960) 110 LJ 10 TrTb Failed appeal against addition to 'A' licence of trailer.

Bell v Ingham [1968] 2 All ER 333; (1968) 118 NLJ 518; [1969] 1 QB 563; (1968) 112 SJ 486; [1968] 3 WLR 401 HC QBD No endorsement of licences unless authorised by Road Traffic Act.

Booth v Director of Public Prosecutions [1993] RTR 379 HC QBD Approval of finding that drawing of empty trailer to test same was drawing of goods vehicle requiring operator's licence.

Boxer v Snelling [1972] Crim LR 441; [1972] RTR 472; (1972) 116 SJ 564 HC QBD On what is a 'vehicle' (here for purposes of Road Traffic Regulation Act 1967, s 31/Folkestone Traffic Regulation Order 1970, Article 15).

Bristow (C), Ltd (1960) 110 LJ 122 TrTb Valid finding that were guilty of entering false weights in licence application forms.

British Engineering Contractors, Ltd v British Transport Commission and others (1960) 110 LJ 176 TrTb 'B' licence restricted to fifteen mile area granted in respect of low-loading vehicle.

British Railways Board v Leinster Ferry Transport, Ltd (1963) 113 LJ 770 TrTb Reference to 'other kinds of transport' in Road Traffic Act 1960, s 173(4), meant other kinds of land transport.

British Road Services Ltd and another v Wurzal [1971] 3 All ER 480; [1972] Crim LR 537; (1971) 135 JP 557; [1972] RTR 45; (1971) 115 SJ 724; [1971] 1 WLR 1508 HC QBD Trailer used to transport goods from England to Continent and vice versa is more than 'temporarily in Great Britain'.

British Transport Commission and others v BH Cecil and Sons (1959) 109 LJ 524 TrTb Reduction in number of vehicles made subject to 'A' licence where had been illegal user of same.

British Transport Commission and others v McKelvie and Co, Ltd (1960) 110 LJ 255 TrTb Are 'providing' transport facility if are eager/prepared to do so.

British Transport Commission v Baker's Transport (Southampton), Ltd (1960) 110 LJ 286 TrTb On form of evidence required when objecting to variation of ('B') licence.

British Transport Commission v C and J (Agencies), Ltd (1958) 108 LJ 43 TrTb Associated companies treated as separate.

British Transport Commission v Charles Alexander and Partners (Transport), Ltd (1957) 107 LJ 810 TrTb On procedure as regards application to vary 'A' licence in respect of fleet of vehicles.

British Transport Commission v Griffiths (1962) 112 LJ 818 TrTb Successful appeal against granting of 'B' licence for two vehicles to be used for carrying home pigeons.

British Transport Commmission v R Hodge, Ltd; WTJ Eastmond, Ltd and others v R Hodge, Ltd (1958) 108 LJ 458 TrTb Objector could argue that granting of 'B' licence would be more appropriate than the 'A' licence sought.

Broadway Haulage and others v Fleming (1965) 115 LJ 108 TrTb 'B' licence granted in continuation of existing one (and not to expire later than) existing licence.

Brown v Allweather Mechanical Grouting Co, Ltd [1953] 1 All ER 474; (1953) 117 JP 136; [1954] 2 QB 443; (1953) 97 SJ 135; [1953] 2 WLR 402 HC QBD Cannot aid and abet use of vehicle not licensed properly as substantive offence attracts excise penalty not conviction.

Carey and others v Heath [1951] 2 All ER 774; (1951) 115 JP 577; [1952] 1 KB 62; (1951) 101 LJ 595; [1951] 2 TLR 797; (1951) WN (I) 513 HC KBD Towing vehicles for repair (by another) required general trade licence; four-wheeled vehicle not two-wheeled when being towed.

Carman Transport, Ltd v British Railways Board and another (1963) 113 LJ 770 TrTb On seeking of licence on foot of traffic coming from outside area of deciding licensing authority.

Carpenter v Campbell and another [1953] 1 All ER 280; (1953) 117 JP 90 HC QBD Owner of unlicensed vehicle ought to be prosecuted.

Chief Constable of West Mercia Police v Williams [1987] RTR 188 HC QBD Could be producing false instrument where produced forged clean driving licence in court when real licence had several (no longer effective) endorsements.

Clark v British Transport Commission and others (1963) 113 LJ 42 TrTb On exercise of discretion in application for carrier's licence pursuant to the Road Traffic Act 1960, s 174.

Cook v Hobbs (1910) 45 LJ 710; (1910) WN (I) 219 HC KBD Vehicle had been built for purpose of carrying goods/burden (appellant and family were burden) but as could be used for other purposes was breach of Customs and Inland Revenue Act 1888.

Corbett and Miller v Barham [1954] Crim LR 470; [1965] 1 WLR 187 HC QBD Use by 'C' licence-holder of vehicle for carriage of goods for hire/reward (taking materials to fill in pond).

D Cattell and Sons v British Transport Commission (1959) 109 LJ 283 TrTb Proof that return loads required ought not to have been demanded upon application being made for conversion of Contract 'A' licence to ordinary 'A' licence.

Davies, Turner and Co v Brodie (1954) 118 JP 532; (1954) 104 LJ 745 HC QBD Forwarding agents not liable for aiding and abetting driving licence violations of persons whom contracted with to convey goods as took all reasonable steps to ensure legalities observed.

Dent Transport (Spennymoor) Ltd's Appeal (No X15 of 1961) (1961) 111 LJ 486 TrTb On basis on which 'A' licence could be revoked.

Devon County Council v Hawkins [1967] 1 All ER 235; (1967) 131 JP 161; [1967] 2 QB 26; (1966) 110 SJ 893; [1967] 2 WLR 285 HC QBD Suffer from prescribed disease if taking drugs that suppress its effects.

Director of Public Prosecutions v Powell [1993] RTR 266 HC QBD No/was special reason arising to justify non-endorsement of licence of person who was guilty of careless driving of child's motor cycle on road/who did not have insurance for or L-plate on same.

Dudley v Holland [1963] 3 All ER 732; [1963] Crim LR 858; (1964) 128 JP 51; (1963) 113 LJ 805; [1965] 1 QB 31; (1963) 107 SJ 1041; [1963] 3 WLR 970 HC QBD 'Keeping' vehicle on road has time element: mere presence not enough.

East v Bladen [1987] RTR 291 HC QBD Mental effects of serious injuries had previously received were special reasons justifying non-endorsement of licence upon conviction for uninsured use of motor vehicle.

Ex parte Hepworth (1931) 75 SJ 408; (1930-31) XLVII TLR 453 HC KBD Failed appeal against unqualified right of licensing authority to refuse licence.

Ex parte Symes (1911-13) XXII Cox CC 346; (1911) 75 JP 33; (1910-11) 103 LTR 428; (1910-11) XXVII TLR 21; (1910) WN (I) 219 HC KBD Licence can be indorsed for failure to drive with light.

Farrall v Department of Transport [1983] RTR 279 HC QBD Failed application to have Luxembourg driving permit recognised as exchangeable for British driving licence more than one year after return of Briton from Luxembourg.

Flores v Scott [1984] Crim LR 296; [1984] RTR 363; (1984) 128 SJ 319; [1984] 1 WLR 690 HC QBD Foreign postgraduate was resident in, not temporarily in Great Britain and so required to comply with road traffic licensing laws.

Flower Freight Co, Ltd v Hammond [1962] 3 All ER 950; [1963] Crim LR 55; (1963) 127 JP 42; (1962) 112 LJ 817; [1963] 1 QB 275; (1962) 106 SJ 919; [1962] 3 WLR 1331 HC QBD If vehicle not within certain category if originally so built, does not come within it through later adaptation.

Great Western Rail Co v West Midland Traffic Area Licensing Authority [1935] All ER 396; (1935) 80 LJ 323; [1936] 105 LJCL 37; (1936) 154 LTR 39; (1935) 79 SJ 941; (1935-36) LII TLR 44; (1935) WN (I) 209 HL On application for/user under 'B' licence.

Guyll v Bright [1987] RTR 104 HC QBD Burden of proving any exception/exemption/proviso/ excuse/qualification to excise licensng requirements rests on person seeking to establish same.

Hammond v Hall and Ham River, Ltd [1965] 1 All ER 108; [1965] Crim LR 309; (1965) 129 JP 107; (1965) 115 LJ 74; (1965) 109 SJ 53; [1965] 1 WLR 180 HC QBD Though paid was not carriage for hire/reward as done in course of business.

Hammond v Hall and Ham River, Ltd [1965] 2 All ER 811; [1965] AC 1049; (1965) 129 JP 488; (1965) 115 LJ 560; (1965) 109 SJ 557; [1965] 3 WLR 337 HL Material carried in course of business not carried for hire/reward (though were paid to carry material).

Harold Wood and Sons, Ltd v British Transport Commission (1957) 107 LJ 507 TrTb Public 'A' licence holder could apply at any time for grant of maintenance vehicles.

Heidak v Winnett [1981] RTR 445 HC QBD Person with valid German driving permit in United Kingdom for under twelve months could rely on permit: that applied for and got provisional licence did not bring him under provisional licensing legislation.

Hoddell (P and A) v Parker (1910) 74 JP 315; (1910-11) 103 LTR 2; (1910) WN (I) 146 HC KBD Locomotive drawing corn to market not an 'agricultural' locomotive for purposes of Locomotives Act 1898, s 9.

Holland v Perry [1952] 2 All ER 720; (1952) 116 JP 581; (1952) 102 LJ 595; [1952] 2 QB 923; (1952) 96 SJ 696; [1952] 2 TLR 634; (1952) WN (I) 458 HC QBD On maximum penalty for using unlicenced vehicle.

James v Evans Motors (County Garages), Ltd [1963] 1 All ER 7; [1963] Crim LR 58t; (1963) 127 JP 104; (1963) 113 LJ 26; (1963) 107 SJ 650; [1963] 1 WLR 685 HC QBD Use by garage owners of lorry to carry rubble not use as manufacturers/repairers/dealers.

Jelliff v Harrington [1951] 1 All ER 384; (1951) 115 JP 100; (1951) 95 SJ 108; [1951] 1 TLR 324; (1951) WN (I) 74 HC KBD Breach of general trade licence by using to tow non-mechanically propelled vehicles.

Jones Transport Services (Liverpool) Ltd v British Transport Commission and others (1963) 113 LJ 74 TrTb On exercise of discretion as regards granting of an 'A' licence.

Kenyon v Thorley [1973] Crim LR 119; [1973] RTR 60 HC QBD Failure to apply brake on unattended vehicle did not require mandatory disqualification/endorsement.

King v British Transport Commission and others (1957) 107 LJ 492 TrTb On need for contract to be made part of evidence of action concerning contract 'A' customers.

Latham's Transporters, Ltd v British Transport Commission and others (1962) 112 LJ 554 TrTb Failed appeal against refusal of 'A' licence (but variation to existing 'B' licence to be made).

London County Council v Fairbank (1911) WN (I) 96 HC KBD Are only guilty of unlicensed keeping of carriage if possess and use it (mere possession not enough).

Lovell v Archer; Lovell v Ducket [1971] Crim LR 240; (1971) 121 NLJ 128t; [1971] RTR 237; (1971) 115 SJ 157 HC QBD Provisional licence to be endorsed where holder of same does not comply with conditions of licence.

Marriott (AT), Ltd v British Transport Commission (1958) 108 LJ 634 TrTb Sucessful appeal against refusal to allow substitution of new diesel articulated vehicle for old one already held under 'A' licence.

Marshall v Ford (1907-09) XXI Cox CC 731; (1908) 72 JP 480; (1908-09) XCIX LTR 796 HC KBD Evidence of police officer on contents of driving licence not before court was admissible.

McCrone v J and L Rigby (Wigan) Ltd; Same v Same [1951] 2 TLR 911 HC KBD Road dumper a mechanically propelled vehicle on which duty payable as road construction vehicle.

Merchandise Transport, Ltd v British Transport Commission and others; Arnold Transport (Rochester), Ltd v British Transport Commission and others [1961] 3 All ER 495; (1961) 111 LJ 644; [1962] 2 QB 173; (1961) 105 SJ 1104 CA Transport Tribunal's decisions fully reviewable/to be decided on individual merits/to allow parties decide whether application in best financial interest; on granting of 'B' licence.

Mitchell's Appeal (1960) 110 LJ 782 (No W31 of 1960) TrTb Was not use of vehicle for hire/reward in course of criminal action unless licensee of vehicle used same in course of robbery.

Munson v British Railways Board and others [1965] 3 All ER 41; (1965) 115 LJ 709; [1966] 1 QB 813; (1965) 109 SJ 597; [1965] 3 WLR 781 CA Answer in previous licence application on normal user relevant in late licence application.

Newton (G), Ltd v Smith; WC Standerwick, Ltd v Smith [1962] 2 All ER 19; [1962] Crim LR 401; (1962) 126 JP 324; (1962) 112 LJ 274; [1962] 2 QB 278; (1962) 106 SJ 287; [1962] 2 WLR 926 HC QBD Employer liable for employee's non-compliance with road service licence; 'wilful or negligent' failure is one offence.

Nichol v Leach and another [1972] Crim LR 571; [1972] RTR 476 HC QBD That did not intend to use vehicle on road/that vehicle towed along road did not mean was not a motor vehicle.

North West Traffic Area Licensing Authority v Brady [1981] Crim LR 407; [1981] RTR 265 CA Licence application 'made' where completed/mailed in December 1976 though not received by licensing authority until January 1977.

North West Traffic Area Licensing Authority v Brady [1979] Crim LR 397; (1979) 129 (2) NLJ 712; [1979] RTR 500 HC QBD Licence application not 'made' when completed/mailed in December 1976 but when received by licensing authority in January 1977.

Nugent v Phillips [1939] 4 All ER 57; (1939-40) XXXI Cox CC 358; (1939) 103 JP 367; (1939) 88 LJ 253; (1939) 161 LTR 386; (1939) 83 SJ 978; (1939) WN (I) 359 HC KBD That person paid to carry horse in motor horse-box meant vehicle used for hire or reward.

Parsons and others v Weaver (1959) 109 LJ 74 TrTb Contract 'A' licence appropriate licence where would be engaged on day haulage work as required.

Pilgram v Dean [1974] 2 All ER 751; [1974] Crim LR 194; (1974) 138 JP 502; (1974) 124 NLJ 102t; [1974] RTR 299; (1974) 118 SJ 149; [1974] 1 WLR 601 HC QBD Conviction for not having and not displaying licence valid.

Police v Lane [1957] Crim LR 542 Magistrates Harpoon gun a gun for purposes of the Gun Licence Act 1870.

R v Ashford and Tenterden Magistrates' Court, ex parte Wood [1988] RTR 178 HC QBD Passenger-car owner guilty of failure to provide blood specimen could have licence endorsed.

R v Berkshire County Council, ex parte Berkshire Lime Co (Childrey), Ltd [1953] 2 All ER 779; (1953) 117 JP 505; [1953] 1 WLR 1146 HC QBD Spreaders also used as conveyors so were goods vehicles.

R v City of Cardiff Justices, ex parte Cardiff City Council [1962] 1 All ER 751; (1962) 126 JP 175; (1962) 112 LJ 274; [1962] 2 QB 436; (1962) 106 SJ 113 HC QBD Can appeal against refusal of licence on basis that are prescribed illness sufferer.

R v Commissioner of Metropolitan Police, ex parte Holloway (1911) 75 JP 490; (1911) 46 LJ 525; (1912) 81 LJKB 205; (1911-12) 105 LTR 532; (1910-11) 55 SJ 773; (1911) WN (I) 184 CA On licensing responsibilities of Metropolitan Police Commissioner under Metropolitan Public Carriage Act 1869 and Order as to Hackney and Stage Carriages 1907.

R v Commissioner of Metropolitan Police, ex parte Pearce (1911) 75 JP 85; (1910) 45 LJ 809; (1911) 80 LJCL 223; (1911) 104 LTR 135 HC KBD On discretion of Metropolitan Police Commisisoner regarding granting of taxi licences pursuant to Metropolitan Public Carriage Act 1869.

R v Cumberland Justices, ex parte Hepworth [1931] All ER 717; (1931-34) XXIX Cox CC 374; (1931) 95 JP 206; (1932) 146 LTR 5; (1930-31) XLVII TLR 610; (1931) WN (I) 209 CA Failed appeal against unqualified right of licensing authority to refuse licence.

R v Gill, ex parte McKim (1911-13) XXII Cox CC 118; (1909) 73 JP 290; (1909) 100 LTR 858 HC KBD Licence can under Motor Car Act 1903, s 7 be indorsed for non-compliance with number plate regulations.

R v Hurst [1966] Crim LR 683 Sessions Driver legally liable for vehicle construction and use violations but no endorsement of licence as true responsibility for defects rested on co-employees.

R v Justices of West Riding of York (1911-13) XXII Cox CC 280 HC KBD Licence cannot be indorsed upon conviction for allowing car to stand in highway and so cause obstruction.

R v Manners-Astley [1967] 3 All ER 899; (1968) 52 Cr App R 5; [1967] Crim LR 658t; (1968) 132 JP 39; (1967) 111 SJ 853; [1967] 1 WLR 1505 CA Must be intent to defraud for conviction for fraudulent use of licence.

R v Minister of Transport, ex parte Valliant Direct Coaches, Ltd; R v Traffic Commissioners for the South Eastern Traffic Area, ex parte, Valliant Direct Coaches, Ltd [1937] 1 All ER 265; (1937) 83 LJ 58; (1937) 81 SJ 138; (1936-37) LIII TLR 227; (1937) WN (I) 27 HC KBD Role of Minister for Transport upon appeal from traffic commissioners' decision.

R v The Licensing Authority for Goods Vehicles for the Metropolitan Traffic Area, ex parte BE Barrett Ld (1949) 113 JP 202; [1949] 2 KB 17; (1949) 99 LJ 135; [1949] 118 LJR 1522; (1949) LXV TLR 309; (1949) WN (I) 126 HC KBD Refusal of mandamus order requiring licensing authority to allow change of operating centre for holder of public carrier's 'A' licence.

R v Traffic Commissioners for the Northern Traffic Area, ex parte Bee-Line Roadways Ltd [1971] RTR 376 HC QBD On procedure when commissioners deciding whether to grant road service licence.

R v Wood and others, ex parte Anderson and another (1922) 86 JP 64; [1922] 91 LJCL 573; (1922) 126 LTR 522; (1921-22) 66 SJ 453; (1921-22) XXXVIII TLR 269; (1922) WN (I) 38 HC KBD Not liable to conviction under Roads Act 1920, s 8(3) where have general licence under Finance Act 1920 for lorry normally/sometimes used to carry goods/passengers.

Re R Hampton and Sons [1965] 3 All ER 106; (1965) 129 JP 508; [1966] 1 QB; (1965) 109 SJ 477; [1965] 3 WLR 576 CA Post-fining suspension of licence valid though double punishment.

Ready Mixed Concrete (East Midlands) Ltd v Yorkshire Traffic Area Licensing Authority (1970) 134 JP 293; [1970] 2 QB 397; [1970] RTR 141; (1970) 114 SJ 111; [1970] 2 WLR 627 HC QBD Ultimate employer and not person for whom presently performing services liable for Road Traffic Act licensing violation.

Reed Transport, Ltd v British Transport Commission (1957) 107 LJ 492 TrTb Subsidiary could plead needs of parent company as justification for 'A' licence.

Rhodes v British Transport Commission (1959) 109 LJ 219 TrTb Variation of 'A' licence so as to make provision for 'through traffic' to the Continent.

Roberts v Morris [1965] Crim LR 46g HC QBD Need only possess licence for carriage of goods for hire/reward when are actually physcally transporting the goods.

Robertson and Son and others v Highland Haulage, Ltd (1958) 108 LJ 651 TrTb On having standing to challenge application for substitution of vehicles under 'A' licence.

Robinson v Director of Public Prosecutions [1989] RTR 42 HC QBD No special reasons meriting non-endorsement of licence of solicitor speeding to arrive at place and deal with matter that could be (and was) adequately dealt with by his clerk.

Roper v Taylor's Central Garages (Exeter), Limited (1951) 115 JP 445; [1951] 2 TLR 284; (1951) WN (I) 383 HC KBD On degree of knowledge necessary to be guilty of operating return service without road service licence contrary to Road Traffic Act 1930, s 72(10).

Sidery v Evans and Peters [1938] 4 All ER 137; (1939-40) XXXI Cox CC 180; (1938) 102 JP 517; (1939) 160 LTR 12; (1938) 82 SJ 892; (1938-39) LV TLR 54 HC KBD Regularly transporting persons to football matches not special occasion.

Spittle v Thames Grit and Aggregates Limited (1939-40) XXXI Cox CC 6; [1938] 107 LJCL 200; (1938) 158 LTR 374; (1937) 81 SJ 902; (1937) WN (I) 365 HC KBD Unlicensed carriage of goods for payment/reward.

Strutt and another v Clift (1910) WN (I) 212 HC KBD Failed appeal against conviction for unlicensed use of carriage to convey burden.

Swaits v Entwhistle (1929) 93 JP 232; [1929] 2 KB 171; (1929) 67 LJ 380; (1929) 98 LJCL 648; (1930) 142 LTR 22; (1929) 73 SJ 366; (1929) WN (I) 143 HC KBD Road Vehicle (Registration and Licensing) Amendment Regulations 1928, r 2 not ultra vires act of Minister of Transport.

Swift v Norfolk County Council [1955] Crim LR 785 HC QBD On what constitutes a 'sudden attack' for purposes of Motor Vehicles (Driving Licences) Regulations 1950, r 5.

Swires (Isaac) and Sons, Ltd v F and H Croft (Yeadon) (1959) 109 LJ 461 TrTb Applicants declaration as to intended normal user of vehicle cannot be changed without consent of applicant.

Taylor v Emerson [1962] Crim LR 638t; (1962) 106 SJ 552 HC QBD Expired licence was a licence so use of same was improper use of licence.

Taylor v Mead [1961] 1 All ER 626; [1961] Crim LR 411; (1961) 125 JP 286; (1961) 111 LJ 323; (1961) 105 SJ 159; [1961] 1 WLR 435 HC QBD 'Constructed or adapted' for use means later material alteration: here hanging rails inside car to transport samples not changing purpose.

Transport Holding Co and others v Mobile Lifting Services, Ltd (1965) 115 LJ 91 TrTb On need to protect exising businesses when awarding heavy goods licences.

Tynan v Jones [1975] Crim LR 458; [1975] RTR 465 HC QBD Burden on person claiming does have licence (in prosecution for driving without licence) to prove same.

Urey v Lummis [1962] 2 All ER 463; (1962) 126 JP 346; (1962) 112 LJ 473; (1962) 106 SJ 430 HC QBD Visiting force driving permit does not entitle one to supervise provisional driver.

Vincent v Whitehead [1966] 1 All ER 917; [1966] Crim LR 225; (1966) 130 JP 214; (1965-66) 116 NLJ 669; (1966) 110 SJ 112; [1966] 1 WLR 975 HC QBD Structure enabling second person to be carried meant vehicle constructed to carry more than one person.

Walker (HI), Ltd v British Transport Commission and other (1961) 111 LJ 678 TrTb On allowable use of trailer pursuant to 'A' licence.

Wing v TD and C Kelly Ltd [1997] TLR 14 HC QBD On when emergency vehicle owners need not have operator's licence (here found that company ought to have had licence for 'emergency' vehicle).

Woodward v Dykes [1969] Crim LR 33; (1968) 112 SJ 787 HC QBD Valid conviction for knowingly making false statement: stated in driving licence application form that deaf person not suffering from illness/disability that might endanger public.

LICENCE (general trade)

Worgan (TK) and Son, Ltd v Gloucestershire County Council; H Lancaster and Co, Ltd v Gloucestershire County Council [1961] 2 All ER 301; (1961) 125 JP 381; (1961) 111 LJ 304; [1961] 2 QB 123; (1961) 105 SJ 403 CA Tractors altered to carry felled timber were solely for haulage and not goods vehicles.

LICENCE (general trade)

Jones v Argyle Motors (Birkenhead), Ltd [1967] Crim LR 244t; (1967) 111 SJ 279 HC QBD On whether carriage of rubbish in motor dealers' vehicle was breach of terms of general trade licence.

LICENCE (goods vehicle)

Clarke v Cherry [1953] 1 All ER 267; (1953) 117 JP 86; (1953) 97 SJ 80; [1953] 1 WLR 268 HC QBD Goods vehicle is such though goods carried not for sale but for trade use.

James v Davies [1952] 2 All ER 758; (1952) 116 JP 603; (1952) 102 LJ 625; [1953] 1 QB 8; (1952) 96 SJ 729; [1952] 2 TLR 662; (1952) WN (I) 480 HC QBD Unloaded vehicle pulling loaded trailer was a goods vehicle.

Lloyd v E Lee, Ltd [1951] 1 All ER 589; (1951) 115 JP 189; [1951] 2 KB 121; (1951) 95 SJ 206; [1951] 1 TLR 624; (1951) WN (I) 116 HC KBD Licence prohibiting user of vehicle for carriage of goods for hire/reward applies to all; identity certificate required no matter who is using vehicle.

Payne v Allcock (1932) 101 LJCL 775; (1932) 76 SJ 308; (1932) WN (I) 111 HC KBD Offence under Finance Act 1922, s 14 to use vehicle licensed for private use as a goods vehicle.

Pearson v Boyes [1953] 1 All ER 492; (1953) 117 JP 131; (1953) 97 SJ 134; [1953] 1 WLR 384 HC QBD Not using goods vehicle as goods vehicle when simply using to tow caravan.

R v Ipswich Justices, ex parte Robson [1971] 2 All ER 1395; [1971] Crim LR 425t; (1971) 135 JP 462; (1971) 121 NLJ 433t; [1971] 2 QB 340; [1971] RTR 339; (1971) 115 SJ 489; [1971] 3 WLR 102 HC QBD Cannot order licensing authority to do thing it cannot do; person seeking HGV licence not 'aggrieved' if licensing authority cannot issue licence.

R v West Midland Traffic Area Licensing Authority, ex parte Great Western Railway Company [1935] 1 KB 449; (1935) 79 LJ 132; (1935) 152 LTR 501; (1935) 79 SJ 109; (1934-35) LI TLR 227; (1935) WN (I) 19 CA On procedure when applying for goods vehicle ('B') licence.

LICENCE (identity mark)

Caldwell v Hague (1914-15) XXIV Cox CC 595; (1915) 79 JP 152; [1915] 84 LJKB 543; (1915) 112 LTR 502 HC KBD Motor dealer's use of car after had failed to renew expired general identity mark licence was criminal offence.

Phelon and Moore, Limited v Keel (1914-15) XXIV Cox CC 234; (1914) 78 JP 247; [1914] 3 KB 165; [1914] 83 LJKB 1516; (1914-15) 111 LTR 214; (1914) WN (I) 190 HC KBD Non-liability of employer's for worker's unauthorised use of car with employer's identity mark.

LICENCE (locomotive)

Cole Brothers v Harrop [1916] 85 LJKB 494; (1915-16) 113 LTR 1013 HC KBD On what constituted use of locomotive for agricultural purposes under Locomotives Act 1898 (and so obviated need for locomotive licence).

London County Council v Lee (1914-15) XXIV Cox CC 388; [1914] 3 KB 255; [1914] 83 LJKB 1373; (1914-15) 111 LTR 569; (1914) WN (I) 224 HC KBD Conviction for use of locomotive outside county in which licensed.

Morris v Tolman (1921-25) XXVII Cox CC 345; (1922) 86 JP 221; [1923] 92 LJCL 215; (1923) 128 LTR 118 HC KBD Aider and abettor could not be convicted as principal where was excluded by statute.

Williams v Morgan (1921-25) XXVII Cox CC 37; (1921) 85 JP 191 HC KBD Locomotive used by non-farmer without licence/permit was permissible as used for agricultural purpose.

LICENCE (motor cycle)

Brown v Anderson [1965] 2 All ER 1; [1965] Crim LR 313t; (1965) 129 JP 298; [1965] 1 WLR 528 HC QBD Bubble car a 'motor cycle' under Road Traffic Act 1960 but not a motor bicycle under Motor Vehicles (Driving Licences) Regulations 1963.

Keen v Parker [1976] 1 All ER 203; [1976] Crim LR 71t; (1976) 140 JP 137; (1975) 125 NLJ 1088t; [1976] RTR 213; (1975) 119 SJ 761; [1976] 1 WLR 74 HC QBD Motorcycle side-car need not be for carriage of persons to be side-car.

LICENCE (penalty)

Johnson v Finbow (1983) 5 Cr App R (S) 95; [1983] Crim LR 480; (1983) 147 JP 563; (1983) 127 SJ 411; [1983] TLR 203; [1983] RTR 363; [1983] 1 WLR 879 HC QBD Not stopping after accident and not reporting same though at different times were on same occasion for purposes of Transport Act 1981, s 19(1): on appropriate penalty points for same.

R v The Clerk to the Croydon Justices, ex parte Chief Constable of Kent [1989] Crim LR 910; (1990) 154 JP 118; [1991] RTR 257 HC QBD Under Transport Act 1982 fixed penalties are payable by unincorporated bodies.

LICENCE (provisional)

Kinsey v Hertfordshire County Council [1972] Crim LR 564t; (1972) 122 NLJ 704t; [1972] RTR 498; (1972) 116 SJ 803 HC QBD Person granted provisional driving licence but unable to use same until date after that on which driving age raised was not licensed to drive.

Lovell v Archer; Lovell v Ducket [1971] Crim LR 240; (1971) 121 NLJ 128t; [1971] RTR 237; (1971) 115 SJ 157 HC QBD Provisional licence to be endorsed where holder of same does not comply with conditions of licence.

LICENCE (restricted)

McKissock v Rees-Davies [1976] Crim LR 452; [1976] RTR 419; (1976) 120 SJ 422 HC QBD Failed prosecution for non-compliance by one-armed driver with terms of restricted driving licence.

LICENCE (road haulage)

Re An Application by Richard Woodward on behalf of Richard Woodward and Doris Woodward, trading as R and D Transport; Richard Woodward and Doris Woodward, trading as R and D Transport v North Western Traffic Area Licensing Authority [1937] 4 All ER 656; [1938] Ch 331; [1938] 107 LJCh 148; (1938) 158 LTR 111; (1938) 82 SJ 15; (1937-38) LIV TLR 216; (1938) WN (I) 9 HC ChD On transfer of road haulage licence.

LICENCE (road service)

Poole v Ibbotson (1949) 113 JP 466; (1949) LXV TLR 701 HC KBD Successful action for causing motor vehicle to be used as express carriage without road service licence contrary to Road Traffic Act 1934 (club bringing club members to away game by coach).

R v Secretary of State for Transport, ex parte Cumbria County Council [1983] RTR 88 HC QBD Minister could disagree wth factual findings of inspector and (as here) allow appeal against refusal of road service licence.

R v Secretary of State for Transport, ex parte Cumbria County Council [1983] RTR 129 CA Minister's disagreeing with factual findings of inspector and allowing appeal against refusal of road service licence was improper.

R v Traffic Commissioners for the North Western Traffic Area, ex parte British Railways Board [1977] RTR 179 HC QBD Certiorari will not issue to quash refusal of road service licence where have not availed of statutory appeal mechanism.

LICENCE (suspension)

Kidner v Daniels (1911-13) XXII Cox CC 276; (1910) 74 JP 127; (1910) 102 LTR 132 HC KBD
Summary court's suspension of licence runs from date of order (despite pending appeal) unless
order otherwise provides.

LICENCE (tractor)

Henderson v Robson and others (1949) 113 JP 313 HC KBD Using tractor to haul pony not an
agricultural purpose so were liable to higher rate of duty.

LICENCE (trade)

Balch v Beeton [1970] Crim LR 285; [1970] RTR 138 HC QBD Scrap trade different from car
dealership/repair trade so was offence to use vehicle licensed for dealership/repair trade to
transport scrap.

Bowers v Worthington [1982] RTR 400 HC QBD Interpretation and application of Road Vehicles
(Registration and Licensing) Regulations 1971, reg 35(4).

Dark v Western Motor and Carriage Co (Bristol), Ltd [1939] 1 All ER 143; (1939) 83 SJ 195 HC
KBD On eligibility for general trade licence.

Griffiths v Studebakers, Limited (1921-25) XXVII Cox CC 565; (1923) 87 JP 199; [1924] 1 KB
102; (1923) 58 LJ 484; [1924] 93 LJCL 50; (1924) 130 LTR 215; (1923-24) 68 SJ 118;
(1923-24) XL TLR 26; (1923) WN (I) 278 HC KBD Company liable for road traffic offence of
employee acting in course of duty — no mens rea necessary.

Lees v Ravenhill (1921-25) XXVII Cox CC 667; (1924) 88 JP 197; (1924) 59 LJ 681; [1925] 94
LJCL 97; (1925) 132 LTR 201; (1924-25) 69 SJ 177; (1924-25) XLI TLR 36; (1924) WN (I)
300 HC KBD Tourer attached to truck a load for conveyance of which special licence required.

Murphy v Brown [1970] Crim LR 234; [1970] RTR 190; (1969) 113 SJ 983 HC QBD Motor
repairer/dealer not licensed to transport pony (for which had given motor vehicle) to place of sale.

Westover Garage Limited v Deacon (1931-34) XXIX Cox CC 327; (1931) 95 JP 155; (1931) 145
LTR 357; (1930-31) XLVII TLR 509 HC KBD General trade licence did not cover conveyances
of prospective purchasers' goods.

LIGHTING (general)

Blackshaw v Chambers [1942] 2 All ER 678; [1943] KB 44; [1943] 112 LJ 72 HC KBD Cyclist
required only one headlamp on bike.

Brown v Crossley (1911-13) XXII Cox CC 402; (1911) 75 JP 177; [1911] 1 KB 603; (1911) 80
LJCL 478; (1911) 104 LTR 429; (1910-11) XXVII TLR 194; (1911) WN (I) 31 HC KBD
Guilty of offence for failure to produce licence for indorsement after conviction for not having
rear light on car.

Buckoke and others v Greater London Council [1971] Ch 655; (1971) 135 JP 321; (1971) 121 NLJ
154t; [1971] RTR 131; (1971) 115 SJ 174; [1971] 2 WLR 760 CA London Fire Brigade Order
144/8 not unlawful as did not require though did in essence permit violation of red traffic lights
— disciplinary action against persons disputing terms of lawful Order not halted.

Bugge v Taylor (1939-40) XXXI Cox CC 450; (1940) 104 JP 467; [1941] 1 KB 198; [1941] 110
LJ 710; (1941) 164 LTR 312; (1941) 85 SJ 82 HC KBD Valid conviction for keeping unlighted
vehicle on road where 'road' was forecourt of hotel.

Ex parte Symes (1911-13) XXII Cox CC 346; (1911) 75 JP 33; (1910-11) 103 LTR 428; (1910-11)
XXVII TLR 21; (1910) WN (I) 219 HC KBD Licence can be indorsed for failure to drive with
light.

Payne v Harland [1980] RTR 478 HC QBD Headlights and sidelights required on car when driven
by day.

Printz v Sewell (1913-14) XXIII Cox CC 23; (1912) 76 JP 295; [1912] 2 KB 511; (1912) 81 LJKB
905; (1912) 106 LTR 880; (1911-12) XXVIII TLR 396 HC KBD Can plead have taken all
reasonable steps not to obscure illumination of number plate.

Provincial Motor Cab Company v Dunning (Parker's case); The Same v The Same (Kynaston's case) (1911-13) XXII Cox CC 159; [1909] 2 KB 599; (1909-10) 101 LTR 231; (1908-09) XXV TLR 646; [1909] 78 LJKB 822; (1909) 73 JP 387 HC KBD Cab company properly convicted of aiding and abetting identification plate illumination offence by employee.

Swift v Spence [1982] RTR 116 HC QBD Road Vehicles (Use of Lights during Daytime) Regulations 1975, s 2(1), never requires sidelights but not dipped lights to be on.

Webster v Terry (1914) 78 JP 34; [1914] 1 KB 51; (1913) 48 LJ 645; [1914] 83 LJKB 272; (1913-14) 109 LTR 982; (1913-14) XXX TLR 23; (1913) WN (I) 290 HC KBD Motor bicycle must comply with lighting requirements under Motor Cars (Use and Construction) Order 1904.

White v Jackson (1915) 79 JP 447; [1915] 84 LJKB 1900; (1915-16) 113 LTR 783; (1914-15) XXXI TLR 505; (1915) WN (I) 256 HC KBD Person driving car with powerful lights guilty of offence in connection with driving of car: licence therefore to be endorsed.

Witts v Williams [1974] Crim LR 259 HC QBD Not having proper lights a continuing offence: person who drove to house and parked car validly required to take breath test by policeman who had been following/on appproaching driver reasonably suspected him to have been drinking.

LOAD (general)

Bindley v Willett [1981] RTR 19 HC QBD Container meant to be secured to vehicle (though in defective condition) was part of vehicle.

Charman (FE) Ltd v Clow [1974] RTR 543 HC QBD Person who sub-contracts with another to deliver ballast is (for purposes of Weights and Measures Act 1963, s 11(2)) user of tipper lorry employed to perform task.

Cornish v Ferry Masters Ltd and another [1975] Crim LR 241; [1975] RTR 292 HC QBD Requirement that loads be secured so as not to be dangerous creates an absolute offence.

Dent v Coleman and another [1977] Crim LR 753; [1978] RTR 1 HC QBD Were guilty of carrying improperly secured load where straps securing load defective though did not know were so.

Director of Public Prosecutions v Seawheel Limited [1993] Crim LR 707; (1994) 158 JP 444 HC QBD On relevancy of ownership to determining whether person 'using' vehicle in contravention of road traffic legislation.

Friend v Western British Road Services Ltd [1975] Crim LR 521; [1976] RTR 103 HC QBD Could secure load on vehicle by means of weight/positioning of load and not be liable for using dangerous vehicle should slow roll over of articulated vehicle occur (as here).

Lowery (P) and Sons Ltd v Wark [1975] RTR 45 HC QBD Necessary to prove mens rea on part of company charged with permitting use on road of vehicle with inadequately secured load.

Police v Hadelka [1963] Crim LR 706 Magistrates Absolute discharge for person who spent five minutes waiting in restricted area for load which had good reason to expect (but did not actually) receive.

Ross Hillman Ltd v Bond [1974] 2 All ER 287; (1974) 59 Cr App R 42; [1974] Crim LR 261; (1974) 138 JP 428; [1974] QB 435; [1974] RTR 279; (1974) 118 SJ 243; [1974] 2 WLR 436 HC QBD Person to have actual knowledge of facts that are unlawful use to be convicted of causing/permitting same.

Siddle C Cook, Ltd v Arlidge [1962] Crim LR 319; (1962) 106 SJ 154 HC QBD On what constitutes abnormal indivisible load.

Smith and another v North-Western Traffic Area Licensing Authority [1974] Crim LR 193; [1974] RTR 236 HC QBD Vehicle carrying twelve separate beams not carrying abnormal 'indivisible' load but vehicle carrying same found to breach vehicle construction/use regulations.

Turberville v Wyer; Bryn Motor Co Ltd v Wyer [1977] RTR 29 HC QBD Where coil inexplicably fell from lorry was prima facie evidence that lorry insecurely loaded; that part of lorry carrying load possibly mistakenly described as trailer in information was irrelevant.

LOCOMOTIVE (general)

Carpenter v Fox (1931-34) XXIX Cox CC 42; (1929) 93 JP 239; [1929] 2 KB 458; (1929) 68 LJ 8; (1929) 98 LJCL 779; (1928-29) XLV TLR 571; (1929) WN (I) 191 HC KBD Prosecution failed as heavy locomotive was neither motor car nor heavy motor car under relevant Orders.

Cole Brothers v Harrop [1916] 85 LJKB 494; (1915-16) 113 LTR 1013 HC KBD On what constituted use of locomotive for agricultural purposes under Locomotives Act 1898 (and so obviated need for locomotive licence).

Cooper v Hawkins [1904] 2 KB 164; (1902-03) 47 SJ 691; (1903-04) 52 WR 233 HC KBD Local authority restrictions on locomotive speed under Locomotives Act 1865 could not apply to Crown locomotive driven by Crown servant on Crown business.

Director of Public Prosecutions v Yates and another [1989] RTR 134 HC QBD Specially constructed towing vehicle a locomotive.

Dobson v Jennings (1919) 83 JP 259; [1920] 1 KB 243; [1920] 89 LJCL 281; (1919) WN (I) 272 HC KBD No need to specially license locomotive used/intended for use in special capacity contained in Locomotives Act 1898, s 9(1).

Evans v Nicholl (1911-13) XXII Cox CC 70; (1909) 73 JP 154; [1909] 1 KB 778; [1909] 78 LJKB 428; (1909) C LTR 496; (1908-09) XXV TLR 239; (1909) WN (I) 25 HC KBD Duty to have other person present on highway under Heavy Motor Car Order 1904.

Hindle and Palmer v Noblett (1908) 72 JP 373; (1908-09) XCIX LTR 26 HC KBD Valid conviction for using motor engines on highway that did not (contrary to Locomotives on Highways Act 1896) so far as was practicable consume their own smoke.

Hoddell (P and A) v Parker (1910) 74 JP 315; (1910-11) 103 LTR 2; (1910) WN (I) 146 HC KBD Locomotive drawing corn to market not an 'agricultural' locomotive for purposes of Locomotives Act 1898, s 9.

Keeble v Miller [1950] 1 All ER 261; (1950) 114 JP 143; [1950] 1 KB 601; (1950) 94 SJ 163; [1950] 66 (1) TLR 429; (1950) WN (I) 59 HC KBD 'Construction' of car determined on date of charge: here heavy motor car had become light locomotive.

London County Council v Lee (1914-15) XXIV Cox CC 388; [1914] 3 KB 255; [1914] 83 LJKB 1373; (1914-15) 111 LTR 569; (1914) WN (I) 224 HC KBD Conviction for use of locomotive outside county in which licensed.

Mayhew v Sutton (1901-07) XX Cox CC 146; (1901) 36 LJ 580; [1902] 71 LJCL 46; (1901-02) 46 SJ 51; (1901-02) XVIII TLR 52; (1901-02) 50 WR 216 HC KBD Can be guilty of driving light locomotive on highway 'to . . . danger of passengers' (Light Locomotives on Highways Order 1896, s 6) even though no passengers on highway.

Morgan v Ennion (1920) 84 JP 205; (1920) 123 LTR 399; (1920) WN (I) 296 HC KBD Valid conviction for not having communication cord from waggons to locomotive (travelling on road).

Roche v Willis [1934] All ER 613; (1934-39) XXX Cox CC 121; (1934) 98 JP 227; (1934) 151 LTR 154 HC KBD Agricultural tractor driven on farm exempt from licensing requirements of Road Traffic Act 1930, s 9(3).

Smith and Sons v Pickering (1914-15) XXIV Cox CC 570; [1915] 1 KB 326; (1914) 49 LJ 677; [1915] 84 LJKB 262; (1915) 112 LTR 452; (1914) WN (I) 426 HC KBD Locomotive as waggon under Locomotives Act 1898, s 2.

Smith v Boon (1899-1901) XIX Cox CC 698; (1901) LXXXIV LTR 593; (1900-01) 45 SJ 485; (1900-01) XVII TLR 472; (1900-01) 49 WR 480 HC KBD Excessive speed determined by reference to traffic on road not immediately around motor vehicle concerned.

Star Omnibus Company (London) Limited v Tagg (1907-09) XXI Cox CC 519; (1907) 71 JP 352; (1907-08) XCVII LTR 481; (1906-07) 51 SJ 467; (1906-07) XXIII TLR 488 HC KBD Failed prosecution against locomotive owner for non-consumption by locomotive of own smoke.

Williams v Morgan (1921-25) XXVII Cox CC 37; (1921) 85 JP 191 HC KBD Locomotive used by non-farmer without licence/permit was permissible as used for agricultural purpose.

Yorkshire (Woollen District) Electric Tramways Limited v Ellis (1901-07) XX Cox CC 795; (1904-05) 53 WR 303 HC KBD Light railway carriage not an omnibus nor a hackney carriage.

MAKING FALSE STATEMENT (general)

R v Cummerson (1968) 52 Cr App R 519; [1968] Crim LR 395t; [1968] 2 QB 534; (1968) 112 SJ 424; [1968] 2 WLR 1486 CA Making false statement under Road Traffic Act 1960, s 235(2)(a) an absolute offence.

MOTOR MANSLAUGHTER (general)

Andrews v Director of Public Prosecutions [1937] 2 All ER 552; (1936-38) 26 Cr App R 34; (1937) 101 JP 386; (1937) 83 LJ 304; [1937] 106 LJCL 370; (1937) 81 SJ 497; (1936-37) LIII TLR 663; (1937) WN (I) 188 HL Can be convicted of reckless driving where negligence insufficient to sustain manslaughter charge in case where victim dies.

de Clifford, The Trial of Lord (1936) 81 LJ 60 HL Failed prosecution of peer for manslaughter (following motor accident).

Jennings v United States Government [1982] 3 All ER 104 (also HC QBD); [1983] 1 AC 624; (1982) 75 Cr App R 367; [1982] Crim LR 748; (1982) 146 JP 396; (1982) 132 NLJ 881; [1983] RTR 1 (also HC QBD); (1982) 126 SJ 659; [1982] TLR 424; [1982] 3 WLR 450 HL Causing death by reckless driving is manslaughter; character of offence for which sought determines if extradition possible.

R v Andrews (1934-39) XXX Cox CC 576; (1937) 156 LTR 464 HL Death through negligent driving to be treated as death through any form of negligence: law of manslaughter remains the same.

R v Gault [1996] RTR 348 CA Slow running over of person with whom had tempestuous relationship properly found to be manslaughter: six years' imposed.

R v Governor of Holloway Prison, ex parte Jennings [1982] Crim LR 590; (1982) 132 NLJ 488; [1983] RTR 1 (also HL); (1982) 126 SJ 413; [1982] TLR 210; [1982] 1 WLR 949 HC QBD Cannot extradite person for manslaughter where evidence suggests are guilty of death by reckless driving.

R v Leach [1937] 1 All ER 319 CCA Possible that person's negligence might be sufficient to sustain conviction for reckless driving but not for manslaughter where victim died.

R v Seymour [1983] 2 All ER 1058; [1983] 2 AC 493; (1983) 77 Cr App R 215; [1983] Crim LR 742; (1984) 148 JP 530; (1983) 133 NLJ 746; [1983] RTR 455; (1983) 127 SJ 522; [1983] TLR 525; [1983] 3 WLR 349 HL Direction to jury where manslaughter arising from reckless driving.

R v Seymour (Edward John) (1983) 76 Cr App R 211; [1983] Crim LR 260; [1983] RTR 202 CA Direction to jury where manslaughter arising from reckless driving.

R v Shipton, ex parte Director of Public Prosecutions [1957] 1 All ER 206; (1957) 101 SJ 148 HC QBD Whether quarter sessions could try dangerous driving when indictment for manslaughter/ accused acquitted of manslaughter.

R v Stringer (1931-34) XXIX Cox CC 605; (1933) 97 JP 99; [1933] 1 KB 704; (1933) 75 LJ 96; (1933) 102 LJCL 206; (1933) 148 LTR 503; (1933) 77 SJ 65; (1932-33) XLIX TLR 189; (1933) WN (I) 28 CCA Could be tried together for manslaughter/dangerous driving and convicted of latter.

MOTORWAY (general)

Higgins v Bernard [1972] 1 All ER 1037; [1972] Crim LR 242t; (1972) 136 JP 314; (1972) 122 NLJ 153t; [1972] RTR 304; (1972) 116 SJ 179; [1972] 1 WLR 455 HC QBD Drowsiness not an emergency justifying stopping on motorway.

Mawson v Oxford [1987] Crim LR 131; [1987] RTR 398 HC QBD Under motorway regulations (1982) vehicles never entitled to stop on motorway carriageway: endorsement had to be ordered.

Wallwork v Rowland [1972] 1 All ER 53; [1972] Crim LR 52t; (1972) 136 JP 137; (1971) 121 NLJ 1025t; [1972] RTR 86; (1972) 116 SJ 17 HC QBD Motorway hard shoulder not part of motorway.

NEGLIGENT DRIVING (general)

Amos v Glamorgan County Council (1967) 117 NLJ 1243t CA Fire brigade not contributorily negligent where motorcyclist not wearing his glasses crashed into rear of fire engine which was stationary/had its lights flashing while firemen attended fire.

Burgoyne v Phillips [1983] Crim LR 265; (1983) 147 JP 375; [1983] RTR 49; [1982] TLR 559 HC QBD On what constitutes 'driving' a car.

Chisholm v London Passenger Transport Board [1938] 4 All ER 850; [1939] 1 KB 426; (1939) 87 LJ 27; [1939] 108 LJCL 239; (1939) 160 LTR 79; (1938) 82 SJ 1050; (1938-39) LV TLR 284; (1939) WN (I) 15 CA Driver not negligent in striking pedestrian already on crossing: case decided by reference to common law principles.

Dickens v Kay-Green (1961) 105 SJ 949 CA Driver who strayed onto wrong side of road during heavy fog deemed to be negligent.

Grange Motors (Cwmbrian) Ltd v Spencer and another [1969] 1 All ER 340; (1969) 119 NLJ 154t; (1968) 112 SJ 908; [1969] 1 WLR 53 CA On duty of care owed by one driver signalling to another.

James v Audigier (1932) 74 LJ 407; (1932-33) XLIX TLR 36; (1932) WN (I) 250 CA Dismissal of appeal over questions that were asked in running-down case over earlier accident in which defendant involved.

James v Audigier (1931-32) XLVIII TLR 600; (1932) WN (I) 181 HC KBD On appropriate questions which may be put to driver in 'running down' case.

Lewis v Ursell [1983] TLR 282 HC QBD Collision of motor car with gateway an accident that occurred owing to presence of that vehicle on road within meaning of the Road Traffic Act 1972, s 8(2).

R v Baldessare (Cyril) (1930-31) Cr App R 70; (1931) 144 LTR 185; (1930) WN (I) 193 CCA Passenger in recklessly driven car may be guilty of criminal negligence by driving.

Snelling v Whitehead (1975) 125 NLJ 870t HL Decision as to who was negligent in motor accident involving young boy.

Tart v GW Chitty and Co, Ltd [1931] All ER 826; [1933] 2 KB 453; (1933) 102 LJCL 568; (1933) 149 LTR 261 HC KBD Contributory negligence where driving so fast could not avoid collision or were not keeping proper look-out.

NOTICE OF INTENDED PROSECUTION (general)

Archer v Blacker [1965] Crim LR 165t; (1965) 109 SJ 113 HC QBD On onus of proof as regards proving that notice of road traffic prosecution served.

Beer v Davies [1958] 2 All ER 255; (1958) 42 Cr App R 198; [1958] Crim LR 398t; (1958) 122 JP 344; (1958) 108 LJ 345; [1958] 2 QB 187; (1958) 102 SJ 383 HC QBD Information rightly dismissed where non-delivery of notice not accused's fault.

Bentley v Dickinson [1983] Crim LR 403; (1983) 147 JP 526 HC QBD Properly acquitted of driving without due care and attention where had not known of accident and was not served with notice of intended prosecution.

Carr v Harrison [1967] Crim LR 54t; (1965-66) 116 NLJ 1573t; (1967) 111 SJ 57 HC QBD Notice of intended prosecution not invalid though contained wrong time of offence as no injustice resulted from the error.

Clarke v Mould [1945] 2 All ER 551; (1945) 109 JP 175; (1945) 173 LTR 370; (1945) 89 SJ 370; (1945) WN (I) 125 HC KBD In adequate notice of prosecution as served on firm rather than individual member thereof.

Director of Public Prosecutions v Pidhajeckyi [1991] Crim LR 471 CA Allegedly careless driver suffering from post-accident amnesia (could not remember accident) ought nonetheless to have been sent notice of prosecution.

Director of Public Prosecutions v Pidhajeckyj (1991) 155 JP 318; [1990] TLR 743 HC QBD Proscution not required to serve notice within fourteen days on person suffering post-accident amnesia.

Gibson v Dalton [1980] RTR 410 HC QBD On what constitutes an adequate warning to motorist for purposes of the Road Traffic Act 1972, s 179(2).

Groome v Driscoll [1969] 3 All ER 1638; [1970] Crim LR 47t; (1970) 134 JP 83; [1970] RTR 105; (1969) 113 SJ 905 HC QBD On notices of prosecution.

Harris v Dudding [1954] Crim LR 796 Magistrates Notice of prosecution valid where indicated liability to prosecution/category of offence at issue.

Holt v Dyson [1950] 2 All ER 840; (1950) 100 LJ 595; (1950) 114 JP 558; [1951] 1 KB 364; (1950) 94 SJ 743; [1950] 66 (2) TLR 1009; (1950) WN (I) 498 HC KBD Improper notice of prosecution if send notice to address where know accused not going to receive it.

Hosier v Goodall [1962] 1 All ER 30; (1962) 126 JP 53; (1962) 112 LJ 105; [1962] 2 QB 401; (1961) 105 SJ 1085 HC QBD Authorised person receiving notice valid serving thereof (even though police knew person on whom served not at address to which notice sent).

Jollye v Dale [1960] 2 All ER 369; [1960] Crim LR 565; (1960) 124 JP 333; (1960) 110 LJ 415; [1960] 2 QB 258; (1960) 104 SJ 467; [1960] 2 WLR 1027 HC QBD One and a half hour delay before notice of intended prosecution given was justified in circumstances.

Layton v Shires [1959] 3 All ER 587; (1960) 44 Cr App R 18; [1959] Crim LR 854; (1960) 124 JP 46; (1959) 109 LJ 652; [1960] 2 QB 294; (1959) 103 SJ 856; [1959] 3 WLR 949 HC QBD Notice of prosecution properly served if registered letter containing notice signed for by authorised person.

Lund v Thompson [1958] 3 All ER 356; (1959) 43 Cr App R 9; [1958] Crim LR 816t; (1958) 122 JP 489; (1958) 108 LJ 793; [1959] 1 QB 283; (1958) 102 SJ 811 HC QBD Post-notice of prosecution indication by police that would not prosecute did not make notice inoperative.

Milner v Allen [1933] All ER 734; (1933) 97 JP 111; [1933] 1 KB 698; (1933) 102 LJCL 395; (1933) 149 LTR 16; (1933) 77 SJ 83; (1932-33) XLIX TLR 240; (1933) WN (I) 30 HC KBD Notice of prosecution adequate though offence mentioned therein not offence with which ultimately charged.

Nicholson v Tapp [1972] 3 All ER 245; [1972] Crim LR 510t; (1972) 136 JP 719; (1972) 116 SJ 527; [1972] 1 WLR 1044 HC QBD Notice to be delivered to addressee within 14 days to be valid.

Percival v Ball (1937) WN (I) 106 HC KBD Notice of intended prosecution for careless driving deemed adequate.

Phipps v McCormick [1972] Crim LR 540; (1971) 115 SJ 710 HC QBD Last known address for correspondence (rather than hospital address) was appropriate place to which to send notice of intended prosecution.

Pope v Clarke [1953] 2 All ER 704; (1953) 117 JP 429; (1953) 103 LJ 525; (1953) 97 SJ 542; [1953] 1 WLR 1060 HC QBD Notice of prosecution valid though time of offence charged incorrectly stated therein.

R v Bilton [1964] Crim LR 828g CCA Notice of intended prosecution served on branch (not registered) office of company in respect of motor offence involving company car was valid.

R v Bolkis [1932] All ER 836; (1932-34) 24 Cr App R 19; (1933) 97 JP 10; (1932) 74 LJ 424; (1933) 148 LTR 358; (1933) 77 SJ 13; (1932-33) XLIX TLR 128; (1933) WN (I) 13 CCA On proper notice of intended prosecution.

R v Edmonton Justices, ex parte Brooks [1960] 2 All ER 475; [1960] Crim LR 564; (1960) 124 JP 409; (1960) 110 LJ 462; (1960) 104 SJ 547 HC QBD Need not contend as preliminary point that no notice of intended prosecution: can wait until later.

R v Jennings [1956] 3 All ER 429; (1956) 40 Cr App R 147; [1956] Crim LR 698; (1956) 100 SJ 861; [1956] 1 WLR 1497 CMAC Could convict of dangerous driving though not notified of prosecution fourteen days beforehand.

R v Rowbotham [1960] Crim LR 429 Sessions Notice of road traffic prosecution void as did not state precise locus of offence.

Rogerson v Edwards (1951) 95 SJ 172; (1951) WN (I) 101 HC KBD Inappropriate for justices to dismiss informations outright, without even hearing them just because notice of intended prosecution mis-addressed.

Sanders v Scott [1961] 2 All ER 403; [1961] Crim LR 554; (1961) 125 JP 419; (1961) 111 LJ 342; [1961] 2 QB 326; (1961) 105 SJ 383; [1961] 2 WLR 864 HC QBD Must prove notice of prosecution not served on driver or on registered owner to establish defence.

Sandland v Neale [1955] 3 All ER 571; (1955) 39 Cr App R 167; [1956] Crim LR 58; (1955) 119 JP 583; (1955) 105 LJ 761; [1956] 1 QB 241; (1955) 99 SJ 799; [1955] 3 WLR 689 HC QBD Notice of prosecution properly served when sent to place most likely to receive it even if know is elsewhere.

Shield v Crighton [1974] Crim LR 605 HC QBD Successful appeal against dismissal of information, originally dismissed because notice of intended prosecution misstated place of offence (despite mistake no confusion could have arisen).

Springate v Questier [1952] 2 All ER 21; (1952) 116 JP 367 HC QBD Notice of prosecution to limited company, notice not containing 'Limited' in name valid.

Stanley v Thomas [1939] 2 All ER 636; (1939-40) XXXI Cox CC 268; (1939) 103 JP 241; [1939] 2 KB 462; [1939] 108 LJCL 906; (1939) 160 LTR 555; (1939) 83 SJ 360; (1938-39) LV TLR 711 HC KBD Notice of prosecution sent to person (in knowledge would not be received by them) valid notice.

Stewart v Chapman [1951] 2 All ER 613; (1951-52) 35 Cr App R 102; [1951] 2 KB 792; (1951) 115 JP 473; (1951) 101 LJ 482; (1951) 95 SJ 641; [1951] 2 TLR 640; (1951) WN (I) 474 HC KBD Time from which notice to be served excludes day offence committed.

Sulston v Hammond [1970] 2 All ER 830; [1970] Crim LR 473t; (1970) 134 JP 601; [1970] RTR 361; (1970) 114 SJ 533; [1970] 1 WLR 1164 HC QBD Notice of intended prosecution not needed for pelican crossing violations.

Taylor v Campbell [1956] Crim LR 342 HC QBD On form of notice of intended prosecution when serving same on registered owner where identity of actual driver of vehicle unknown.

Young v Day [1959] Crim LR 460t HC QBD Valid finding by justices that place where alleged offence committed was inadequately specified in notice of prosecution.

OBSTRUCTION OF HIGHWAY (general)

Absalom v Martin [1973] Crim LR 752; (1973) 123 NLJ 946t HC QBD Billposter's part-parking of van on pavement while putting up bill did not merit conviction for wilful obstruction of highway.

Arrowsmith v Jenkins [1963] Crim LR 353; (1963) 127 JP 289; (1963) 113 LJ 350; (1963) 107 SJ 215; [1963] 2 WLR 856 HC QBD Person who does act that leads to obstruction of highway which did not intend is nonetheless guilty of obstruction of highway.

Attorney General v Gastonia Coaches Ltd (1976) 126 NLJ 1267t; [1977] RTR 219 HC ChD Obstruction of highway by coaches a public nuisance; smell/noise from premises of motor coach operator along with interference in access to private premises a private nuisance.

Baxter v Matthews [1958] Crim LR 263 Magistrates Valid conviction under Highway Act 1835, s 78, for negligent interruption of free passage of carriage on the highway.

Bostock v Ramsey Urban District Council (1899-1900) XVI TLR 18 HC QBD Not malicious prosecution to bring proceedings for obstruction of highway by employee against employer.

Bryant v Marx (1931-34) XXIX Cox CC 545; (1932) 96 JP 383; (1932) 147 LTR 499; (1932) 76 SJ 577; (1931-32) XLVIII TLR 624 HC KBD Obstructing road under Motor Vehicles (Construction and Use) Regulations 1931, r 74(1) includes obstructing footpath.

Carey v Chief Constable of Avon and Somerset [1995] RTR 405 CA Removal by police of converted coach parked by kerb but not obstructing road-users was unlawful.

Divito v Stickings [1948] 1 All ER 207; (1948) 112 JP 166 HC KBD Cannot 'pitch' a van under Highways Act 1935, s 72.

Evans v Barker [1972] Crim LR 53; (1971) 121 NLJ 457t; [1971] RTR 453 HC QBD On when parking of car in road not an obstruction but an unnecessary obstruction.

Gill v Carson and Nield (1917) 81 JP 250; [1917] 2 KB 674; (1917-18) 117 LTR 285 HC KBD On what constitutes wilful obstruction of highway (none here).

Lambeth London Borough Council v Saunders Transport Ltd [1974] Crim LR 311 HC QBD Person who hired out skip not liable for its being unlighted at night, having agreed with person hiring that responsibility for lighting skip would fall on the latter.

Mounsey v Campbell [1983] RTR 36 HC QBD Bumper to bumper parking by van driver occasioned unnecessary obstruction.

Palastanga v Solman [1962] Crim LR 334; (1962) 106 SJ 176 HC QBD Prosecution failure to produce Stationery Office copy of regulations in court and so prove same did not merit dismissal of summons.

Police v O'Connor [1957] Crim LR 478 Sessions That large vehicle properly parked on road did not constitute unreasonable user adequate to support conviction for obstruction of road.

R v Adler [1964] Crim LR 304t CCA Obstruction of highway to be public nuisance must involve unlawful user.

R v Bartholomew (1907-09) XXI Cox CC 556; (1908) 72 JP 79; [1908] 1 KB 554; [1908] 77 LJCL 275; (1907-08) XXIV TLR 238 CCR Coffee stall on highway not common nuisance.

R v Justices of West Riding of York (1911-13) XXII Cox CC 280 HC KBD Licence cannot be indorsed upon conviction for allowing car to stand in highway and so cause obstruction.

R v Lyndon and another, ex parte Moffat (1908) 72 JP 227 HC KBD Causing obstruction by leaving unattended car on highway not offence in connection with driving for which licence can be endorsed.

R v Moule [1964] Crim LR 303t; (1964) 108 SJ 100 CCA Valid conviction for causing unlawful obstruction of highway of person who incited others to sit down and block highway.

R v The Justices of Yorkshire (West Riding), ex parte Shackleton (1910) 74 JP 127; [1910] 1 KB 439; (1910) 79 LJKB 244; (1910) 102 LTR 138; (1910) WN (I) 23 HC KBD Obstructing of highway with car not offence in connection with driving of car for purposes of Motor Car Act 1903, s 4.

Solomon v Durbridge (1956) 120 JP 231 HC QBD Car left in certain place for unduly long period could be (and here was) obstruction of highway.

Torbay Borough Council v Cross and another (1995) 159 JP 682 HC QBD On application of de minimis doctrine to wilful obstruction of highway cases: here inapplicable (shopping display outside shop).

Wade v Grange [1977] RTR 417 HC QBD Doctor blocking road by way in which parked car when answering emergency call was guilty of causing unnecessary obstruction.

Worthy v Gordon Plant (Services) Ltd; WR Anderson (Motors), Ltd v Hargreaves [1989] RTR 7; [1962] 1 All ER 129; [1962] Crim LR 115t; (1962) 126 JP 100; (1962) 112 LJ 154; [1962] 1 QB 425; (1961) 105 SJ 1127 HC QBD Not guilty of obstruction where park vehicles in parking place during operative hours, whatever intent.

OVERTAKING (general)

Burton v Nicholson (1909) 73 JP 107; [1909] 1 KB 397; [1909] 78 LJKB 295; (1909) C LTR 344; (1908-09) XXV TLR 216; (1909) WN (I) 19 HC KBD Must overtake tramcar on the outside.

Clark v Wakelin (1965) 109 SJ 295 HC QBD Delay in road traffic accident actions unhelpful; overtaker not negligent where person overtaken behaved in manner that was not reasonably forseeable.

Connor v Paterson [1977] 3 All ER 516; [1977] Crim LR 428; (1978) 142 JP 20; (1977) 127 NLJ 639t; [1977] RTR 379; (1977) 121 SJ 392; [1977] 1 WLR 1450 HC QBD Was unlawful under 'Zebra' Pedestrian Crossings Regulations 1971/1524 to overtake car which had stopped at crossing to let people cross even though all people crossing had crossed at moment of overtaking.

Davison v Leggett (1969) 133 JP 552; (1969) 119 NLJ 460t CA On burden of proof (as regards establishing negligence) where there is collision in central traffic lane between two overtaking cars coming from opposite directions.

Dilks v Bowman-Shaw [1981] RTR 4 HC QBD Decision that overtaking in left-hand lane of motorway not inconsiderate driving was valid.

Gullen v Ford; Prowse v Clarke [1975] 2 All ER 24; [1975] Crim LR 172; (1975) 139 JP 405; (1975) 125 NLJ 111t; [1975] RTR 302; (1975) 119 SJ 153; [1975] 1 WLR 335 HC QBD Overtaking of person stopped because sees person about to cross crossing is offence.

Squire v Metropolitan Police Commissioner [1957] Crim LR 817t HC QBD Was prima facie case of dangerous driving against person simultaneously overtaking bus and two preceding cars while was oncoming traffic.

PARKING (general)

Arthur and another v Anker and another [1996] 3 All ER 783; (1996) 146 NLJ 86; [1997] QB 564; [1996] RTR 308; [1995] TLR 632; [1996] 2 WLR 602 CA Unauthorised parking on land (consequences of which were posted) meant took risk of clamping/having to pay for removal of clamp.

Beames v Director of Public Prosecutions [1989] Crim LR 659; [1990] RTR 362 HC QBD Driver could not use unexpired time of another driver and (once expired) add coins to meter up to maximum time (Parking Places and Controlled Zone (Manchester) Order 1971).

Boulton v Pilkington [1981] RTR 87 HC QBD Authority to wait in restricted area when loading/unloading meant for trade purposes not while collecting take-away.

Bowman v Director of Public Prosecutions [1990] Crim LR 600; (1990) 154 JP 524; [1991] RTR 263; [1990] TLR 52 HC QBD Justices can rely on local knowledge in determining whether particular car park a public place.

Bulman v Godbold [1981] RTR 242 HC QBD Not unlawful waiting for frozen fish deliverer to halt on restricted street but was not attending van (with motor still running) where entered premises to re-load fish into hotel refrigerator.

Bunting v Holt [1977] RTR 373 HC QBD Point where curving kerb joined adjacent road, not nominal intersection point, was point from which distance of parked car from junction to be measured.

Capell v Director of Public Prosecutions (1991) 155 JP 361 HC QBD Off-public road parking bay deemed to be public place.

Clifford-Turner v Waterman [1961] 3 All ER 974; [1962] Crim LR 51; (1961) 111 LJ 757; (1961) 105 SJ 932 HC QBD Waiting and loading regulations allow to stop for as long as takes someone to immediately get in/out.

Cooper v Hall [1968] 1 All ER 185; [1968] Crim LR 116t; (1968) 132 JP 152; (1967) 117 NLJ 1243t; (1967) 111 SJ 928; [1968] 1 WLR 360 HC QBD Not defence to parking in restricted area that was no yellow line.

Coote and another v Stone (1970) 120 NLJ 1205; [1971] RTR 66 CA Parked car on clearway not a common law nuisance.

Cran and others v Camden London Borough Council [1995] RTR 346 HC QBD Parking designation order quashed where council had engaged in inadequate consultation regarding same.

Crossland v Chichester District Council [1984] RTR 181 CA Punitive amount payable if failed to display 'pay and display' parking ticket was a charge so local government measure complied with Road Traffic Regulation Act 1967, s 31(1) (as amended).

Derrick v Ryder [1972] Crim LR 710; [1972] RTR 480 HC QBD Person to be convicted for parking on single yellow line on Sunday (even though reading of Highway Code might indicate restriction did not operate on Sunday).

Elkins v Cartlidge [1947] 1 All ER 829; (1947) 177 LTR 519; (1947) 91 SJ 573 HC KBD Parking enclosure a public place.

Evans v Barker [1972] Crim LR 53; (1971) 121 NLJ 457t; [1971] RTR 453 HC QBD On when parking of car in road not an obstruction but an unnecessary obstruction.

Funnell v Johnson [1962] Crim LR 488g HC QBD On burden of proof as regards proving that person had been waiting in restricted street.

George v Garland [1980] RTR 77 HC QBD Police officer exempted from disabled parking space requirements where acting in course of duty.

Goode v Hayward [1976] Crim LR 578 CrCt Successful appeal against conviction for illegal parking where double yellow lines at issue had been wiped out by road works.

Hassan v Director of Public Prosecutions (1992) 156 JP 852; [1992] RTR 209 HC QBD Single yellow line ineffective absent accompanying warning sign.

Houghton and another v Scholfield [1973] Crim LR 126; [1973] RTR 239 HC QBD Cul de sac validly found to be road to which public had access.

Keene v Muncaster [1980] Crim LR 587; (1980) 130 NLJ 807; [1980] RTR 377; (1980) 124 SJ 496 HC QBD Police officer could not give himself permission not to leave left/near side of parked vehicle turned to carriageway at night (Motor Vehicles (Construction and Use Regulations) 1973, reg 115 interpreted).

Kelly v WRN Contracting, Ltd and another (Burke, third party) [1968] 1 All ER 369; (1968) 112 SJ 465; [1968] 1 WLR 921 Assizes Action for breach of statutory duty if car parked contrary to regulations (but not negligently) helped cause accident.

Levinson v Powell [1967] 3 All ER 796; [1967] Crim LR 542t; (1968) 132 JP 10; (1967) 117 NLJ 784t; (1967) 111 SJ 871; [1967] 1 WLR 1472 HC QBD Taxi not a goods vehicle: waiting restrictions apply; taxi driving to place requested to observe law.

Mounsey v Campbell [1983] RTR 36 HC QBD Bumper to bumper parking by van driver occasioned unnecessary obstruction.

Norton v Hayward [1969] Crim LR 36; (1968) 112 SJ 767 HC QBD That part of land to which local authority prohibited waiting order extended was private did not mean order did not apply in respect of same.

O'Neill v George [1969] Crim LR 202; (1969) 113 SJ 128 HC QBD Valid to convict for parking without payment in bay where was free parking on bank holidays but not on day when banks closed because of special royal proclamation.

Oxford v Austin [1981] RTR 416 HC QBD On determining whether car park a 'road'.

Parish v Judd [1960] 3 All ER 33; (1960) 124 JP 444; (1960) 110 LJ 701; (1960) 104 SJ 644 HC QBD Prima facie negligent to leave unlit car in unlit place on road at night (not here as lamp/reasonable care); nuisance if unlit car in unlit place on road at night is dangerous obstruction.

Police v Hadelka [1963] Crim LR 706 Magistrates Absolute discharge for person who spent five minutes waiting in restricted area for load which had good reason to expect (but did not actually) receive.

Pratt v Hayward [1969] 3 All ER 1094; [1969] Crim LR 377t; (1969) 133 JP 519; (1969) 119 NLJ 414t; (1969) 113 SJ 369; [1969] 1 WLR 832 HC QBD Waiting and loading to happen together if waiting prohibition not to apply.

R v Derby Crown Court, ex parte Sewell [1985] RTR 251 HC QBD Car parked under 15 yards from first paving stone of curving kerb at junction ought to have had its lights lit.

R v Parking Adjudicator, ex parte Wandsworth London Borough Council (1996) 140 SJ LB 261; (1996) TLR 26/11/96 CA Absent rebutting evidence person who keeps car is presumed to be registerd owner of same (and so liable where car has been illegally parked).

R v Parking Adjudicator, ex parte Wandsworth London Borough Council (1996) TLR 22/7/96 CA Issue of fact to be resolved by Parking Adjudicator as to who is owner of car at moment of illegal parking.

Rawlinson v Broadley (1969) 113 SJ 310 HC QBD Valid conviction for not acting as required in 'pay-and-display' car park.

Richards v McKnight [1977] 3 All ER 625; [1976] Crim LR 749; (1977) 141 JP 693; (1976) 126 NLJ 1091t; [1977] RTR 289; (1976) 120 SJ 803; [1977] 1 WLR 337 HC QBD If use of vehicle not necessary to performance of task waiting prohibition applies.

Riley v Hunt [1980] Crim LR 382; [1981] RTR 79 HC QBD Requirement that parking charge be paid 'on leaving . . . vehicle in . . . parking place' meant before left car park.

Roberts v Powell [1966] Crim LR 225; (1965-66) 116 NLJ 445g; (1966) 110 SJ 113 HC QBD Successful appeal against conviction for unlawful parking where parked in parking bay (albeit one where meter had been removed).

Rodgers v Taylor [1987] RTR 86 HC QBD Exception to restricted waiting provisions applied to hackney carriages awaiting hire not to taxis left parked and unattended on street.

Sandy v Martin [1974] Crim LR 258; (1975) 139 JP 241; [1974] RTR 263 HC QBD Inn car park a public place during hours that persons invited to use it but absent contrary evidence not thereafter (so were not drunk and in charge of motor vehicle in public place).

PASSENGER VEHICLE (general)

Startin v Solihull Metropolitan Borough Council (1978) 128 NLJ 1072t; [1979] RTR 228 HC QBD Charge of £10 for late-leaving of car in car park (reduced to £1.50 if paid in certain time) not unreasonable.

Strong v Dawtry [1961] 1 All ER 926; [1961] Crim LR 319t; (1961) 125 JP 378; (1961) 111 LJ 240; (1961) 105 SJ 235 HC QBD Are guilty of not paying parking meter if park then dash for change and return immediately.

Wade v Grange [1977] RTR 417 HC QBD Doctor blocking road by way in which parked car when answering emergency call was guilty of causing unnecessary obstruction.

Wilson v Arnott [1977] 2 All ER 5; [1977] Crim LR 43; (1977) 141 JP 278; (1976) 126 NLJ 1220t; [1977] RTR 308; (1976) 120 SJ 820; [1977] 1 WLR 331 HC QBD Permissible to wait in parking bay whose use suspended.

PASSENGER VEHICLE (general)

R v South Wales Traffic Licensing Authority, ex parte Ebbw Vale Urban District Council [1951] 1 All ER 806; (1951) 115 JP 278; [1951] 2 KB 366; [1951] 1 TLR 742; (1951) WN (I) 192 CA Licensing authority could hear/decide application by company wholly owned by the British Transport Commission.

Smith v London Transport Executive [1951] 1 All ER 667; (1951) 115 JP 213; [1951] 1 TLR 683; (1951) WN (I) 157 HL On rôle of British Transport Commission.

Smith v London Transport Executive [1948] 2 All ER 306; [1948] Ch 652; [1948] 117 LJR 1483; (1948) LXIV TLR 426; (1948) WN (I) 292 HC ChD On rôle of British Transport Commission.

Smith v London Transport Executive [1949] Ch 685; (1949) 99 LJ 401; (1949) LXV TLR 538; (1949) WN (I) 315 CA On rôle of British Transport Commission.

PETROL (general)

Appleyard v Bingham [1914] 1 KB 258; (1913) 48 LJ 660; [1914] 83 LJKB 193; (1913) WN (I) 300 HC KBD Breach of petrol storage regulations.

Chapman v O'Hagan; Chapman v Smith [1949] 2 All ER 690; (1949) 113 JP 518; (1949) 99 LJ 611; (1949) LXV TLR 657; (1949) WN (I) 399; (1949) 113 JP 464 HC KBD Need not seek to put analyst's certificate in evidence where respondent admitted had commercial petrol in tank.

Grandi and others v Milburn [1966] 2 All ER 816; [1966] Crim LR 450t; [1966] 2 QB 263; (1966) 110 SJ 409; [1966] 3 WLR 90 HC QBD Absent licence is offence for occupier/tanker owner to sell petrol directly from tanker on occupier's property.

Simmonds v Pond (1918-21) XXVI Cox CC 365; (1919) 83 JP 56; [1919] 88 LJCL 857; (1919) 120 LTR 124; (1918-19) XXXV TLR 187; (1919) WN (I) 25 HC KBD Conviction for unauthorised use of petrol was one connected with use of motor car.

Spicer v Clark (1950) 100 LJ 288; (1949) LXV TLR 674 HC KBD Taking petrol from tank in single flow but squirting it into three bottles was taking single sample divided into three parts as required under Motor Spirit (Regulation) Act 1948, s 10(3).

Taylor v Ciecierski [1950] 1 All ER 319; (1950) 114 JP 162; (1950) 94 SJ 164; (1950) WN (I) 80 HC KBD Commercial petrol in private car.

PROTECTIVE HEADGEAR (general)

Losexis Ltd v Clarke [1984] RTR 174; [1983] TLR 323 HC QBD Successful prosecution for sale of crash helmet that did not comply with British Standard specifications: no defence that helmet for off-road use only.

R v Aylesbury Crown Court, ex parte Cholal [1976] Crim LR 635; [1976] RTR 489 HC QBD Sikh person was obliged to wear motor cycle helmet (though that required removal of turban which could not do for religious reasons).

'PUBLIC PLACE' (general)

R v Collinson (Alfred Charles) (1931-2) 23 Cr App R 49; (1931) 75 SJ 491 CCA Admission of public to field (not normally open to public) made field a public place under Road Traffic Act 1930, s 15(1).

PUBLIC SERVICE VEHICLE (general)

Appleby (RW) Ltd and another v Vehicle Inspectorate [1994] RTR 380 HC QBD Non-liability of parent company for non-provision of tachograph records by subsidiary to employee; tachograph necessary when driving from depot to docks to collect passengers (no notion of 'positioning' journey under the tachograph legislation).

Armstrong v Ogle (1926-30) XXVIII Cox CC 253; (1926) 90 JP 146; [1926] 2 KB 438; (1926) 61 LJ 474; [1926] 95 LJCL 908; (1926) 135 LTR 118; (1925-26) XLII TLR 553; (1926) WN (I) 156 HC KBD Conviction of omnibus driver for unlicensed plying for hire.

Attorney-General v Premier Line Limited (1931) 72 LJ 425; (1932) 101 LJCh 132; (1932) 146 LTR 297; (1931) 75 SJ 852; (1931) WN (I) 270 HC ChD Construction of licensing of motor-coach service provisions under Road Traffic Act 1930.

Attorney-General v Sharp (1930) 70 LJ 62; (1928-29) XLV TLR 628 HC ChD Picking-up passengers who had bought tickets in garage outside area where bus company allowed ply for hire was plying for hire in that area by bus company.

Attorney-General v Sharp (1930) 94 JP 234; (1930) 99 LJCh 441; (1930) 143 LTR 367; (1929-30) XLVI TLR 554 CA Because statute creating offence provided only for fine did not preclude granting of injunction where appropriate upon proof of offence.

Barnett v French (1981) 72 Cr App R 272; [1981] Crim LR 415; (1981) 131 NLJ 447; [1981] RTR 173; (1981) 125 SJ 241; [1981] 1 WLR 848 HC QBD Non-endorsement of convict's licence appropriate as was not personally culpable but nominated by Government Department as person responsible for offence involving public service vehicle.

Birmingham and Midland Motor Omnibus Company v Thompson (1918) 82 JP 213; [1918] 2 KB 105; [1918] 87 LJCL 915; (1918-19) 119 LTR 140; (1918) WN (I) 131 HC KBD Omnibus must always have licence when plying for hire.

Dennis v Miles (1921-25) XXVII Cox CC 649; (1924) 88 JP 105; [1924] 2 KB 399; [1924] 93 LJCL 1115; (1924) 131 LTR 146; (1923-24) 68 SJ 755; (1923-24) XL TLR 643; (1924) WN (I) 173 HC KBD Properly convicted for overloaded motor omnibus under Railway Passenger Duty Act 1842.

Director of Public Prosecutions v Sikondar [1993] Crim LR 76; (1993) 157 JP 659; [1993] RTR 90; [1992] TLR 251 HC QBD Any carriage beyond pure generosity (whether or not pursuant to contract) was use of car for hire/reward.

Edwards v Rigby [1980] RTR 353 HC QBD Bus driver who had girl take actions which resulted in her being dragged by bus did not fail to take all reasonable steps to ensure passenger safety.

Ellis v Smith [1962] 3 All ER 954; [1963] Crim LR 128; (1963) 127 JP 51; (1963) 113 LJ 26; (1962) 106 SJ 1069 HC QBD Until relief (bus) driver assumes charge last (bus) driver is person in charge of vehicle.

Evans v Dell [1937] 1 All ER 349; (1934-39) XXX Cox CC 558; (1937) 101 JP 149; (1937) 156 LTR 240; (1937) 81 SJ 100; (1936-37) LIII TLR 310 HC KBD Owner's innocent unawareness that vehicle used as unlicensed stage carriage precluded liability therefor.

Gough v Rees (1931-34) XXIX Cox CC 74; (1930) 94 JP 53; (1930) 142 LTR 424; (1929-30) XLVI TLR 103 HC KBD Absent owner could be convicted of aiding/abetting/counselling/procuring overcrowding offence.

Greyhound Motors Limited v Lambert (1926-30) XXVIII Cox CC 469; (1927) 91 JP 198; [1928] 1 KB 322; (1927) 64 LJ 358; (1928) 97 LJCL 122; (1928) 138 LTR 269; (1927) 71 SJ 881; (1927) WN (I) 271 HC KBD Conviction possible for plying for hire without notice where possible that passengers collected not all pre-paid passengers.

Griffin v Grey Coaches Limited and another (1926-30) XXVIII Cox CC 576; (1929) 93 JP 61; (1929) 67 LJ 32; (1929) 98 LJCL 209; (1929) 140 LTR 194; (1928) 72 SJ 861; (1928-29) XLV TLR 109; (1928) WN (I) 313 HC KBD Motor coach owner plying for hire though no motor coach present.

Kelani Valley Motor Transit Company, Limited v Colombo-Ratnapura Omnibus Company, Limited [1946] 115 LJ 76; (1946) 90 SJ 599; (1945-46) LXII TLR 459 PC 'Route' under Motor Car Ordinance 1938 not same as 'highway' under Omnibus Service Licensing Ordinance 1942: effect of same on awarding of road service licences.

Leonard v Western Services Limited (1926-30) XXVIII Cox CC 294; (1927) 91 JP 18; [1927] 1 KB 702; (1926) 62 LJ 449; [1927] 96 LJCL 213; (1927) 136 LTR 308; (1926-27) XLIII TLR 131 HC KBD Omnibus owner issuing return tickets in area where licensed to ply for hire but collecting passengers in area where not licensed not committing unlawful plying for hire.

Miller v Pill; Pill v Furse; Pill v J Mutton and Son [1933] All ER 34; (1931-34) XXIX Cox CC 643; [1933] 2 KB 308; (1933) 97 JP 197; (1933) 102 LJCL 713; (1933) 149 LTR 404; (1933) 77 SJ 372; (1932-33) XLIX TLR 437 HC KBD Annual summer outing/weekly market/weekly cattle mart not special occasions so not contract carriage so road service licence required.

Neal v Guy (1926-30) XXVIII Cox CC 515; (1928) 92 JP 119; [1928] 2 KB 451; (1928) 65 LJ 515; (1928) 97 LJCL 644; (1928) 139 LTR 237; (1928) 72 SJ 368; (1928) WN (I) 141 HC KBD On applicability of by-laws to motor omnibuses (rather than horse-drawn omnibuses).

Police v Okoukwo [1954] Crim LR 869 Magistrates Person may at own risk board stationary bus at any point along route.

R v Brighton Corporation, ex parte Thomas Tilling, Limited (1916) 80 JP 219 HC KBD Mandamus ordered/refused in respect of properly/improperly refused omnibus/char-a-banc licence.

R v Fletcher, ex parte Ansonia (1907-09) XXI Cox CC 578; (1908) 72 JP 249; (1908) XCVIII LTR 749 HC KBD Omnibus wrongfully standing/plying for hire.

R v Minister of Transport, ex parte HC Motor Works, Ltd [1927] 96 LJCL 686 HC KBD Fare requirement as precondition to issuing omnibus licence was valid.

Robinson v Secretary of State for the Environment [1973] 3 All ER 1045; [1973] RTR 511; (1973) 117 SJ 603; [1973] 1 WLR 1139 HC QBD Suspension of public service vehicle licence for fixed period reasonable.

Rout v Swallow Hotels, Limited [1993] Crim LR 77; (1993) 157 JP 771; [1993] RTR 80; [1992] TLR 439 HC QBD Hotel courtesy coach was being used for hire/reward as money spent on accommodation/food at hotel was reason coach provided.

Sales v Lake and others [1922] All ER 689; (1921-25) XXVII Cox CC 170; (1922) 86 JP 80; [1922] 1 KB 553; [1922] 91 LJCL 563; (1922) 126 LTR 636; (1921-22) 66 SJ 453; (1922) WN (I) 66 HC KBD Tour bus picking up clients at pre-arranged points not 'plying for hire' contrary to statute.

Secretary of State for the Environment v Hooper [1981] RTR 169 HC QBD Crown not Secretary of State owned Government department vehicle so latter improperly convicted of loading offence in respect of vehicle.

Sheffield Corporation v Kitson (1929) 93 JP 135; (1929) 67 LJ 401; (1929) 98 LJCL 561; (1930) 142 LTR 20; (1929) 73 SJ 348; (1928-29) XLV TLR 515 HC KBD Action against local authority for operation of unlicensed omnibus failed as was not brought by aggrieved party.

Spires v Smith [1956] 2 All ER 277; [1956] Crim LR 493; (1956) 120 JP 363; (1956) 106 LJ 347; (1956) 100 SJ 400; [1956] 1 WLR 601 HC QBD Bus conductor cannot be guilty of permitting bus to carry more than the maximum number of standing passengers.

Steff v Beck [1987] RTR 61 HC QBD Driver failed to take all reasonable precautions regarding safety of passengers where pulled off from stop before pensioner had seated herself.

Thomson v Birmingham Motor Omnibus Co (1917-18) 62 SJ 683 HC KBD Valid conviction for unlicensed use of omnibus as omnibus plying for hire.

Traffic Commissioners for the South Wales Traffic Area v Snape and another [1977] Crim LR 427; (1977) 127 NLJ 664t; [1977] RTR 367 HC QBD Minibus altered so as to be capable of seating eight people was adapted for use as public service vehicle.

Vickers v Bowman [1976] Crim LR 77; [1976] RTR 165 HC QBD Vehicle was used as express carriage carrying passengers at separate fares even though money collected weekly by one passenger from all and then handed to driver.

Warwickshire County Council and others v British Railways Board [1969] 1 WLR 1117 CA Construction of clause 'until . . . time for . . . appeal . . . has expired' in Road Traffic Commissioners granting of licence (esentially meant licences granted not operative until appeal procedure exhausted).

Westacott v Centaur Overland Travel Ltd and another [1981] RTR 182 HC QBD Valid decision by justices that minibus with eleven seats (four blocked) not adapted for carriage of eight or more passengers.

Williams v Bond and others [1986] Crim LR 564; [1986] RTR HC QBD Was breach of excess hour regulations where four drivers drove four vehicles in which an extra driver sat at different times.

Yorkshire (Woollen District) Electric Tramways Limited v Ellis (1901-07) XX Cox CC 795; (1904-05) 53 WR 303 HC KBD Light railway carriage not an omnibus nor a hackney carriage.

RECKLESS DRIVING (general)

Andrews v Director of Public Prosecutions [1937] 2 All ER 552; (1936-38) 26 Cr App R 34; (1937) 101 JP 386; (1937) 83 LJ 304; [1937] 106 LJCL 370; (1937) 81 SJ 497; (1936-37) LIII TLR 663; (1937) WN (I) 188 HL Can be convicted of reckless driving where negligence insufficient to sustain manslaughter charge in case where victim dies.

Director of Public Prosecutions v Hastings (1994) 158 JP 118; [1993] RTR 205 HC QBD Not reckless driving for passenger to grab steering wheel so as to swerve towards friend (whom inadvertently hit).

Hand v Director of Public Prosecutions [1991] Crim LR 473; (1992) 156 JP 211; [1991] RTR 225 HC QBD On what constitutes 'reckless' driving — on relevance of level of alcohol consumption to determination of same.

Hargreaves v Baldwin (1905) 69 JP 397; (1905-06) XCIII LTR 311; (1904-05) XXI TLR 715 HC KBD Speed relevant to reckless driving charge even though speeding a different offence (Motor Car Act 1903, s 1(1)).

Houghton v Manning (1904-05) 49 SJ 446 HC KBD Was not criminally reckless/negligent driving for driver to drive car along road with toll-keeper who had sought toll hanging onto car until fell off.

Jarvis v Norris [1980] RTR 424 HC QBD Vengefulness not a defence to/could afford evidence of requisite intent for reckless driving.

Jennings v United States Government [1982] 3 All ER 104 (also HC QBD); [1983] 1 AC 624; (1982) 75 Cr App R 367; [1982] Crim LR 748; (1982) 146 JP 396; (1982) 132 NLJ 881; [1983] RTR 1 (also HC QBD); (1982) 126 SJ 659; [1982] TLR 424; [1982] 3 WLR 450 HL Causing death by reckless driving is manslaughter; character of offence for which sought determines if extradition possible.

R v Austin (Howard) (1980) 2 Cr App R (S) 203; [1981] RTR 10 CA Consecutive sentences merited where in course of single occasion were guilty of reckless driving and causing bodily harm by wanton/furious driving.

R v Baldessare (Cyril) (1930-31) Cr App R 70; (1931) 144 LTR 185; (1930) WN (I) 193 CCA Passenger in recklessly driven car may be guilty of criminal negligence by driving.

R v Bell (David) [1984] Crim LR 685; [1985] RTR 202 CA On what constitutes reckless driving.

R v Bennett [1991] Crim LR 788; [1992] RTR 397 CA Convictions quashed where inadequate direction as regards recklessness.

R v Clarke (Andrew) (1990) 91 Cr App R 69; [1990] RTR 248 CA On what constitutes reckless driving: adequacy of Lawrence direction where drinking of alcohol involved.

R v Conway [1988] 3 All ER 1025; (1989) 88 Cr App R159; [1989] Crim LR 74; (1988) 152 JP 649; [1989] QB 290; [1989] RTR 35; (1988) 132 SJ 1244; [1988] 3 WLR 1238 CA Necessity a defence to reckless driving if duress of circumstances present.

R v Cox [1993] 2 All ER 19; (1993) 96 Cr App R 452; (1993) 157 JP 114; [1993] RTR 185; [1993] 1 WLR 188 CA Custodial sentence justifiable if to right-thinking members of the public justice would not otherwise be done; prevalence of offence/public concern relevant to seriousness.

R v Crossman (Richard Alan) (1986) 82 Cr App R 333; [1986] Crim LR 406; [1986] RTR 49; (1986) 130 SJ 89 CA Was causing death by reckless driving to drive along road with unsecured load that knew could fall off and injure another and which did fall and kill pedestrian.

R v Denton (Stanley Arthur) (1987) 85 Cr App R 246; [1987] RTR 129; (1987) 131 SJ 476 CA Defence of necessity unavailable where person charged with reckless driving claimed had driven carefully at relevant time.

R v Fairbanks (John) (1986) 83 Cr App R 251; [1986] RTR 309; (1970) 130 SJ 750; [1986] 1 WLR 1202 CA Was material irregularity not to have left alternative verdict of driving without due care and attention where person charged with causing death by reckless driving.

R v Fisher [1992] Crim LR 201; [1993] RTR 140 CA Best that R v Lawrence direction on recklessness be given ipissima verba.

R v Governor of Holloway Prison, ex parte Jennings [1982] Crim LR 590; (1982) 132 NLJ 488; [1983] RTR 1 (also HL); (1982) 126 SJ 413; [1982] TLR 210; [1982] 1 WLR 949 HC QBD Cannot extradite person for manslaughter where evidence suggests are guilty of death by reckless driving.

R v Griffiths (Gordon Rupert) (1989) 88 Cr App R 6; [1990] RTR 244 CA On what constitutes recklessness in charge of causing death by reckless driving.

R v Jeavons (Phillip) (1990) 91 Cr App R 307; [1990] RTR 263 CA Circumstances did not merit jury being given opportunity of convicting accused of careless driving where was charged with reckless driving.

R v Jones and others, ex parte Thomas (1921) 85 JP 112; [1921] LJCL 543; (1921) WN (I) 35 HC KBD Conviction for reckless driving at dangerous speed not duplicitous.

R v Lamb (Charles Roland) (1990) 91 Cr App R 181; [1990] RTR 284; [1990] TLR 94 CA On what constitutes reckless driving; on adequacy of R v Lawrence direction.

R v Lawrence (Stephen Richard) (1980) 71 Cr App R 291; [1980] RTR 443 CA Death by reckless driving quashed where was inadequate direction on meaning of recklessness.

R v Lawrence [1981] 1 All ER 974; [1982] AC 510; (1981) 73 Cr App R 1; [1981] Crim LR 409; (1981) 145 JP 227; (1981) 131 NLJ 339; [1981] RTR 217; (1981) 125 SJ 241; [1981] 2 WLR 524 HL Actus reus of reckless driving: driving without due care/attention, serious risk to other road user/property. Mens rea: No thought to/disregard of risk.

R v Lawrence (Justin) [1990] RTR 45 CA On imposition of concurrent/consecutive sentences for offences arising from same facts.

R v Leach [1937] 1 All ER 319 CCA Possible that person's negligence might be sufficient to sustain conviction for reckless driving but not for manslaughter where victim died.

R v Madigan (Anthony Brian) (1982) 75 Cr App R 145; [1982] Crim LR 692; [1983] RTR 178 CA On what constitutes 'reckless' driving (Road Traffic Act 1972, s 2).

R v Murphy [1980] 2 All ER 325; (1980) 71 Cr App R 33; [1980] Crim LR 309; (1980) 144 JP 360; (1980) 130 NLJ 474; [1980] QB 434; [1980] RTR 145; (1980) 124 SJ 189; [1980] 2 WLR 743 CA Failure to drive with due care/attention or recklessness as to same is essence of reckless driving; mens rea is attitude leading to driving — need not show intent to take risk.

R v Powell (James Thomas); R v Elliott (Jeffrey Terence); R v Daley (Frederick); R v Rafferty (Robert Andrew) (1984) 79 Cr App R 277 CA On sentencing for death by reckless driving; when disqualifying should have reference to accused's driving record.

R v Reid [1992] 3 All ER 673; (1992) 95 Cr App R 391; [1992] Crim LR 814; (1994) 158 JP 517; [1992] RTR 341; (1992) 136 SJ LB 253; [1992] 1 WLR 793 HL Reckless driving occurs where accused disregards serious risk of injury to another/does not give that prospect consideration.

R v Reid (John Joseph) (1990) 91 Cr App R 263; [1990] RTR 276 CA On necessary mens rea for reckless driving; on adequacy of R v Lawrence direction.

R v Renouf [1986] 2 All ER 449; (1986) 82 Cr App R 344; [1986] Crim LR 408; [1986] RTR 191; (1986) 130 SJ 265; [1986] 1 WLR 522 CA That was part of reasonable use of force in making arrest could excuse reckless driving.

R v Seymour [1983] 2 All ER 1058; [1983] 2 AC 493; (1983) 77 Cr App R 215; [1983] Crim LR 742; (1984) 148 JP 530; (1983) 133 NLJ 746; [1983] RTR 455; (1983) 127 SJ 522; [1983] TLR 525; [1983] 3 WLR 349 HL Direction to jury where manslaughter arising from reckless driving.

R v Seymour (Edward John) (1983) 76 Cr App R 211; [1983] Crim LR 260; [1983] RTR 202 CA Direction to jury where manslaughter arising from reckless driving.

R v Thompson (Calvin) [1980] Crim LR 188; [1980] RTR 386 CA Person indicted acquitted of reckless driving having withdrawn unaccepted guilty plea to careless driving could not be convicted of latter offence.

R v Willer [1987] RTR 22 CA Possible duress and slow speed at which car driven at relevant time rendered conviction for reckless driving unsafe.

Troughton v Manning (1901-07) XX Cox CC 861; (1905) 69 JP 207; (1905) XCII LTR 855; (1904-05) XXI TLR 408; (1904-05) 53 WR 493 HC KBD Not reckless driving to drive at reasonable speed with person clinging to side of car.

RECOVERY VEHICLE (general)

Gibson v Nutter [1984] RTR 8; [1983] TLR 261 HC QBD On what constitutes 'disabled' vehicle; goods vehicle test certificate necessary where using recovery vehicle in connection with more than one disabled vehicle.

Pearson (E) and Son (Teesside) Ltd v Richardson [1972] 3 All ER 277; [1972] Crim LR 444t; (1972) 136 JP 758; [1972] RTR 552; (1972) 116 SJ 416; [1972] 1 WLR 1152 HC QBD Vehicle that can raise but not tow disabled vehicle not 'recovery vehicle'.

Seeney v Dean and another [1972] Crim LR 545; [1972] RTR 25 HC QBD On what constitutes a recovery vehicle (Land Rover with towing buoy attached a recovery vehicle).

REGISTRATION (general)

Beverley Acceptances Ltd v Oakley and others [1982] RTR 417 CA Appliction of Factors Act 1889, s 2(1); car registration document not 'document of title' for purposes of Factors Act 1889, s 1(4); pledgee who had not breached duty of care towards holders of bills of sale not estopped by negligence from taking priority over their rights.

Bowers v Worthington [1982] RTR 400 HC QBD Interpretation and application of Road Vehicles (Registration and Licensing) Regulations 1971, reg 35(4).

R v Gill, ex parte McKim (1911-13) XXII Cox CC 118; (1909) 73 JP 290; (1909) 100 LTR 858 HC KBD Licence can under Motor Car Act 1903, s 7 be indorsed for non-compliance with number plate regulations.

Swaits v Entwhistle (1929) 93 JP 232; [1929] 2 KB 171; (1929) 67 LJ 380; (1929) 98 LJCL 648; (1930) 142 LTR 22; (1929) 73 SJ 366; (1929) WN (I) 143 HC KBD Road Vehicle (Registration and Licensing) Amendment Regulations 1928, r 2 not ultra vires act of Minister of Transport.

REGISTRATION/LICENSING (general)

Scutt v Luxton (1950) 94 SJ 33 HC KBD Use of towing lorry as regular carrier's lorry was improper even though it occurred in the course of business.

ROAD (general)

Adams and another v Commissioner of Police of the Metropolis and another; Aberdeen Park Maintenance Co Ltd (Third Party) [1980] RTR 289 HC QBD Failed application to have 'road' declared such (where police commissioner had been refusing to prosecute road users on basis that was not road).

Blackmore v Chief Constable of Devon and Cornwall [1984] TLR 699 HC QBD On determining whether certain road a road or other public place for purposes of the Road Traffic Act 1972, s 6(1).

Borthwick v Vickers [1973] Crim LR 317; [1973] RTR 390 HC QBD Justices could rely on own knowledge of local area when deciding if certain trip must have involved using public road.

Botwood v Phillips [1976] Crim LR 68; [1976] RTR 260 HC QBD On what constitutes 'dock road'.

Bowman v Director of Public Prosecutions [1990] Crim LR 600; (1990) 154 JP 524; [1991] RTR 263; [1990] TLR 52 HC QBD Justices can rely on local knowledge in determining whether particular car park a public place.

Buchanan v Motor Insurers' Bureau [1955] 1 All ER 607; [1955] 1 WLR 488 HC QBD 'Road' not road under Road Traffic Act 1930, s 121(1) as no public access by right/tolerance.

Cutter v Eagle Star Insurance Co Ltd (1996) TLR 3/12/96; [1997] 1 WLR 1082 CA Use of car in multi-storey car park deemed to have been use of car on road.

Deacon v AT [1976] Crim LR 135; [1976] RTR 244 HC QBD Council housing estate road not a 'road' for purpose of Road Traffic Act 1972, s 196(1).

Director of Public Prosecutions v Coulman [1993] Crim LR 399; [1993] RTR 230; [1992] TLR 609 HC QBD Inward Freight Immigration Lanes at Dover were public place.

Griffin v Squires [1958] 3 All ER 468; [1958] Crim LR 817; (1959) 123 JP 40; (1958) 108 LJ 829; (1958) 102 SJ 828 HC QBD Car park not a road.

Havell v Director of Public Prosecutions [1993] Crim LR 621; (1994) 158 JP 680 HC QBD Private members' club car park not a public place.

Hawkins v Phillips and another [1980] Crim LR 184; [1980] RTR 197 HC QBD Filter lane/slip road was 'main carriageway'.

Kreft v Rawcliffe [1984] TLR 306 HC QBD On determining whether a road to which public had access for purposes of the Road Traffic Act 1972, s 196(1).

Lang v Hindhaugh [1986] RTR 271 HC QBD Was driving on road to drive on footpath that was highway.

Lock v Leatherdale [1979] Crim LR 188; [1979] RTR 201 HC QBD Named but unadopted road on uncompleted housing estate not a 'road to which . . . public has access'. (Road Traffic Act 1972, s 196(1)).

O'Brien v Trafalgar Insurance Company, Limited (1945) 109 JP 107; (1944-45) LXI TLR 225 CA Roads around factory deemed not to fall within definition of road in Road Traffic Act 1930, s 121(1).

Price v Director of Public Prosecutions [1990] RTR 413 HC QBD Pavement outside shop which was part pavement, part shop frontage was 'road'.

Pugh v Knipe [1972] Crim LR 247; [1972] RTR 286 HC QBD Was not driving in public place to drive on land which belonged to private club/public did not use.

R v Beaumont [1964] Crim LR 665; (1964) 114 LJ 739 CCA Occupation road deemed not to road on which could be guilty of offence of drunk driving.

R v Miller [1975] 2 All ER 974; (1975) 61 Cr App R 182; [1975] Crim LR 723; (1975) 139 JP 613; [1975] RTR 479; (1975) 119 SJ 562; [1975] 1 WLR 1222 CA Not defence to driving on road while disqualified that believed place driving on not road.

R v Murray (Gerrard) [1984] RTR 203 CA Were driving on road while disqualified where drove on Mersey dock road.

R v Waters (John James) (1963) 47 Cr App R 149 CCA On what constitutes a public place for the purpose of the Road Traffic Act 1960, s 6.

Thomas v Dando [1951] 1 All ER 1010 HC KBD Unpaved private forecourt not adjoining road not a 'road' under Road Transport Lighting Act 1927, s 1(1).

White and another v Richards [1993] RTR 318 CA On extent of private right to re-/pass along track: determined by reference to state of track when right granted.

SEAT BELT (general)

Director of Public Prosecutions v Shaw (David) (1993) 157 JP 1035; [1993] RTR 200 HC QBD On what constitutes a lawful seat belt.

Webb v Crane [1988] RTR 204 HC QBD Newsagent/newspaper distributor driving to collect newspapers not engaged in round/delivery and so not exempt from seat belt requirement.

SENTENCE ('totting up')

Maynard v Andrews [1973] RTR 398 HC QBD No 'totting up' where person only convicted of single offence inside three years preceding commission of offence for which being sentenced here.

R v Brentwood Magistrates' Court, ex parte Richardson (1992) 95 Cr App R 187; (1992) 156 JP 839; [1993] RTR 374; [1992] TLR 1 HC QBD Construction of 'totting-up' provisions of Road Traffic Act 1988 so as to determine whether mandatory disqualification of defendant required.

R v Jones (Robert Thomas) (1971) 55 Cr App R 32 CA 'Totting up' disqualification to be subsequent to any other disqualification period; similar offences may be taken into account when imposing disqualification for particular offence.

R v Sibthorpe (John Raymond) (1973) 57 Cr App R 447 CA Road traffic disqualifications can only be made consecutive under 'totting-up' provisions of Road Traffic Act 1962, s 5(3).

SENTENCE (aggravated vehicle-taking)

Attorney-General's Reference No 68 of 1995 (Paul Thomas Dawes) [1996] 2 Cr App R (S) 358 CA Six years' young offender detention for manslaughter arising from aggravated vehicle taking.

R v Bird (Simon Lee) (1993) 14 Cr App R (S) 343; [1993] Crim LR 85 CA On sentencing for aggravated vehicle taking (here twelve months' young offender detention imposed).

R v Carroll (Stephen) (1995) 16 Cr App R (S) 488; [1995] Crim LR 92 CA Eighteen months' young offender detention for person who pleaded guilty to aggravated vehicle-taking.

R v Evans (Stephen) (1994) 15 Cr App R (S) 137 CA Two years' young offender detention for aggravated vehicle taking.

R v Gostkowski [1995] RTR 324 CA Disqualification imposed on passenger party to aggravated vehicle-taking reduced in light of disqualification imposed on driver.

R v Marron (Paul James) (1993) 14 Cr App R (S) 615 CA Six months' imprisonment for aggravated vehicle-taking.

R v Ore (Martin); R v Tandy (Stuart) (1994) 15 Cr App R (S) 620; [1994] Crim LR 304 CA Four years' young offender detention for causing death by aggravated vehicle taking.

R v Robinson (Kristian Paul); R v Scurry (Lee Patrick) (1994) 15 Cr App R (S) 452 CA Eighteen/twenty-one months' young offender detention for aggravated vehicle-taking.

R v Sealey (Mark) (1994) 15 Cr App R (S) 189 CA Nine months' imprisonment for aggravated vehicle-taking.

R v Sharkey (Bernard Lee); R v Daniels (Andrew Anthony) (1995) 16 Cr App R (S) 257; [1994] Crim LR 866 CA Inappropriate that sixteen year old be sentenced to twelve months' young offender detention after guilty plea to aggravated vehicle-taking resulting in injury to another.

R v Sherwood (Carl Edward); R v Button (Nigel John) (1995) 16 Cr App R (S) 513; [1995] Crim LR 176 CA Inappropriate to impose over two years' imprisonment on offender who caused death by aggarvated vehicle taking unless that offence expressly indicted.

R v Timothy (Stephen Brian) (1995) 16 Cr App R (S) 1028 CA Nine months' imprisonment for aggravated vehicle-taking.

SENTENCE (careless driving)

Director of Public Prosecutions v Powell [1993] RTR 266 HC QBD No/was special reason arising to justify non-endorsement of licence of person who was guilty of careless driving of child's motor cycle on road/who did not have insurance for or L-plate on same.

SENTENCE (causing bodily harm by wanton driving)

R v Austin (Howard David) (1980) 2 Cr App R (S) 203; [1981] RTR 10 CA Consecutive sentences merited where in course of single occasion were guilty of reckless driving and causing bodily harm by wanton/furious driving.

SENTENCE (causing death by dangerous driving)

Attorney-General's Reference No 1 of 1994 (Ian Campbell Kenneth Day) (1995) 16 Cr App R (S) 193 CA Community service order/five years' disqualification for person guilty of causing death by dangerous driving allowed stand unchanged.

Attorney-General's Reference No 22 of 1994 (Christopher Charles Nevison) (1995) 16 Cr App R (S) 670; [1995] Crim LR 255 CA Five years' (not fifteen months') imprisonment for causing two deaths by dangerous driving (after drinking).

Attorney-General's Reference No 34 of 1994 (Colin Francis Vano) (1995) 16 Cr App R (S) 785; [1995] Crim LR 346; [1996] RTR 15 CA Six months' imprisonment generally merited for causing death by dangerous driving of person on crossing but here twenty-eight days' appropriate.

Attorney-General's Reference No 34 of 1994 (1995) 159 JP 237 CA R v Boswell applied: 28 days' custody and three years' disqualification inappropriate for causing death by dangerous driving — six month custodial sentence merited.

Attorney-General's Reference No 37 of 1994 (Troy Stewart Sergeant) (1995) 16 Cr App R (S) 760 CA Four years' imprisonment for causing death by dangerous driving allowed stand.

Attorney-General's Reference No 38 of 1994 (Jamie Archer) (1995) 16 Cr App R (S) 714; [1995] Crim LR 257 CA Four years' (not thirty months') detention for nineteen year old guilty of causing death by dangerous driving (in course of overtaking).

Attorney-General's Reference No 42 of 1994 (Kevin Norman Vickers) (1995) 16 Cr App R (S) 742 CA Five (not three) years' imprisonment for causing death by dangerous driving.

Attorney-General's Reference No 46 of 1994 (Paul Michael Antonsen) (1995) 16 Cr App R (S) 914 CA Four (not three) years' imprisonment for causing death by dangerous driving one morning after heavy drinking previous evening.

Attorney-General's Reference No 30 of 1995 (Richard Law) [1996] 1 Cr App R (S) 364 CA Twelve (not six) months' imprisonment for causing death by dangerous driving.

Attorney-General's Reference No 67 of 1995 (David Russell Lloyd) [1996] 2 Cr App R (S) 373 CA Seven (not five) years' imprisonment for causing death by dangerous driving (when racing).

Attorney-General's Reference No 6 of 1996 (Adam Kousourous) [1997] 1 Cr App R (S) 79 CA Twenty-one (not nine) months' imprisonment for causing death by dangerous driving (speeding).

Attorney-General's Reference No 20 of 1996 (Salim Abdulla Omer) [1997] 1 Cr App R (S) 285 CA Three years' imprisonment for causing death by dangerous driving (when racing) allowed stand.

Attorney General's Reference No 28 of 1996 (Mark Douglas Hysiak) [1997] 2 Cr App R (S) 79 CA Three years' young offender detention for eighteeen year old guilty of causing death of passenger by dangerous driving.

Attorney-General's Reference No 44 of 1996 (Richard French) [1997] 1 Cr App R (S) 375 CA Three and a half years' young offender detention for causing death by dangerous driving.

Attorney General's Reference No 48 of 1996 (Paul Swain) [1997] 2 Cr App R (S) 76 CA Four years' imprisonment/five years' disqualification for causing death by dangerous driving after drinking.

Attorney-General's Reference No 69 of 1996 (Anthony Paul Jackson) [1996] 2 Cr App R (S) 360 CA Five (not three) years' imprisonment for causing death by dangerous driving (speeding wrong way up one way road).

R v Angell [1977] Crim LR 682 CA £250 fine/five years' disqualification for causing death by dangerous driving, plus £50 fine/one years' disqualification for driving with excess alcohol fully merited.

R v Bailey (Paul John) [1996] 1 Cr App R (S) 129 CA Three months' imprisonment for causing death by dangerous driving.

R v Barber (Simon) [1997] 1 Cr App R (S) 65 CA Five years' imprisonment following guilty plea to charge of causing death by dangerous driving.

R v Bevan (Paul Martin) [1996] 1 Cr App R (S) 14 CA Eighteen months' imprisonment for causing death by dangerous driving (overtaking).

R v Bruin [1979] RTR 95 CA Three year disqualification and costs but immediate release from imprisonment (after serving three weeks of four week sentence) for person guilty of death by dangerous driving where were no aggravating circumstances.

R v Burns (Michael) (1995) 16 Cr App R (S) 821 CA Four years' imprisonment for causing death by dangerous driving (speeding after drinking).

R v Carr (Simon) [1996] 1 Cr App R (S) 107 CA Six months' imprisonment for causing death by dangerous driving (night-time motorway speeding).

R v Challoner (1964) 108 SJ 1049 CCA Nine months' imprisonment plus ten years' disqualification for person with history of road traffic offences who was guilty here of causing death by dangerous driving.

R v Clarke-Sutton (Damon Matthew) (1995) 16 Cr App R (S) 937 CA Twelve months' imprisonment for causing death by dangerous driving (drove with person standing on bonnet).

R v Day (Attorney General's Reference No 1 of 1994) [1994] Crim LR 764; [1995] RTR 184; [1995] TLR 24 CA 240 hours of community service/five year disqualification satisfactory for causing death by dangerous driving: that ignored speed signs/knew area to pose hazards did not aggravate excessive speeding.

R v De Meersman (Eddy Louis) [1997] 1 Cr App R (S) 106 CA Three years' imprisonment for causing death by dangerous driving (driver who fell asleep at the wheel).

R v Divisi (James Christopher) (1995) 16 Cr App R (S) 23; [1994] Crim LR 699 CA Twelve months' young offender detention for person who caused death by dangerous driving when was sixteen.

R v Dutton [1972] Crim LR 321t; (1972) 122 NLJ 128; [1972] RTR 186 CA Nine months' imprisonment merited for person guilty of causing death by dangerous driving as result of taking risk on purpose; on imprisonment for causing death by dangerous driving.

R v Gisbourne [1977] Crim LR 299 CA £50 fine plus twelve months' disqualification for causing death by dangerous driving.

R v Groves (Scott Paul) (1995) 16 Cr App R (S) 769 CA Four years' imprisonment for causing death by dangerous driving (speeding after drinking).

R v Hudson (Brian) [1979] RTR 401 CA Imprisonment inappropriate where (five) deaths by dangerous driving resulted from single instant of inattention: £100 fine and three year disqualification.

R v Jerrum (Maurice Albert) (1967) 51 Cr App R 251 CA Eighteen months' imprisonment and eleven year disqualification appropriate for three death by dangerous driving convictions.

R v Jolliffe [1970] Crim LR 50; [1972] RTR 188 CA On imprisonment for causing death by dangerous driving: here three months and twenty-five days substituted for nine months.

R v Kang (Balvinder Singh) [1997] 1 Cr App R (S) 306 CA Thirty months' imprisonment/four years' disqualification for causing death by dangerous driving (had been driving articulated vehicle with defective brakes).

R v Le Mouel (Andre Marcel Marie) [1996] 1 Cr App R (S) 42 CA Four months' imprisonment for causing death by dangerous driving (foreigner who drove on wrong side of road).

R v Lowry (Thomas Gordon) [1996] 2 Cr App R (S) 416 CA Eighteen months' imprisonment for causing death by dangerous driving (through overtaking).

R v Lundt-Smith [1964] Crim LR 543; (1964) 128 JP 534; (1964) 108 SJ 424; [1964] 2 WLR 1063 HC QBD Ambulance driver involved in fatal collision after crashed red light en route to hospital in emergency had special reasons justifying non-disqualification upon conviction of causing death by dangerous driving.

R v Lyons (Terence Patrick) (1971) 55 Cr App R 565 CA On sentencing under-twenty-one year old of previously good character for causing death by dangerous driving.

R v Mallone (Patrick) [1996] 1 Cr App R (S) 221 CA Four (not five) years' imprisonment for causing death by dangerous driving — reduction inter alia because of grave injuries offender suffered.

R v Milburn [1974] Crim LR 434t; (1974) 124 NLJ 387t; [1974] RTR 431 CA Dangerous driving sentence of appellant reduced to twelve months' imprisonment and ten years' disqualification (with need to re-take test) where had been severe disparity between his sentence and that of co-accused.

R v Moon (Paul); R v Moon (David) [1997] 2 Cr App R (S) 44 CA Five years' disqualification plus six years' detention in young offender institution/three years' detention under Children and Young Persons Act 1933 for nineteen/fifteen year olds guilty of causing death by dangerous driving.

R v Morgan (James Bernard) (1968) 52 Cr App R 235; [1968] Crim LR 226 CA Custodial sentence necessary in serious death by dangerous driving action.

R v Nunn (Adam John) [1996] 2 Cr App R (S) 136; [1996] Crim LR 210 CA On relevance of opinions of deceased's survivors to sentencing of person who caused deceased's death by dangerous driving.

R v Preston [1971] Crim LR 438 CA £200 fine (six months' imprisonment in default) for person guilty of causing death (of passenger-friend) by dangerous driving.

R v Pring (Marcus Anthony); R v Pring (Harold William John) [1996] 2 Cr App R (S) 53 CA Eight months' imprisonment for causing causing death by dangerous driving in light of medical status.

R v Pritchard (Terry) (1995) 16 Cr App R (S) 666 CA Thirty months' imprisonment for motorcyclist who caused death by dangerous driving.

R v Rayner; R v Wing [1994] TLR 535 CA On sentencing for causing death by dangerous driving.

R v Robinson (Dorothy) [1975] RTR 99 CA £50 fine/twelve months' disqulification for person guilty of death by dangerous driving as result of momentary inattention.

R v Rowbotham (Kevin) [1997] 1 Cr App R (S) 187 CA Four years' imprisonment for causing death by dangerous driving (drove at person).

R v Rowe [1975] Crim LR 245 CA Quashing of prosecution costs order where had been too high.

R v Rumbold [1977] Crim LR 682 CA Six months' imprisonment/four years' disqualification for person guilty of causing death by dangerous driving while unfit to drive through drink or drugs.

R v Sergeant (Attorney General's Reference (No 37 of 1994)) [1995] RTR 309 CA Four years' imprisonment/ten year disqualification/ten penalty points merited for previously convicted road traffic offender guilty here of several traffic offences including causing death to another after lengthy piece of very bad driving after which sought to evade apprehension/frustrate investigation.

R v Severn (Kevin) (1995) 16 Cr App R (S) 989 CA Eighteen months' imprisonment for causing death by dangerous driving (did not see person on crossing in darkness).

R v Shaw (Bradley) (1995) 16 Cr App R (S) 961 CA Four years' young offender detention for causing death by dangerous driving (speeding).

R v Sweeney (Barrie) [1996] 2 Cr App R (S) 148 CA Five years' young offender detention for late-teenager guilty of causing death by dangerous driving.

R v Vano (Attorney General's Reference (No 34 of 1994)) (1995) 16 Cr App R (S) 785; [1995] Crim LR 346; [1996] RTR 15 CA Double jeopardy element/mental health of defendant resulted in his non-return to prison despite inadequacy of 28 day imprisonment imposed where caused death by dangerous driving.

R v Vickers (Attorney General's Reference (No 42 of 1994)) [1995] Crim LR 345; [1996] RTR 9 CA Five years' imprisonment for drunken driver whose wanton driving caused death of pedestrian.

R v Walton (Stephen) [1996] 2 Cr App R (S) 220 CA Two years and three months' imprisonment for causing death by dangerous diving (unsecurely attached tow-load).

R v Wheatley [1984] RTR 327 CA Two (not three) years' imprisonment, half the sentence suspended, for young man guilty of causing death by dangerous driving.

R v Wright (Ernest) [1979] RTR 15 CA Three years' imprisonment (concurrent) for each of three deaths caused by dangerous driving/twelve months' imprisonment (concurrent) for driving while unfit to drink justified even in light of offender's having been 'sent to Coventry' by small community in which lived.

R v Yarnold [1978] RTR 526 CA Fine plus disqualification appropriate sentence for person causing death by dangerous driving where was speeding but not recklessly so.

SENTENCE (causing death by reckless driving)

Attorney General's Reference No 3 of 1989; Attorney General's Reference No 5 of 1989 [1989] RTR 337 CA Appropriate sentencing for causing death by reckless driving.

Attorney-General's Reference No 3 of 1989 (Peter Anthony Sumner) (1989) 11 Cr App R (S) 486 CA Fifteen months' imprisonment (not probation order) for causing death by reckless driving.

Attorney-General's Reference No 5 of 1989 (Mark Charles Hill-Trevor) (1989) 11 Cr App R (S) 489 CA Twenty-one months' detention in young offender institution for causing death by dangerous driving.

Attorney General's Reference No 5 of 1990 (James William Bain) (1990-91) 12 Cr App R (S) 514 CA Four years' imprisonment/five years' disqualification for particularly serious case of causing death by reckless driving.

Attorney General's Reference No 15 of 1990 (Roy Francis Lambert) (1990-91) 12 Cr App R (S) 510 CA Two years' imprisonment/four years' disqualification for causing death by reckless driving.

Attorney-General's Reference No 24 of 1990 (Lisa Veronica Ashley) (1990-91) 12 Cr App R (S) 686; [1991] Crim LR 566 CA Eighteen (not nine) months' detention for eighteen year old female guilty of causing death by reckless driving (after had drunk considerable quantites of alcohol).

Attorney General's Reference No 2 of 1991 (Edward Peter Campbell Dillon) (1992) 13 Cr App R (S) 337; [1992] Crim LR 316 CA Twelve months' (not twenty-eight days') imprisonment for causing death by reckless driving (after drinking).

Attorney-General's Reference No 17 of 1991 (Anthony Hart) (1992) 13 Cr App R (S) 656 CA Two years' imprisonment/four year (not twelve month) disqualification from driving for causing two deaths by reckless driving.

Attorney-General's Reference No 6 of 1992 (Jaon Andrew Ewing) (1993) 14 Cr App R (S) 70; [1992] Crim LR 601 CA Three years' imprisonment for causing two deaths by reckless driving (after drinking) plus nine months' consecutive for dishonesty offences merited.

Attorney-General's References Nos 17 and 18 of 1992 (William Thomason and Martin Douglas) (1993) 14 Cr App R (S) 428; [1993] Crim LR 226 CA Three years' (not twenty-one months') imprisonment for causing deaths by reckless driving (speeding after drinking).

Attorney-General's References Nos 30 and 31 of 1992 (Garry Roger James Godden and Steven John Boosey) (1993) 14 Cr App R (S) 386 CA Eighteen/twelve (not six/four) months' imprisonment for causing death by reckless driving.

Attorney-General's Reference No 37 of 1992 (Andrew John Hayton) (1994) 15 Cr App R (S) 71; [1993] Crim LR 632 CA Eighteen (not three) months' imprisonment for causing death by reckless driving (after drinking).

R v Abraham (Mansel) [1980] RTR 471 CA Imprisonment/fine improper/proper for police officer causing death by reckless driving in course of police driving exercise.

R v Adams (John) (1990-91) 12 Cr App R (S) 393 CA Four years' imprisonment for causing two deaths by reckless driving (after drinking).

R v Ashley (Attorney General's Reference (No 24 of 1990)) [1991] RTR 113 CA Eighteen months in young offender institution/five year disqualification for young first offender who after drinking drove on wrong side of road and hit person causing them to die.

R v Bain (Attorney General's Reference (No 5 of 1990)) [1991] Crim LR 312; [1991] RTR 169 CA Four years' imprisonment/five year disqualification for offender who contrived situation in which drove too close to pedestrians and thereby became guilty inter alia of causing death by dangerous driving.

R v Beeby (Leslie Albert) (1983) 5 Cr App R (S) 56 CA Twelve month disqualification/£750 fine for person who caused death by reckless driving when fell asleep at the wheel.

R v Bekka (Janet Angela Helen) (1992) 13 Cr App R (S) 520 CA Two/four years' imprisonment/ disqualification for causing death by dangerous driving (after drinking).

R v Bennett [1991] Crim LR 788; [1992] RTR 397 CA Convictions quashed where inadequate direction as regards recklessness.

R v Bennett (Paul) (1989) 153 JP 317 CA Four years' imprisonment appropriate for experienced criminal guilty of two death by reckless driving charges but disqualification reduced from seven to four years where great length likely to result in trouble.

R v Boswell; R v Elliott (Jeffrey); R v Daley (Frederick); R v Rafferty (1984) 6 Cr App R (S) 257; [1984] Crim LR 502; [1984] RTR 315; [1984] TLR 395; [1984] 1 WLR 1047 CA On sentencing (disqualification) for death by reckless driving.

R v Boxell (Michael John) (1989) 11 Cr App R (S) 269 CA Four years' imprisonment for causing three deaths by reckless driving (speeding after drinking).

R v Brown [1985] Crim LR 611 CA Twenty-eight days' imprisonment for reckless driving (drove car into lamp post, then wall) following domestic wrangle.

R v Brown (Kenneth Michael) (1981) 3 Cr App R (S) 361; [1982] Crim LR 242 CA Twenty-eight days' imprisonment/five years' disqualification for person guilty of causing death by reckless driving (speeding after drinking).

R v Brown (Victor) (1985) 7 Cr App R (S) 97; [1985] Crim LR 607 CA Six months' imprisonment for causing death by reckless driving (pedestrian crossing collision after which failed to stop).

R v Burder (Paul John) (1993) 14 Cr App R (S) 111 CA £1,500 fine/two years' disqualification for fire engine driver guilty of causing death by reckless driving in course of answering emergency call.

R v Chadwick (Peter) (1990-91) 12 Cr App R (S) 349; [1991] Crim LR 216; [1991] RTR 176 CA Fifteen months' imprisonment/five years' disqualification for causing deaths of three persons by reckless driving.

R v Clancy [1979] RTR 312 CA On what is meant by 'reckless' driving.

R v Craig (1994) 158 JP 449; [1993] TLR 294 CA More severe sentences to be expected for drunken persons guilty of causing death by reckless driving: three and a half years' imprisonment plus five years' disqualification justified.

R v Crossman (Richard Alan) (1986) 82 Cr App R 333; [1986] Crim LR 406; [1986] RTR 49; (1986) 130 SJ 89 CA Was driving recklessly to bring articulated vehicle on road knowing there was grave risk that load thereon could fall and injure another.

R v Daniels [1986] Crim LR 484 CA Twelve months' imprisonment/three year disqualification for driver guilty of causing death by reckless driving.

R v Davis [1979] Crim LR 259 CA Three months' suspended sentence/five year disqualification for causing death by reckless driving.

R v Davis (Mark Lawrence) (1987) 9 Cr App R (S) 211 CA Twelve months' imprisonment for causing death by reckless driving (speeding while driving with excess alcohol).

R v Devall (Richard) (1992) 13 Cr App R (S) 598 CA Four years' imprisonment/eight years' (not life) disqualification for causing death by reckless driving (after drinking).

R v Dunwoody (Christopher Michael) (1983) 5 Cr App R (S) 76 CA Three months' imprisonment for young man of good character guilty of causing death by reckless (high speed) driving.

R v Eastwood (Mark) (1985) 7 Cr App R (S) 77 CA Six months' youth custody for fifteen year old guilty of causing death by reckless driving.

R v Farrugia [1979] Crim LR 791; [1979] RTR 422 CA Long disqualification periods inappropriate for young men: five year disqualification imposed as part of sentence for causing death by reckless driving.

R v Godden (Attorney General's Reference (No 30 of 1992)); R v Boosey (Attorney General's Reference (No 31 of 1992)) [1993] RTR 259 CA Eighteen/twelve months' imprisonment for two drivers whose aggressive racing with each other along road ended with collision/death of a passenger.

R v Gooch (Shaun Lee) (1994) 15 Cr App R (S) 390 CA Five years' imprisonment for causing deaths of five people by reckless driving (speeding).

R v Goodman (Ian Peter) (1988) 10 Cr App R (S) 438 CA Six months' imprisonment/two years' disqualification for driver guilty of causing death by reckless driving after fell asleep at wheel.

R v Gorman (John Paul) (1989) 11 Cr App R (S) 560 CA Thirty months' imprisonment for causing death by reckless driving (overtaking).

R v Greatbanks (David Patrick) (1986) 8 Cr App R (S) 478 CA Thirty months' youth custody for seventeen year old guilty of causing death by reckless driving in course of police chase of car taken without consent.

R v Griffiths [1990] RTR 244 CA On what constitutes reckless driving: that consumed excessive alcohol and drove very dangerously was prima facie evidence of reckless driving.

R v Hammett [1993] RTR 275 CA Judge's introducing possibility of careless driving conviction in direction at trial for causing death by reckless driving a material irregularity.

R v Harrington-Griffin (Melvin) (1989) 153 JP 274; [1989] RTR 138 CA Thirty months' imprisonment/seven year disqualification appropriate for person guilty of death by reckless driving where had been drunk, driven for some time at excessive speed, and run from scene of accident.

R v Hayton (Attorney-General's Reference (No 37 of 1992)) [1993] RTR 310 CA Eighteen months' imprisonment/eight year disqualification for person repeatedly guilty of driving with excess alcohol.

R v Heppinstall (Kenneth John) (1985) 7 Cr App R (S) 20; [1985] Crim LR 452 CA Eighteen months' imprisonment for causing death by dangerous driving (victim a passenger; driver had been drinking).

R v Hives [1991] RTR 27 CA Was reckless driving where struck woman crossing road (who was only a footstep away from the pavement) and thought had hit something like a lamp post.

R v Holmes (Ronald) (1990-91) 12 Cr App R (S) CA Two hundred hours' community service/four years' disqualification for causing death by reckless driving (brief instant of recklessness).

R v Hudson (Alan James) (1989) 89 Cr App R 51; (1989) 11 Cr App R (S) 65; [1989] RTR 206 CA Two years' imprisonment and five years' disqualification for person convicted of death by reckless driving who had previously good character/work record.

R v Janes [1985] Crim LR 684 CA Two years' imprisonment plus five years' disqualification for person guilty of causing death by reckless driving.

R v Jewell [1982] Crim LR 52 CA Six months' imprisonment/three years' disqualification/licence endorsement/requirement that re-take driving test for person guilty of causing death by reckless driving.

R v Jones (Lee Adam) (1985) 7 Cr App R (S) 170 CA Two years' imprisonment/five years' disqualification for causing two deaths by reckless driving.

R v Lambert (Attorney General's Reference (No 15 of 1990)) (1991) 92 Cr App R 194; [1991] Crim LR 312; [1991] RTR 195 CA Two years' imprisonment/four year disqualification for speeding drunk driver whose recklessness resulted in death of another.

R v Lawrence (Peter Frederick) (1993) 14 Cr App R (S) 20 CA Six months' imprisonment/six year disqualification for causing death by reckless driving (approached crossing of which had unclear view at too great a speed).

R v Lawson (Wayne Peter) (1985) 7 Cr App R (S) 165; [1985] Crim LR 685 CA Twelve months' youth custody/three years' disqualification for eighteen year old guilty of causing two deaths by reckless driving.

R v Lemings (Colin William) (1982) 4 Cr App R (S) 384; [1983] Crim LR 268 CA Nine months' imprisonment (three suspended)/three years' disqualification for person who had been drinking and caused death by dangerous driving.

R v Manders (Richard John) (1989) 11 Cr App R (S) 442 CA Eighteen months in young offender institution/five years' disqualification for eighteen year old guilty of causing three deaths by dangerous driving (speeding).

R v Marshall (Gordon Kane) (1988) 10 Cr App R (S) 246; [1991] RTR 201 CA Two and a half years' imprisonment/seven years' disqualification for off-duty police officer (with drink taken) guilty of causing death by reckless driving.

R v Matthews (Melvin) (1981) 3 Cr App R (S) 217; [1981] Crim LR 789 CA Roughly three months' imprisonment/three years' disqualification for causing death by dangerous driving through careless overtaking.

R v Mawson (Gary) (1992) 13 Cr App R (S) 218; [1992] Crim LR 68; [1991] RTR 418; [1991] TLR 356 CA Eight months' imprisonment for causing death by reckless driving (through crash occasioned after drove lorry into stationary lorry).

R v McLaren (Alexander) (1983) 5 Cr App R (S) 332; [1984] RTR 126 CA £500 fine (minimum merited) plus three years' disqualification for causing death by reckless driving (offence not of the worst kind but nor was it due to momentary inattention).

R v Miah (Mohammed) (1992) 13 Cr App R (S) 278 CA Six months' imprisonment (suspended for twelve months)/three year disqualification for causing death by reckless driving after crashed traffic lights.

R v Midgley [1979] Crim LR 259; [1979] RTR 1 CA Six months' imprisonment/four years' disqualification for young man with excess alcohol who struck sign causing death of a passenger; appeal against sentence is appeal against entirety of sentence.

R v Morgan (James Percival) (1988) 10 Cr App R (S) 192 CA Eighteen months' imprisonment for causing death by reckless driving (speeding after drinking).

R v Murphy (Anthony John) (1989) 89 Cr App R 176; [1989] RTR 236 CA Six years' disqualification and requirement that sit new driving test substituted for eight years' disqualification for reckless driving resulting in serious injury/death.

R v Muscroft (Wayne Brian) (1990-91) 12 Cr App R (S) 41 CA Two years in young offender institution for young man guilty of causing three deaths by reckless driving (speeding after drinking).

R v Nash (Ronald David) (1988) 10 Cr App R (S) 99 CA Four years' imprisonment for causing death by reckless driving (death of pursuing policeman in car chase).

R v Noden [1991] RTR 32 CA Nine months' imprisonment/three year disqualification appropriate for person who caused death by driving recklessly in foggy conditions.

R v O'Sullivan (Patrick John); R v Burtoft (Duncan Paul) (1983) 5 Cr App R (S) 283; [1983] Crim LR 827 CA On appropriate senences for reckless driving/causing death by reckless driving.

R v Owens (Ralph) (1981) 3 Cr App R (S) 311 CA Five years' imprisonment for causing death by reckless driving (struck and killed person chasing car a fled from accident scene).

R v Peters [1993] Crim LR 519; [1993] RTR 133 CA That had drunk too much alcohol was admissible to explain nature of driving.

R v Pettipher (1989) 11 Cr App R (S) 321; [1991] RTR 183 CA Two years' imprisonment/three year disqualification merited for person with excess alcohol taken guilty of speeding which resulted in three deaths.

R v Plant (Barnett) (1987) 9 Cr App R (S) 241 CA Eight months' imprisonment for causing death by reckless driving (speeding).

R v Price (Peter Patrick) (1983) 5 Cr App R (S) 42 CA Eighteen months' imprisonment/five years' disqualification for causing three deaths by reckless driving.

R v Reardon (Stuart) (1993) 14 Cr App R (S) 275 CA Six months' imprisonment for causing two deaths by reckless driving (drove too fast on narrow road).

R v Robson (James Keith) (1989) 11 Cr App R (S) 78 CA Three years' imprisonment for causing death by reckless driving (speeding after drinking).

R v Rodenhurst (Ian Henry) (1989) 11 Cr App R (S) 219; [1989] RTR 333 CA Roughly three months in young offender institution for twenty year old guilty of causing three deaths by reckless driving through brief instant of recklessness.

R v Rowcroft (Philip) (1984) 6 Cr App R (S) 112; [1984] Crim LR 431 CA Nine months' custody for causing death by reckless driving.

R v Scott (Alexander) (1990-91) 12 Cr App R (S) 684 CA Suspended sentence quashed/two years' disqualification ordered/£500 fine approved for causing death by reckless driving (travelled with rear of long lorry jutting into fast lane of dual carriageway).

R v Smith [1980] RTR 460 HC QBD Three (not five) years' disqualification appropriate in serious case of causing death by reckless driving.

R v Smith (Gary Michael) [1991] RTR 109; [1990] TLR 678 CA On sentencing for causing death by reckless driving.

R v Staddon (1991) 141 NLJ 1004t; [1992] RTR 42 CA On distinction in sentencing for reckless driving/causing death by reckless driving.

R v Stewart (Robert) (1988) 10 Cr App R (S) 123 CA Three and a half years' imprisonment for causing death by reckless driving while racing car.

R v Taylor (Kevin Thomas) (1982) 4 Cr App R (S) 346; [1981] Crim LR 422 CA Twelve months' imprisonment for causing death by reckless driving (drunk learner driver).

R v Taylor (Miah) [1992] Crim LR 69 CA Twelve month suspended sentence/three years' disqualification for causing death by reckless driving: custodial sentence not always necessary for this offence.

R v Thomas (Jeffrey William) (1979) 1 Cr App R (S) 86 CA Custodial sentence may be merited where very high speed driving resulted in death of (here two) others.

R v Turner (Michael John) (1988) 10 Cr App R (S) 234 CA Two years' imprisonment/seven year disqualification (given occupation) for lorry driver guilty of causing death by reckless driving (drove too fast in fog).

R v Turner (Nigel David) (1990-91) 12 Cr App R (S) 472 CA Eighteen months' imprisonment/ three years' disqualification for causing death by reckless driving.

R v Upchurch (Edward Henry) (1992) 13 Cr App R (S) 476 CA Conditional discharge (following two and ahalf months in custody) for causing death by reckless driving (did not see pedestrian using crossing).

R v Webster (Richard) (1992) 13 Cr App R (S) 615 CA Probation/three years' disqualification for offender who when newly licensed seventeen year old driver caused death by reckless driving.

R v West (William John) (1986) 8 Cr App R (S) 5 CA Six months' imprisonment for sixty year old man who caused death by reckless driving (continued driving though knew was dangerous because of his epilepsy).

R v Whitmore (Wayne) (1985) 7 Cr App R (S) 193; [1985] Crim LR 685 CA One hundred hours' community service for causing death by reckless driving (no aggravating factors).

R v Willetts (1993) 14 Cr App R (S) 592; [1993] RTR 252 CA Concurrent twelve month imprisonment sentences for person guilty of causing two deaths by reckless driving who had been drinking slightly/had been careless on approach to pedestrian crossing.

R v Winterton (Alexander David) (1993) 14 Cr App R (S) 529; [1993] Crim LR 322 CA Four months' detention in young offender institution for causing death by reckless driving.

Reference by the Attorney-General under Section 36 of the Criminal Justice Act 1988 (No 3 of 1989); Reference by the Attorney-General under Section 36 of the Criminal Justice Act 1988 (No 5 of 1989) (1990) 90 Cr App R 358; [1990] Crim LR 277 CA Death by reckless driving merited custodial sentence of at least 15 months' duration.

SENTENCE (concurrent/consecutive sentences)

R v Berry (1976) 63 Cr App R 44; [1978] RTR 111 CA Sentence for road traffic offence to be consecutive to other unconnected offences.

R v Wheatley [1984] RTR 273 CA On legitimacy of consecutive sentences for road traffic offences.

SENTENCE (conspiracy to obstruct/pervert course of justice)

R v Moynes; R v Dawson [1976] RTR 303 CA Nine months' imprisonment for each of two offenders who conspired to defeat justice by agreeing to one impersonating other if stopped by police.

SENTENCE (criminal damage)

R v Cartwright [1990] RTR 409 CA Two years' probation (with requirement that undergo medical treatment) substituted for twelve months' imprisonment imposed on manic depressive for inter alia reckless driving, criminal damage and drunk driving.

SENTENCE (dangerous driving)

R v Adeniyi [1978] Crim LR 634 CA Borstal training plus two year disqualification for dangerous driving by Nigerian 'A'-level student.

R v Ball [1974] RTR 296 CA Four-month disqualification merited in case where sentencing court acted under mistaken view that twelve month disqualification mandatory for dangerous driving.

R v Bignall (1968) 52 Cr App R 10; (1967) 117 NLJ 1061t CA Can impose disqualification order alone.

R v Collier [1973] Crim LR 188 CA Determination of appropriate factual basis on which to sentence person found guilty of dangerous driving.

R v Considine (1965) 109 SJ 78 CCA Four months' imprisonment merited by dangerous driver who had appealed sentence on basis that fact that was drinking ought to have been mitigating factor.

R v Frow (John) (1995) 16 Cr App R (S) 609 CA Nine months' imprisonment for dangerous driving.

R v Moore (Martin David) (1995) 16 Cr App R (S) 536 CA Nine months' imprisonment for heavy goods vehicle driver guilty of dangerous driving.

R v Offer (Harry John Miles) [1996] 1 Cr App R (S) 143 CA Twenty-eight days' imprisonment for dangerous driving (motorway speeding).

R v Sherwood [1995] RTR 60 CA Seven years in young offender institution for motor manslaughter driver concurrent with eighteen months for dangerous driving merited by seriousness of manslaughter; maximum of two years' imprisonment for passenger guilty of theft where indictment had not spelt out facts meriting greater sentence.

R v Storey [1973] Crim LR 189 CA Nine months' imprisonment plus three years' disqualification merited by first time road traffic offender guilty of seriously dangerous driving.

R v Templeton (Graham) [1996] 1 Cr App R (S) 380 CA Nine months' imprisonment for dangerous driving.

R v Wood (1971) 121 NLJ 749t CA On when life disqualification merited (not here).

SENTENCE (disqualification)

Adams v Bradley [1975] Crim LR 168; [1975] RTR 233 HC QBD That drank alcoholic beverage in mistaken belief that was beverage of lower alcohol content not special reason justifying non-disqualification.

Alexander v Latter [1972] Crim LR 646; (1972) 122 NLJ 682t; [1972] RTR 441 HC QBD Person drinking particularly strong beer at another's prompting without warning as to strength had special reason meriting non-disqualification.

Ambrose v Jamison [1967] Crim LR 114 CrCt Disqualified driver not disqualified where (acting under instructions from police officer) he drove while disqualified.

Ashworth v Johnson; Charlesworth v Johnson [1959] Crim LR 735 Sessions Successful appeals against disqualifications with requirement that re-take driving test: on when such a form of disqualification merited.

Attorney General's Reference No 48 of 1996 (Paul Swain) [1997] 2 Cr App R (S) 76 CA Four years' imprisonment/five years' disqualification for causing death by dangerous driving after drinking.

Baker v Cole [1971] 3 All ER 680; [1966] Crim LR 453; (1971) 135 JP 592; [1972] RTR 43; [1971] 1 WLR 1788 HC QBD Factors to be considered as mitigating when sentencing for third traffic conviction.

Bolliston v Gibbons (1984) 6 Cr App R (S) 134 CA Where court deciding if person guilty of second drunk driving offence in ten years should face three year disqualification only special reasons pertaining to second offence can be raised as justifying non-disqualification.

Bolliston v Gibbons [1985] RTR 176; [1984] TLR 200 HC QBD Where sentencing person to second compulsory disqualification in ten years cannot take into account special reasons extant at time of original disqualification.

Brewer v Metropolitan Police Commissioner [1969] Crim LR 149t; (1969) 133 JP 185; (1968) 112 SJ 1022; [1969] 1 WLR 267 HC QBD Worker who unknowingly inhaled alcohol fumes at factory (and had no constructive knowledge of same) had special reasons justifying non-disqualification.

Brewer v Metropolitan Police Commissioner (1969) 53 Cr App R 157 CA Lacing of drinks/ inadvertent ingestion of alcoholic fumes could be special reason justifying non-disqualification.

Bullen v Keay [1974] Crim LR 371; [1974] RTR 559 HC QBD Non-disqualification not merited by fact that accused had attempted to commit suicide by taking drugs and had not intended to drive after taking same.

Coombs v Kehoe [1972] 2 All ER 55; [1972] Crim LR 560; (1972) 136 JP 387; (1972) 122 NLJ 153t; [1972] RTR 224; (1972) 116 SJ 486; [1972] 1 WLR 797 HC QBD Driving merely to park not a special reason justifying non-disqualification.

Crampsie v Director of Public Prosecutions [1993] RTR 383; [1993] TLR 82 HC QBD On whether discretionary/mandatory disqualification could be/to be imposed where evidence proved accused in charge of vehicle/did not prove was not driving or attempting to drive.

Damer v Davison (1975) 61 Cr App R 232; [1975] Crim LR 522 CA On removal of disqualification order.

De Munthe v Stewart [1982] RTR 27 HC QBD Person (with excess alcohol) requested by constable to re-park car just parked could not plead re-park request as special reason meriting non-disqualification.

Delaroy-Hall v Tadman; Watson v Last; Earl v Lloyd (1969) 53 Cr App R 143; [1969] Crim LR 93t; (1969) 133 JP 127; (1968) 118 NLJ 1173t; [1969] 2 QB 208; (1968) 112 SJ 987; [1969] 2 WLR 92 HC QBD Amount by which exceeded blood-alcohol limit not special reason justifying non-disqualification.

Dennis v Tame [1954] 1 WLR 1338 HC QBD Unless there are special reasons justifying non-disqualification (here for driving uninsured motor-vehicle) disqualification must be ordered.

Director of Public Prosecutions v Bristow (1997) 161 JP 35; (1996) TLR 28/10/96 HC QBD Objective test as to whether emergency which led person to drive after drinking excess alcohol a special reason justifying non-disqualification.

Director of Public Prosecutions v Corcoran (1991) 155 JP 597; [1991] RTR 329 HC QBD Non-disqualification merited upon conviction for driving with excess alcohol given short length intended to drive/fact that did not pose danger to public.

Director of Public Prosecutions v Cox [1996] RTR 123 HC QBD Non-disqualification of drunk driver merited where had only driven in response to emergency (alarm/burglary at club of which was key-holder).

Director of Public Prosecutions v Doyle [1993] RTR 369; [1992] TLR 651 HC QBD Could not find special reasons meriting non-disqualification where offender had wilfully elected to drink and drive.

Director of Public Prosecutions v Enston [1996] RTR 324; [1995] TLR 89 HC QBD Woman's threat to accuse man who had been drinking of raping her unless he drove her to cash-point was special reason justifying non-disqualification.

Director of Public Prosecutions v Feeney (1989) 89 Cr App R 173 HC QBD Disqualification justified: were special reasons justifying conveyance of passenger to her home but not justifying continuing driving on to own home.

Director of Public Prosecutions v Kinnersley [1993] RTR 105; (1993) 137 SJ LB 12; [1992] TLR 651 HC QBD Fear of contracting HIV (though not mentioned at time refused to give breath specimen) could be special reason meriting non-disqualification: police not under duty to explain breathalyser is sterile.

Director of Public Prosecutions v Kinnersley (1993) 14 Cr App R (S) 516 CA Could (as here) find special reasons justifying non-disqualification though did not accept plea of reasonable excuse for non-provision of breath specimen.

Director of Public Prosecutions v Knight [1994] RTR 374 HC QBD Bona fide fear of attack on baby-sitter and baby (plus fact that driving not notably bad/were only slightly over limit) meant had special reasons meriting non-disqualification.

Director of Public Prosecutions v O'Connor (Michael Dennis) (1992) 13 Cr App R (S) 188 HC QBD On non-disqualification for drink driving offences arising from 'spiking' of drinks unbeknownst to driver.

Director of Public Prosecutions v O'Meara [1989] RTR 24 HC QBD That did not know that drink from previous night could result in one being over prescribed level next morning not special reason meriting non-disqualification.

Director of Public Prosecutions v Upchurch [1994] RTR 366 HC QBD Were special reasons meriting non-disqualification where offender had driven injured friends to hospital when no ambulance/taxi available.

Director of Public Prosecutions v Waller [1989] RTR 112 HC QBD Driving fiancee from restaurant where she faced threat of attack not special reason meriting non-disqualification as though began as emergency must have been other means of remedying situation.

Director of Public Prosecutions v Whittle [1996] RTR 154 HC QBD Person's panicked taking over of driving after driver-wife stated was feeling unwell not emergency meriting non-disqualification.

Director of Public Prosecutions v Younas [1990] RTR 22 HC QBD Were special reasons meriting non-disqualification where consumed one and three quarter pints of lager each glass of which was (unbeknownst to drinker) 'spiked' with rum.

Donahue v Director of Public Prosecutions [1993] RTR 156 HC QBD Disqualification valid despite finding of special reasons which would justify non-disqualification.

Dyson v Ellison [1975] 1 All ER 276; (1975) 60 Cr App R 191; [1975] Crim LR 48; (1975) 139 JP 191; (1974) 124 NLJ 1132t; [1975] RTR 205; (1975) 119 SJ 66; [1975] 1 WLR 150 HC QBD Court to consider previous endorsements in deciding if diqualification necessary; appeal against obligatory endorsement is by way of re-hearing.

Evans v Bray [1976] Crim LR 454; [1977] RTR 24 HC QBD Disqualification merited where driver with excess alcohol bringing tablets to wife in emergency had not duly considered possible alternatives such as calling emergency services.

Fearson v Sydney [1966] 2 All ER 694; [1966] Crim LR 396t; (1966) 130 JP 329; (1966) 110 SJ 449; [1966] 1 WLR 1003 HC QBD Previous disqualifications were aggravating, not mitigating factors when deciding whether disqualification necessary on foot of new offence.

Flewitt v Horvath [1972] Crim LR 103; [1972] RTR 121 HC QBD Case remitted to justices who accepted inadmissible hearsay evidence of 'spiking' of drink with alcohol and then (oddly) imposed £50 fine.

Fraser v Barton (1974) 59 Cr App R 15 HC QBD On special circumstances justifying non-diqualification.

Gardner v Director of Public Prosecutions (1989) 89 Cr App R 229; (1989) 153 JP 357; [1989] RTR 384 HC QBD Disqualification discretionary where person in charge of (ie, not driving) vehicle refused to provide breath specimen upon request.

George v Director of Public Prosecutions [1989] RTR 217 HC QBD On whether charge one of being in charge of/driving vehicle (and so whether disqualification discretionary/mandatory).

Glendinning v Batty [1973] Crim LR 763; (1973) 123 NLJ 927t; [1973] RTR 405 HC QBD Inappropriate to stretch the law so as to find 'special reasons' justifying non-disqualification.

Goldsmith v Laver [1970] RTR 162; [1970] Crim LR 286 HC QBD Was not special reason justifying non-disqualification that person convicted was (unknown to himself) a diabetic and so more affected by alcohol.

Gordon v Smith [1971] Crim LR 173; [1971] RTR 52; (1971) 115 SJ 62 HC QBD Public interest (in soldier going to Northern Ireland as driver) did not merit special limited disqualification for drunk driving: twelve months' disqualification ordered.

Gosling v Paul (1961) 125 JP 389 HC QBD Shortness of remaining period of disqualification not reason justifying non-imprisonment for driving while disqualified.

Haime v Walklett (1983) 5 Cr App R (S) 165 CA Not special reason justifying non-disqualification that person only drove car from roadside to car park so as to leave it there overnight.

Harding v Oliver [1973] Crim LR 764; [1973] RTR 497 HC QBD Not special reason justifying non-disqualification that drink-and-driving offender lost portion of blood specimen given to him by police.

Hockin v Weston [1972] Crim LR 541t; (1971) 121 NLJ 690t; [1972] RTR 136; (1971) 115 SJ 675 HC QBD Could not be reasonable excuse for failure to provide specimen on part of person pleading guilty of said failure.

Holland v Phipp [1983] RTR 123; (1982) 126 SJ 466; [1982] TLR 262; [1982] 1 WLR 1150 HC QBD Court bound by periods of disqualification appearing on court register.

Holroyd v Berry [1973] Crim LR 118; (1972) 122 NLJ 1155t; [1973] RTR 145 HC QBD Doctor pleading was essential he be allowed continue drive in area without enough doctors and feeling obliged to resign if disqualified was nonetheless disqualified.

Hosein v Edmonds [1970] RTR 51; (1969) 113 SJ 759 HC QBD Request to speak with solicitor at same moment that refused specimen not special reason justifying non-disqualification.

Hughes v Challes (1984) 148 JP 170; [1984] RTR 283; [1983] TLR 658 HC QBD Disqualification of reckless driver who was normally competent but for single occasion when was ill was unmerited.

Hunter v Coombs [1962] 1 All ER 904; (1962) 112 LJ 321; (1962) 106 SJ 287 HC QBD Conviction quashed where bad information (despite prosecution being so notified).

Jacobs v Reed [1973] Crim LR 531; (1973) 123 NLJ 568t; [1974] RTR 81 HC QBD On when sudden emergency (that led to person driving with excess alcohol) can justify non-disqualification.

James v Morgan [1988] RTR 85 HC QBD Successful appeal by prosecutor against successful 'spiking of drinks' claim by defendant before trial court as special reason justifying non-disqualification.

Jones v Nicks [1977] Crim LR 365; [1977] RTR 72 HC QBD That believed road subject to certain speed limit/that disqualification would cause hardship did not merit non-disqualification for speeding.

Jones v Powell [1965] 1 All ER 674; [1965] Crim LR 311; (1965) 129 JP 187; (1965) 115 LJ 213; [1965] 2 QB 216; [1965] 2 WLR 683 HC QBD Disqualification for uninsured driving; disqualification periods for any reason to be consecutive.

Kerr v Armstrong (1973) 123 NLJ 638t; [1974] RTR 139; [1973] Crim LR 532 HC QBD That person involved in accident (which did not involve another) telephoned police to inform them of same was not special reason justifying non-disqualification.

Knight v Baxter [1971] Crim LR 368; [1971] RTR 270; (1971) 115 SJ 350 HC QBD That had drunk alcohol on empty stomach not special reason justifying non-disqualification.

Knowler v Rennison [1947] 1 All ER 302; [1947] KB 488 HC KBD Whether special reasons justifying non-disqualification a question of law; hardship/that would not re-offend not a special reason.

Lambie v Woodage [1972] 2 All ER 462; [1972] Crim LR 442t; (1972) 136 JP 554; (1972) 122 NLJ 426t; [1972] RTR 396; (1972) 116 SJ 376; [1972] 1 WLR 754 HL In considering reasonable grounds for not disqualifying judges may consider anything reasonable.

Learmont v Director of Public Prosecutions [1994] RTR 286; [1994] TLR 109 HC QBD Disqualification issue returned to justices for re-consideration after mistakenly counted earlier double disqualification upon single conviction as two disqualifications.

Lodwick v Brow [1984] RTR 394 HC QBD Non-disqualification not merited by loss of specimen of person who because of post-traumatic amnesia could not remember giving specimen.

Lodwick v Brow (1984) 6 Cr App R (S) 38 CA That person lost portion of blood specimen given to them by police not special reason jstifying non-disqualification.

MacLean v Cork [1968] Crim LR 507t; (1968) 118 NLJ 638t; (1968) 112 SJ 658 HC QBD That needed to drive in pursuit of business not a special reason justifying non-disqualification.

Malin v Cavey [1967] Crim LR 712; (1967) 111 SJ 744 HC QBD Failed appeal against disqualification; on what is meant by 'special reasons'.

Milliner v Thorne [1972] Crim LR 245; [1972] RTR 279 HC QBD That were on straight road/could see well both ways/did not put another at risk not special reasons meriting non-disqualification upon conviction for driving with excess alcohol.

Mullarkey v Prescott [1970] RTR 296 HC QBD That had artificial legs/did not drive far along quiet road in Winter/might need state assistance if were disqualified not special reasons justifying non-disqualification for driving with excess alcohol.

Newnham v Trigg [1970] RTR 107 HC QBD That did not know how much whisky had drunk (served for medicinal purposes) not special reason justifying non-disqualification; could be that greater the time between original drink and provision of specimen, the greater the offence.

Nicholson v Brown [1974] Crim LR 187; [1974] RTR 177; (1974) 118 SJ 259 HC QBD That careless driving not very careless or culpability not extreme not special reasons justifying non-disqualification.

Owen v Jones (1987) 9 Cr App R (S) 34 CA Exceptionally proof of exceptional circumstances justifying non-disqualification need not be supported by evidence where magistrates can rely on own personal knowledge.

Owen v Jones [1988] RTR 102 HC QBD Proof of exceptional circumstances justifying non-disqualification need not be supported by evidence where magistrates can rely on own personal knowledge.

Owens v Imes [1973] Crim LR 60 HC QBD Requirement that person guilty of driving offence under the Road Safety Act 1967, s 1(1) be disqualified remained even though offender being given absolute discharge.

Park v Hicks [1979] Crim LR 57; [1979] RTR 259 HC QBD Failed attempt to rely upon medical emergency as special reason for non-disqualification following driving with excess alcohol as defence too flimsy.

Petherick v Buckland (1955) 119 JP 82; (1955) 99 SJ 78; [1955] 1 WLR 48 HC QBD Disqualification can be limited to particular vehicle.

Pilbury v Brazier [1950] 2 All ER 835; (1950) 114 JP 548; [1950] 66 (2) TLR 763 HC KBD That insurers considered themselves bound a special reason justifying non-disqualification.

Piridge v Gant [1985] RTR 196 HC QBD On exercising discretion not to disqualify where defendant pleads as special reason that non-alcoholic drink was 'spiked' with alcohol.

Powell v Gliha [1979] Crim LR 188; [1979] RTR 126 HC QBD Disqualification merited where emergency requiring person to drive after had been drinking arose in part through her own positive actions.

Pugsley v Hunter [1973] 2 All ER 10; [1973] Crim LR 247; (1973) 137 JP 409; (1973) 123 NLJ 225t; [1973] RTR 284; (1973) 117 SJ 206 HC QBD Defendant must prove special circumstances.

Punshon v Rose (1968) 118 NLJ 119tt; (1969) 113 SJ 39 HC QBD That blood level in part attributable to drinking on night before apprehension (defendant also having drunk on day in question) not a special reason meriting non-disqualification.

R v Agnew [1969] Crim LR 152; (1969) 113 SJ 58 CA Valid refusal not to disqualify despite short distance driven while over the limit/fact that were only person available to drive car that had been asked to move.

R v Anderson [1972] Crim LR 245; [1972] RTR 113; (1972) 116 SJ 103 CA Defendant's destruction of own part of specimen after was told would not be (but later was) prosecuted was special reason meriting non-dsqualification.

R v Arif (Mohammed) (1985) 7 Cr App R (S) 92; [1985] Crim LR 523 CA Can under Powers of Criminal Courts Act 1973, s 44, disqualify person from driving for period exceeding that for which are imprisoned.

R v Aspden [1975] RTR 456 CA Bona fide actions of policeman that may/may not have breached manufacturer instructions on delay between drinking and testing breath did not render breathalyser evidence inadmissible; one year disqualification for cooperative (repeat) offender who required car for business.

R v Bain [1973] RTR 213 CA On whether/when imposition of consective disqualification periods permissible.

R v Baines (1970) 54 Cr App R 481; [1970] Crim LR 590; (1970) 120 NLJ 733t; [1970] RTR 455; (1970) 114 SJ 669 CA That person who had been drinking only drove to lend assistance to person stranded without petrol in remote area after latter telephoned for help not special reason justifying non-disqualification.

R v Ball [1974] RTR 296 CA Four-month disqualification merited in case where sentencing court acted under mistaken view that twelve month disqualification mandatory for dangerous driving.

R v Bentham (William) (1981) 3 Cr App R (S) 229; [1982] RTR 357 CA Order of CrCt where seek removal of disqualification is a sentence; improper for CrCt to require person to re-sit driving test when ordering disqualification to be lifted.

R v Bignall (1968) 52 Cr App R 10; (1967) 117 NLJ 1061t CA Can impose disqualification order alone.

R v Bond [1968] 2 All ER 1040; (1968) 52 Cr App R 505; [1968] Crim LR 622; (1968) 118 NLJ 734t; (1968) 112 SJ 908; [1968] 1 WLR 1885 CA Life disqualification unmerited for driving while disqualified.

R v Boswell; R v Elliott (Jeffrey); R v Daley (Frederick); R v Rafferty (1984) 6 Cr App R (S) 257; [1984] Crim LR 502; [1984] RTR 315; [1984] TLR 395; [1984] 1 WLR 1047 CA On sentencing (disqualification) for death by reckless driving.

R v Bradfield and Sonning Justices, ex parte Holdsworth [1971] 3 All ER 755; [1972] Crim LR 542t; (1971) 135 JP 612; (1971) 121 NLJ 664t; [1972] RTR 108; (1971) 115 SJ 608 HC QBD Lifting of disqualification.

R v Brentwood Magistrates' Court, ex parte Richardson (1992) 95 Cr App R 187; (1992) 156 JP 839; [1993] RTR 374; [1992] TLR 1 HC QBD Construction of 'totting-up' provisions of Road Traffic Act 1988 so as to determine whether mandatory disqualification of defendant required.

R v Brown (Edward) [1975] RTR 36 CA Borstal training/two and a half year disqualification/ deprivation of vehicle order all appropriate for person guilty of handling who used van to convey stolen goods.

R v Brown and Taylor, ex parte Metropolitan Police Commissioner (1962) 46 Cr App R 218 HC QBD Disqualification order not sentence unless stated to be so — when justices commit young offender to quarter sessions with view to disqualification is not sentence.

R v Buckley (Karl Peter) (1994) 15 Cr App R (S) 695; [1994] Crim LR 387 CA Life disqualification for driver guilty here and previously of numerous driving offences.

R v Buckley (Nicholas) (1988) 10 Cr App R (S) 477; [1989] Crim LR 386 CA Driving disqualification appropriate only where competency as driver in issue.

R v Camfield [1971] RTR 449 CA On passing consecutive sentences of disqualification.

R v Cockermouth Justices and others, ex parte Patterson [1971] Crim LR 287; [1971] RTR 216 HC QBD Pre-Theft Act 1968 endorsement of licence/disqualification for taking and driving away car without owner's consent was improper.

R v Cottrell [1955] 3 All ER 817; (1955) 39 Cr App R 198; [1956] Crim LR 127; (1956) 120 JP 45; (1956) 100 SJ 55; [1956] 1 WLR 70 HC QBD Disqualification cannot be partly lifted.

R v Cottrell [1956] 1 All ER 751; (1956) 40 Cr App R 46; [1956] Crim LR 339 CCA On removal of disqualification.

R v Cottrell (No 2) (1956) 120 JP 163; (1956) 100 SJ 264; [1956] 1 WLR 342 CCA Disqualification cannot be partly lifted.

R v Davegun (Surbjeet Singh) (1985) 7 Cr App R (S) 110; [1985] Crim LR 608 CA Court may when imposing disqualification under Powers of Criminal Courts Act 1973 seek not to unduly jeopardise livelihood of offender who makes his living by way of driving.

R v Davies (Paul Neil) (1993) 157 JP 820 CA Improper to impose two consecutive four month sentences/improper to impose consecutive disqualifications for driving while disqualifed/driving with excess alcohol convictions: two three month consecutive sentences and concurrent disqualification imposed instead.

R v Davis [1979] Crim LR 259 CA Three months' suspended sentence/five year disqualification for causing death by reckless driving.

R v Devine (Michael) (1990-91) 12 Cr App R (S) 235; [1990] Crim LR 753 CA Driving disqualification may (as here) be imposed where used same to avoid being caught for conspiracy offence.

R v Donnelly [1975] 1 All ER 785; (1975) 60 Cr App R 250; [1975] Crim LR 178T; (1975) 139 JP 293; (1975) 125 NLJ 111; [1975] RTR 243; (1975) 119 SJ 138; [1975] 1 WLR 390 CA Re-taking of driving test unjustified unless solid ground for believing incompetent driver.

R v Dunbar [1970] Crim LR 52; (1969) 113 SJ 856 CA Ten years' disqualification inadvertently imposed reduced to five.

R v Ealing Justices, ex parte Scrafield [1994] RTR 195 HC QBD Ten year disqualification (in addition to three months' imprisonment) merited by individual repeatedly guilty of drink driving offences.

R v Earle [1976] RTR 33 CA Disqualification quashed for person guilty of various offences (including taking car) who had been sentenced to three years.

R v Farnes [1983] RTR 441; [1982] TLR 468 CA Inappropriate that disqualification on young person being imprisoned should extend long beyond imprisonment.

R v Farrugia [1979] Crim LR 791; [1979] RTR 422 CA Long disqualification periods inappropriate for young men: five year disqualification imposed as part of sentence for causing death by reckless driving.

R v Fenwick (1979) 129 (2) NLJ 681; [1979] RTR 506 CA Two (not five) years' disqualification for doctor who drove straight across junction at which was signed to give way (disqualification causing particular hardship).

R v Ford [1982] RTR 5 CA Three year disqualification merited where overtook another on hill-crest.

R v Gibbons (Laurence) (1987) 9 Cr App R (S) 21; [1987] Crim LR 349 CA Disqualification exceeding period of imprisonment by three years upheld.

R v Gibson [1980] RTR 39 CA One-year disqualification merited where pleaded guilty to being in charge of vehicle with blood-alcohol concentration above prescribed limit.

R v Godfrey (Leonard Peter) (1967) 51 Cr App R 449; (1967) 117 NLJ 810t CA That unreasonably refused urine sample not ground for increasing duration of disqualification.

R v Gostkowski [1995] RTR 324 CA Disqualification imposed on passenger party to aggravated vehicle-taking reduced in light of disqualification imposed on driver.

R v Graham [1955] Crim LR 319 CCA Disqualification (unlike here) must run from date of conviction.

R v Hansel (Rudyard Lloyd) (1982) 4 Cr App R (S) 368; [1983] Crim LR 196; [1983] RTR 445 CA Need not be chronological correspondence between period of disqualification and period of imprisonment where both imposed.

R v Hazell (1965) 109 SJ 112 CCA Three months' imprisonment plus five years' disqualification for twenty year old guilty of dangerous driving.

R v Heslop [1978] Crim LR 566; [1978] RTR 441 CA Two year disqualification/re-testing requirement where unclear whether death by dangerous driving result of momentary inattention of inexperienced driver or very blameworthy driving.

R v Higgins [1973] RTR 216 CA On whether/when consecutive disqualification periods allowed.

R v Hollier [1973] Crim LR 584; [1973] RTR 395 CA No 'totting up' where person only convicted of single offence inside three years preceding commission of offence for which being sentenced here.

R v Holt [1962] Crim LR 565t CCA No special reasons meriting non-disqualification of driver for driving under influence of drink/drugs even though his doctor did not warn him of adverse effects of drinking small amount of alcohol in tandem with taking prescription drug.

R v Ireland (Anthony Albert) (1988) 10 Cr App R (S) 474; [1989] Crim LR 458 CA Judge minded to impose non-mandatory driving disqualification must inform counsel so that latter can make submission to court on same.

R v Jackson; R v Hart [1969] 2 All ER 453; (1969) 53 Cr App R 341; [1969] Crim LR 321t; (1969) 113 SJ 310; (1969) 133 JP 358; [1970] 1 QB 647; [1970] RTR 165; [1969] 2 WLR 1339 CA That disabled, have liver complaint and unimpaired ability to drive despite drinking not special reasons justifying non-disqualification.

R v Johnson [1969] Crim LR 443t CA On multiple disqualification.

R v Johnston [1972] Crim LR 647; [1973] RTR 403 CA On ordering consecutive disqualification.

R v Johnston (Malcolm Victor) (1972) 56 Cr App R 859 CA Unless is statutory provision allowing same disqualification order not to begin from future date.

R v Jones (Michael) [1977] RTR 385 CA Period of disqualification reduced as public interest (in having skilled computer technicians available) was served thereby.

R v Jones (Robert Thomas) (1971) 55 Cr App R 32 CA 'Totting up' disqualification to be subsequent to any other disqualification period; similar offences may be taken into account when imposing disqualification for particular offence.

R v Kent [1983] 3 All ER 1; (1983) 77 Cr App R 120; [1983] Crim LR 553; [1983] RTR 393; (1983) 127 SJ 394; [1983] TLR 335; [1983] 1 WLR 794 CA Disqualification coupled with endorsement of penalty points not permissible.

R v Kent (Kenneth Gordon); R v Tanser (Herbert Paul) (1983) 5 Cr App R (S) 16; [1983] Crim LR 406 CA Court when disqualifying driver may take into account that offender makes his living by way of driving.

R v Kent (Michael Peter) (1983) 5 Cr App R (S) 171 CA On disqualification from driving.

R v King (Phillip) (1992) 13 Cr App R (S) 668; [1993] RTR 245 CA On imposing disqualification (here life disqualification not merited).

R v Kingston [1980] RTR 51 CA Fining/costs adequate for coach driver who used coach for own ends without employer's consent — disqualfication unnecessary.

R v Krebs [1977] RTR 406 CA Person unwittingly drinking alcohol stronger than that which had led to believe by another was consuming had special reason meriting non-disqualification.

R v Lake (Brian Edward) and others (1986) 8 Cr App R (S) 69 CA Counsel to be told by sentencer that has disqualification in mind so that counsel may make submissions on that point.

R v Lambeth Metropolitan Magistrate, ex parte Everett [1967] 3 All ER 648; (1967) 51 Cr App R 425; [1967] Crim LR 543t; (1968) 132 JP 6; (1967) 117 NLJ 730t; [1968] 1 QB 446; (1967) 111 SJ 545; [1967] 3 WLR 1027 HC QBD New disqualification begins when earlier disqualification ends.

R v Lane [1986] Crim LR 574 CA Judge minded to make disqualification (in light of car being used in commission of crime) ought to forewarn counsel so that latter might make representations on matter.

R v Lark (David Gordon) (1993) 14 Cr App R (S) 196 CA Inappropriate to impose life disqualification on presumption that offender will eventually apply for it to be lifted: ten year disqualification substituted.

R v Lazzari (George Thomas) (1984) 6 Cr App R (S) 83 CA Requirement to re-sit driving test may only be imposed where driver's competency to drive is suspect.

R v Lee [1971] RTR 30 CA Inappropriate that young first time road traffic offender guilty of taking and uninsured driving of car be sentenced to borstal training and ten years' disqualfication (borstal plus three years' disqualification substituted).

R v Lobley (1974) 59 Cr App R 63; [1974] Crim LR 373; [1974] RTR 550 CA Two year disqualification plus requirement to re-take driving test imposed on person guilty of death by dangerous driving as result of brief lack of attention.

R v Lundt-Smith [1964] 3 All ER 225; [1964] Crim LR 543; (1964) 128 JP 534; [1964] 2 QB 167; (1964) 108 SJ 424; [1964] 2 WLR 1063 HC QBD Ambulance driver involved in fatal collision after crashed red light en route to hospital in emergency had special reasons justifying non-disqualification upon conviction of causing death by dangerous driving.

R v Maidstone Crown Court, ex parte Litchfield [1992] TLR 322 HC QBD Fine ought not to be increased fourfold to make up for quashing of disqualification order.

R v Mallender [1975] Crim LR 725; [1975] RTR 246 CA Order to re-take driving test quashed (but three year disqualification continued) for driver guilty of driving with excess alcohol but whose competency as a driver was not in issue.

R v Manchester Justices, ex parte Gaynor [1956] 1 All ER 610; (1956) 100 SJ 210; [1956] 1 WLR 280 HC QBD Justices having ordered disqualification removed from deferred date could hold another hearing to see if immediate removal justified.

R v Marshall [1954] Crim LR 386 CCA Twenty-seven months' disqualification for person guilty of using motor car in commission of crime: disqualification generally merited in such cases.

R v Marshall (John) [1976] RTR 483 CA Five year disqualification appropriate for young man driving too fast so as to impress others.

R v Mathews [1975] RTR 32 CA Disqualification inappropriate where five years' imprisonment also imposed for various offences of dishonesty at different banks to which had driven.

R v Matthews (Douglas David) (1987) 9 Cr App R (S) 1; [1987] Crim LR 348 CA Reduction in period of disqualification from driving as considerably surpassed that of period of imprisonment.

R v McCluskie (Alexander) (1992) 13 Cr App R (S) 334; [1983] Crim LR 273 CA Ten years' (not life) disqualification for repeated road traffic offender guilty of careless driving.

R v McIntyre [1976] Crim LR 639; [1976] RTR 330 CA Person with excess alcohol who genuinely thought police officer had instructed him to move car had special reason meriting non-disqualification.

R v McLaughlin [1978] Crim LR 300; [1978] RTR 452 CA Thirty months' imprisonment/twenty year disqualification merited by repeated drunk driving offender who had alcohol problem.

R v Messom (Samuel Andrew) (1973) 57 Cr App R 481; [1973] Crim LR 252t; (1972) 122 NLJ 1134t; [1973] RTR 140 CA That drink was laced may be special reason justifying non-disqualification from driving.

R v Middleton Justices, ex parte Collins; R v Bromley Justices, ex parte Collins; R v Bexley Justices, ex parte Collins [1969] 3 All ER 800; [1970] 1 QB 216; [1969] 3 WLR 632 HC QBD Quashing of disqualifications.

R v Miller (Jason Mark) (1994) 15 Cr App R (S) 505; [1994] Crim LR 231 CA Disqualification until re-sit driving test a punishment to be imposed where offender's competency to drive an issue.

R v Miller (John) [1978] RTR 98 CA Twelve month disqualification appropriate for death by dangerous driving offender whose offence arose from momentary inattention.

R v Mills [1974] RTR 215 CA Two year disqualification for driving with excess alcohol was not excessive.

R v Money (Marc Sean) (1988) 10 Cr App R (S) 237; [1988] Crim LR 626 CA Defence counsel to be allowed make submission to court before disqualification order imposed on party who used motor vehicle as part of committing offence.

R v Mullarkey [1970] Crim LR 406dt HC QBD Disqualification of legless driver from using invalid carriage (after drunk driving episode) would have harsh consequences but was merited.

R v Mulroy [1979] RTR 214 CA Disqualification merited where sentencing for taking conveyance without lawful authority/robbery but reduced to fifteen months given difficulties offender might otherwise face upon twelve months' imprisonment imposed for same offences.

R v Muncaster [1974] Crim LR 320t; (1974) 124 NLJ 103t CA Justices exhorted to reduce unmerited twenty year disqualification which had led to offender being brought before court on several occasions.

R v Murphy (Anthony John) (1989) 89 Cr App R 176; [1989] RTR 236 CA Six years' disqualification and requirement that sit new driving test substituted for eight years' disqualification for reckless driving resulting in serious injury/death.

R v Newton (David) [1974] Crim LR 321; [1974] RTR 451 CA Unconvincing claim that driver's drinks had been 'spiked' with alcohol at party each time put down and left glass did not justify non-disqualification.

R v Nuttall [1972] Crim LR 485; [1971] RTR 279; (1971) 115 SJ 489 Assizes Removal of five-year disqualification.

R v O'Connor (David) [1979] RTR 467 CA Fine/three year disqualification merited by provisional driver who caused death by dangerous driving when ignored supervising driver's instructions.

R v O'Toole (Robert John) (1971) 55 Cr App R 206; [1971] Crim LR 294 CA On disqualification as a sentence; on privileges and responsibilities of driver of emergency vehicle answering emergency call.

R v Olarinde (Tunde) [1967] 2 All ER 491; (1967) 51 Cr App R 249; (1967) 131 JP 323; (1967) 111 SJ 117; [1967] 1 WLR 911 CA Life disqualification was valid.

R v Parrington (Steven) (1985) 7 Cr App R (S) 18; [1985] Crim LR 452 CA Vehicle must have been involved in commission of offence for which convicted to justify disqualification order under Powers of Criminal Courts Act 1973, s 44.

R v Pashley [1974] RTR 149 CA Twelve months' imprisonment and twelve months' disqualification for person who drove with blood-alcohol levels well above prescribed limit.

R v Patel (Rajesh) (1995) 16 Cr App R (S) 756; [1995] Crim LR 440; [1995] RTR 421 CA Person who drove after another to assault him could (under Powers of Criminal Courts Act 1973) be disqualified from driving.

R v Peat (Alan Michael) (1984) 6 Cr App R (S) 311; [1985] Crim LR 110 CA Long disqualifications generally inappropriate; requierment to re-sit driving test merited only when competency to drive is in issue.

R v Powell (James Thomas); R v Elliott (Jeffrey Terence); R v Daley (Frederick); R v Rafferty (Robert Andrew) (1984) 79 Cr App R 277 CA On sentencing for death by reckless driving; when disqualifying should have reference to accused's driving record.

R v Preston [1986] RTR 136 CA Cannot impose licence endorsement and discretionary disqualification; must be circumstances relating to facts making up offence to justify non-endorsement.

R v Raynor [1982] RTR 286 CA Three year disqualification/re-testing requirement imposed on young man of previously good character guilty of causing death by dangerous driving through incompetence/inexperience.

R v Reading Justices, ex parte Bendall [1982] RTR 30 HC QBD Disqualification/endorsement of licence quashed where person eligible for provisional licence drove unlicensed but in compliance with provisional licence conditions.

R v Recorder of Leicester (1946) 175 LTR 173; (1946) 90 SJ 371 HC KBD On removal of disqualification pursuant to Summary Jurisdiction Act 1879.

R v Riley (Terence) (1983) 5 Cr App R (S) 335; [1984] Crim LR 48; [1983] TLR 609 CA On disqualification where motor vehicle used in commission of crime.

R v Rivano (Frank Sean) (1993) 14 Cr App R (S) 578; (1994) 158 JP 288 CA On when life disqualification merited (not merited here).

R v Russell (Ian) [1993] RTR 249 CA Disqualification period reduced where recorder had paid inadequate attention to prospect of offender being rehabilitated.

R v Sandbach Justices, ex parte Pescud (1983) 5 Cr App R (S) 177 CA On procedure as regards pleading mitigating circumstances justifying non-disqualification where have twelve or more driving licence endorsements.

R v Sandwell (David Anthony) (1985) 80 Cr App R 78; [1985] RTR 45; [1984] TLR 481 CA Cannot impose consecutive disqualification periods — sentence varied in light of entirety of case.

R v Schofield [1964] Crim LR 829; (1964) 108 SJ 802 CCA Five (not thirty) years' disqualification for person who drove while unfit to do so through drink or drugs.

R v Scott [1969] 2 All ER 450; (1969) 53 Cr App R 319; (1969) 133 JP 369; [1970] 1 QB 661; [1970] RTR 173; (1969) 113 SJ 470; [1969] 2 WLR 1350 CA Prescriptive drug-taking not special reason justifying non-disqualification.

R v Scott (Stephen Anthony James) (1989) 11 Cr App R (S) 249; [1989] Crim LR 920 CA Judge minded to make life disqualification order to inform counsel of same so that counsel may make submissions on the matter.

R v Scurry (Alan) (1992) 13 Cr App R (S) 517 CA Disqualification by reason of age does not count as disqualification when later determining whether extended disqualification merited in light of previous record.

R v Seaman [1971] RTR 456; (1971) 115 SJ 741 CA Desire of Australian to contact Australian officials not reasonable excuse for refusing specimen nor special reason justifying non-disqualification.

R v Sharman [1974] Crim LR 129; [1974] RTR 213 CA Two year disqualification for driving with excess alcohol was not excessive.

R v Shippam [1971] Crim LR 434t; (1971) 121 NLJ 361t; [1971] RTR 209; (1971) 115 SJ 429 CA 'Spiking' of drink with alcohol unknown to person consuming same a special reason justifying non-disqualification.

R v Shirley [1969] 3 All ER 678; (1969) 53 Cr App R 543; [1969] Crim LR 497t; (1969) 133 JP 691; (1969) 113 SJ 721; [1969] 1 WLR 1357 CA On lengthy disqualification sentences.

R v Shoreham-by-Sea Justices, ex parte Pursey [1955] Crim LR 383 HC QBD Disqualification order quashed as duration exceeded maximum permissible: preferable that court be able to effect downward variation of sentence.

R v Sibthorpe (John Raymond) (1973) 57 Cr App R 447 CA Road traffic disqualifications can only be made consecutive under 'totting-up' provisions of Road Traffic Act 1962, s 5(3).

R v Sixsmith Stipendiary Magistrate, ex parte Morris [1966] 3 All ER 473; [1966] Crim LR 560; (1966) 130 JP 420; [1968] 1 QB 438; [1966] 3 WLR 1200 HC QBD On making disqualification orders.

R v Smith [1980] RTR 460 HC QBD Three (not five) years' disqualification appropriate in serious case of causing death by reckless driving.

R v Southend-on-Sea Justices, ex parte Sharp and another [1980] RTR 25 HC QBD Person's licence wrongfully endorsed as his age/weight of vehicle brought him within exempting provisions of Road Traffic (Drivers' Ages and Hours of Work) Act 1976, Schedule 2.

R v Stimpson [1977] Crim LR 114 CA Twelve months' imprisonment, order to resit driving test plus three (not five) years' disqualification for driving with excess alcohol.

R v Tantrum (Neil) (1989) 11 Cr App R (S) 348 CA Four years' imprisonment/seven years' (not life) disqualification for drunk who caused death by reckless driving.

R v Thomas [1983] 3 All ER 756; (1984) 78 Cr App R 54; [1983] RTR 437; (1983) 127 SJ 749; [1983] 1 WLR 1490 CA Non-extension of disqualification after custodial sentence a 'ground for mitigating normal consequences of conviction'.

R v Thomas (Derek) [1975] RTR 38 CA Five years' imprisonment plus five year disqualification merited for person convicted of theft who used car in commisison of theft offences.

R v Thomas (Kevin) (1983) 5 Cr App R (S) 354; [1984] Crim LR 49; [1983] TLR 599 CA That road traffic offender sentenced to imprisonment a mitigating factor as regards determination of appropriate period of disqualification.

R v Tupa [1974] RTR 153 CA £250 fine plus five year disqualification for driving with blood alcohol concentration three times above prescribed limit.

R v Ward [1971] RTR 503 CA On when life disqualification merited.

R v West (Alan Richard) (1986) 8 Cr App R (S) 266 CA Two years' disqualification/twelve months' imprisonment (six months suspended) for driving while disqualified.

R v West London Magistrate, ex parte Anderson [1973] RTR 232 CA Failed appeal against disqualification where sought/failed to prove order not made before offender in court.

R v West London Magistrates, ex parte Anderson [1973] Crim LR 311 HC QBD Failed appeal against disqualification where sought/failed to prove order not made before offender in court.

R v Weston (Paul Kenneth) (1982) 4 Cr App R (S) 51 CA Six years' imprisonment/two years' disqualification for youth guilty of inter alia various car-related offences.

R v Williams [1962] 3 All ER 639; (1962) 46 Cr App R 463; [1963] Crim LR 205; (1962) 126 JP 523; (1962) 112 LJ 699; (1962) 106 SJ 819; [1962] 1 WLR 1268 CCA No disqualification upon conviction for stealing motor car.

R v Wilson [1969] Crim LR 158t CA Two years and nine months' imprisonment plus two and a half years' disqualification for persistent road traffic offender guilty of drunk driving/driving while disqualified/using driving licence with intent to deceive/making false statement so as to acquire insurance.

R v Wintour [1982] RTR 361 CA Entire sentence invalid where CrCt judge had sought to impose additional element (requirement that re-sit test) when removing disqualification.

R v Wright (Desmond Carl) (1979) 1 Cr App R (S) 82 CA Disqualification (for use of motor vehicle in course of offence) improper if will impede person getting job after released from prison.

Redmond v Parry [1986] RTR 146 HC QBD That only drove a few feet (and intended to drive no further) could amount to special reason justifying non-disqualification.

Reynolds v Roche [1972] RTR 282 HC QBD Rushing home late in case babysitter left child alone/that disqualification would adversely affect offender's family not special reasons justifying non-disqualification.

Scoble v Graham [1970] RTR 358 HC QBD Person stunned after collision might (but did not here) have reasonable excuse for refusing specimen.

Smith v Geraghty [1986] RTR 222 HC QBD On establishing 'spiking' of drink with alcohol as special reason justifying non-disqualification.

Soanes v Ahlers (1981) 145 JP 301; [1981] RTR 337 HC QBD 'Totting-up' under the Road Traffic Act 1972, s 92(3).

Stone v Bastick [1965] 3 All ER 713; (1966) 130 JP 54; (1965-66) 116 NLJ 189; [1967] 1 QB 74; (1965) 109 SJ 877; [1965] 3 WLR 1233 HC QBD Format of disqualification certificates: only offence for which disqualified to appear.

Surtees v Benewith [1954] 1 WLR 1335 HC QBD Unless there are special reasons justifying non-disqualification (here for driving uninsured motor-vehicle) disqualification must be ordered.

Taylor v Austin [1969] Crim LR 152; (1969) 133 JP 182; (1968) 112 SJ 1024; [1969] 1 WLR 264 HC QBD That excess alcohol did not affect driving ability/were not responsible for accident/ would suffer greatly financially did not justify non-disqualification.

Taylor v Rajan (1974) 59 Cr App R 11 HC QBD On special circumstances justifying non-disqualification.

Thompson v Diamond [1985] RTR 316 HC QBD Mother's emergency admission to hospital not an emergency justifying non-disqualification for driving with excess alcohol (to be with mother).

Vaughan v Dunn [1984] Crim LR 365; [1984] RTR 376 HC QBD That had been engaged in undercover police work which necessitated visiting public houses not special reason justifying non-disqualification.

Weatherson v Connop [1975] Crim LR 239 HC QBD Spiking of drink did not here constitute special reason justifying non-disqualification: on when could do so.

West v Jones [1956] Crim LR 423t; (1956) 120 JP 313 HC QBD When disqualifying person from driving disqualification must be general and cannot be qualified.

Williams v Neale [1972] Crim LR 598; [1971] RTR 149 HC QBD Non-disqualification of driver whose drink had (unknown to him) been 'spiked' with alcohol.

Woodage v Lambie [1971] 3 All ER 674; (1971) 135 JP 595; (1971) 121 NLJ 665t; [1972] RTR 36; [1972] Crim LR 536t; (1971) 115 SJ 588; [1971] 1 WLR 1781 HC QBD That earlier offences not serious not mitigating circumstance when sentencing for third traffic offence.

SENTENCE (driving while disqualified)

Davidson-Houston v Lanning [1955] 2 All ER 737; [1955] Crim LR 311; (1955) 119 JP 428; (1955) 99 SJ 510; [1955] 1 WLR 858 HC QBD Cannot imprison under-twenty one year old (for driving while disqualified) unless no other option.

Gosling v Paul (1961) 125 JP 389 HC QBD Shortness of remaining period of disqualification not reason justifying non-imprisonment for driving while disqualified.

Patterson v James [1962] Crim LR 406t HC QBD On deciding whether person to be fined/imprisoned for driving while disqualified.

R v Cornish (1965) 109 SJ 434 CCA Three years' probation order for young man (guilty of previous offences and here guilty of taking car without consent/driving while disqualified) on condition that attend Langley Trust house.

R v Craven [1979] Crim LR 324 CA Nine months' disqualification (inter alia) for driving while disqualified.

R v Isaacs [1990] RTR 240 CA Justices' consecutive six-month imprisonment sentences (for driving while disqualified) made concurrent as was no jurisdiction in justices to pass former sentence.

R v Jones [1970] Crim LR 693 CA Further offence of driving while disqualified may be taken into account when sentencing for similar offence for which convicted before later offence.

R v Jordan (William) [1996] 1 Cr App R (S) 181; [1995] Crim LR 906; [1996] RTR 221 CA Separate sentences for drunk driving/driving while disqualified offences arising from same facts were justified (five months' imprisonment each, consecutive).

R v Lynn (Frederick John) (1971) 55 Cr App R 423; [1971] RTR 369 CA Disqualification order approved where person driving while disqualified did not actually know was disqualified.

R v Mitchell (Thomas Christopher) (1982) 4 Cr App R (S) 277; [1982] Crim LR 837 CA Nine months' imprisonment for driving while disqualified.

R v Pegrum (Leslie David John) (1986) 8 Cr App R (S) 27 CA Twelve months' imprisonment for driving while disqualified (not first time did so).

R v Phillips (Gordon William) (1955) 39 Cr App R 87 CCA Where imprisonment and disqualification imposed together period of disqualification to be longer than that of imprisonment.

R v Skinner (Roger Martin) (1986) 8 Cr App R (S) 166 CA Sentences for reckless driving/driving while disqualified convictions arising from same event to be concurrent.

R v Warrior [1984] Crim LR 188 CA Nine months' youth custody for sixteen year old guilty of driving while disqualified/without insurance.

R v Webb [1991] RTR 192 CA Twelve months of suspended eighteen month sentence brought into effect where guilty of driving while disqualified.

R v Wilson [1969] Crim LR 158t CA Two years and nine months' imprisonment plus two and a half years' disqualification for persistent road traffic offender guilty of drunk driving/driving while disqualified/using driving licence with intent to deceive/making false statement so as to acquire insurance.

SENTENCE (drugged/drunk driving)

Chatters v Burke (1986) 8 Cr App R (S) 222 CA On special reasons for (as here) not disqualifying driver for driving with excess alcohol.

Davies (Gordon Edward) v Director of Public Prosecutions [1990] Crim LR 60; (1990) 154 JP 336 HC QBD Failed appeal against conviction for refusing specimens by individual who had offered medical reasons for refusal (objective test applied as to credibility of reasons preferred).

Denny v Director of Public Prosecutions (1990) 154 JP 460; [1990] RTR 417 HC QBD Not improper for police officer to seek breath specimens on properly working brethalyser after failed attempts on defective breathalyser.

Director of Public Prosecutions v Beech [1992] Crim LR 64; (1992) 156 JP 31; (1991) 141 NLJ 1004t; [1992] RTR 239; [1991] TLR 342 HC QBD Not reasonable excuse for non-provision of breath specimen that had made self so drunk could not understand what were instructed to do.

Director of Public Prosecutions v Corcoran (1991) 155 JP 597; [1991] RTR 329 HC QBD Non-disqualification merited upon conviction for driving with excess alcohol given short length intended to drive/fact that did not pose danger to public.

Director of Public Prosecutions v Corcoran (1990-91) 12 Cr App R (S) 652 CA Non-disqualification may be merited if person guilty of driving with excess alcohol did not drive far/pose danger to others.

Director of Public Prosecutions v O'Meara (1988) 10 Cr App R (S) 56 CA Not special reason justifying non-disqualification that excess alcohol result of previous evening's drinking (interrupted by sleep).

Director of Public Prosecutions v Warren (Frank John) [1992] Crim LR 200; (1992) 156 JP 753; [1992] RTR 129 HC QBD Failed prosecution for driving with excess alcohol as accused not given adequate opportunity to comment on what type of specimen would prefer to give (where breathalyser not functioning properly).

Director of Public Prosecutions v Welsh (1997) 161 JP 57; (1996) TLR 15/11/96 HC QBD Rounding down of readings/deduction of 6mg from final alcohol reading are common practices not immutable rules.

Director of Public Prosecutions v White (1988) 10 Cr App R (S) 66; [1988] RTR 267 CA Not special reason justifying non-disqualification that were breathalysed on two separate occasions within half-hour period, proving negative at first and later proving positive.

Park v Hicks [1979] Crim LR 57; [1979] RTR 259 HC QBD Failed attempt to rely upon medical emergency as special reason for non-disqualification following driving with excess alcohol as defence too flimsy.

R v Adames [1997] RTR 110 CA Two (not three) years' imprisonment for sick old man guilty of causing death while driving with excess alcohol taken.

R v Beardsley [1979] RTR 472 CA Eighteen months' imprisonment, fine, order to pay prosecution costs in relation to offence for which convicted (to maximum of £100) for person convicted of driving with excess alcohol.

R v Berry [1976] Crim LR 638 CA On when (as here) imprisonment (here two months) merited for driving with excess alcohol.

R v Burdon (James Heggie) (1927-28) 20 Cr App R 80 CCA Approval of sentence of twenty-two months in second division for causing bodily harm by way of wanton driving following drinking of alcohol.

R v Cambridge Magistrates' Court and the Crown Prosecution Service, ex parte Wong (1992) 156 JP 377; [1992] RTR 382 HC QBD Where taking prescribed cough medicine had unaware to person pushed them slightly over alcohol limit this could be special reason justifying non-disqualification — whether should be non-disqualification a matter for discretion of justices.

R v Cartwright [1990] RTR 409 CA Two years' probation (with requirement that undergo medical treatment) substituted for twelve months' imprisonment imposed on manic depressive for inter alia reckless driving, criminal damage and drunk driving.

R v Cook (Arthur Paul) [1996] 1 Cr App R (S) 350 CA £500 for driving with excess alcohol.

R v Eadie (Eric) (1978) 66 Cr App R 234; [1978] Crim LR 368; (1978) 128 NLJ 162t; [1978] RTR 392 CA On sentencing for drunken driving.

R v Holt [1962] Crim LR 565t CCA No special reasons meriting non-disqualification of driver for driving under influence of drink/drugs even though his doctor did not warn him of adverse effects of drinking small amount of alcohol in tandem with taking prescription drug.

R v Jenkins [1978] RTR 104 CA Disqualification for drunk driving reduced from three years to eighteen months.

R v Jones (David John) (1980) 2 Cr App R (S) 152 CA Concurrent sentences where on foot of same incident were convicted of driving while disqualified/driving with excess blood alcohol.

R v Jordan (William) [1996] 1 Cr App R (S) 181; [1995] Crim LR 906; [1996] RTR 221 CA Separate sentences for drunk driving/driving while disqualified offences arising from same facts were justified (five months' imprisonment each, consecutive).

R v Lavin (John Martin) (1967) 51 Cr App R 378; [1967] Crim LR 481 CA £50 fine for driving under influence of drink (substituted for six months' imprisonment which was too radical a change in sentencing policy).

R v Newman [1976] Crim LR 265 CA £50 fine for first time road traffic offender guity of drunk driving.

R v Newman [1978] RTR 107 CA Imprisonment sentence quashed and fine substituted for person guilty of driving with three times permitted level of alcohol in blood.

R v Nokes (John James Richard Francis) (1978) 66 Cr App R 3; [1978] RTR 101 CA Need not be custodial sentence for first breathalyser conviction.

R v Pashley [1974] RTR 149 CA Twelve months' imprisonment and twelve months' disqualification for person who drove with blood-alcohol levels well above prescribed limit.

R v Redbridge Justices, ex parte Dent (1990) 154 JP 895 HC QBD Person who drafted erroneous Crown Prosecution Service summary of case did not have to be produced before court.

R v Rennie [1978] RTR 109 CA Improper that had activated suspended sentence (relating to different type of offence) when sentencing person for drunk driving.

R v Salters (John William) (1979) 1 Cr App R (S) 116; [1979] RTR 470 CA Imprisonment merited by person guilty of being in charge of motor vehicle with grossly excessive blood-alcohol concentration.

R v Schofield [1964] Crim LR 829; (1964) 108 SJ 802 CCA Five (not thirty) years' disqualification for person who drove while unfit to do so through drink or drugs.

R v Shoult (Alan Edward) [1996] 2 Cr App R (S) 234 CA On imprisoning offenders guilty of driving with surfeit alcohol.

R v Smith [1961] 2 All ER 743; (1961) 45 Cr App R 207; (1961) 111 LJ 501; (1961) 105 SJ 532 CCA Preventive detention possible sentence in drunk driving case but not merited here.

R v Smith [1961] 2 All ER 743; (1961) 125 JP 506 CCA Preventive detention possible sentence in drunk driving case but not merited here.

R v Stimpson [1977] Crim LR 114 CA Twelve months' imprisonment, order to resit driving test plus three (not five) years' disqualification for driving with excess alcohol.

R v Sylvester (John Dennis) (1979) 1 Cr App R (S) 250 CA Three months' immediate imprisonment merited for persistent driving with excess alcohol (albeit by small amount).

R v Thomas (1973) 57 Cr App R 496; [1973] Crim LR 379; [1973] RTR 325 CA £100 fine/two years' disqualification but no custodial sentence for person guilty of driving with excess alcohol.

R v Tupa [1974] RTR 153 CA £250 fine plus five year disqualification for driving with blood alcohol concentration three times above prescribed limit.

R v Wilson [1969] Crim LR 158t CA Two years and nine months' imprisonment plus two and a half years' disqualification for persistent road traffic offender guilty of drunk driving/driving while disqualified/using driving licence with intent to deceive/making false statement so as to acquire insurance.

SENTENCE (endorsement of licence)

Agnew v Director of Public Prosecutions (1990-91) 12 Cr App R (S) 523; (1991) 155 JP 927; [1991] RTR 144; [1990] TLR 732 HC QBD Endorsement of licence of police driver in police training execrcise who was involved in accident after deviated from his instructions and was found guilty of driving without due care/attention.

Barnes v Gevaus (1980) 2 Cr App R (S) 258 CA On practice as regards finding special reasons justifying non-endorsement.

Barnett v Fieldhouse [1987] RTR 266 HC QBD Father's being away on holiday at relevant time was special reason for not endorsing licence where was convicted of permitting uninsured use of car by son.

Beighton v Brown [1965] 1 All ER 793; [1965] Crim LR 375; (1965) 129 JP 199; (1965) 115 LJ 247; (1965) 109 SJ 295; [1965] 1 WLR 553 HC QBD No endorsement for dangerously secured load.

Bradburn v Richards (1975) 125 NLJ 998t; [1976] RTR 275 HC QBD Justices could not extend period for variation of sentence allowed under Road Traffic Act 1972, s 101.

Burgess v West [1982] Crim LR 235; [1982] RTR 269 HC QBD Non-endorsement of licence of speeding offender where was no sign indicating decrease in speed limit.

Chief Constable of West Mercia Police v Williams [1987] RTR 188 HC QBD Could be producing false instrument where produced forged clean driving licence in court when real licence had several (no longer effective) endorsements.

Director of Public Prosecutions v Fruer; Director of Public Prosecutions v Siba; Director of Public Prosecutions v Ward [1989] RTR 29 HC QBD Where motorwy not in ordinary use could be that there were special reasons justifying driving on central reservation.

Director of Public Prosecutions v Powell [1993] RTR 266 HC QBD No/was special reason arising to justify non-endorsement of licence of person who was guilty of careless driving of child's motor cycle on road/who did not have insurance for or L-plate on same.

Dyson v Ellison [1975] 1 All ER 276; (1975) 60 Cr App R 191; [1975] Crim LR 48; (1975) 139 JP 191; (1974) 124 NLJ 1132t; [1975] RTR 205; (1975) 119 SJ 66; [1975] 1 WLR 150 HC QBD Court to consider previous endorsements in deciding if diqualification necessary; appeal against obligatory endorsement is by way of re-hearing.

East v Bladen [1987] RTR 291 HC QBD Mental effects of serious injuries had previously received were special reasons justifying non-endorsement of licence upon conviction for uninsured use of motor vehicle.

East v Bladen (1986) 8 Cr App R (S) 186 CA Reasonable belief that were insured while driving (when were not) could as here be special reason justifying non-disqualification.

Hawkins v Roots; Hawkins v Smith [1975] Crim LR 521; [1976] RTR 49 HC QBD Non-endorsement not justified just because of slightness of incident prompting careless/inconsiderate driving charge.

Marks v West Midlands Police [1981] RTR 471 HC QBD Slightness of speeding/driver's concern for unwell elderly passenger were special reasons justifying non-endorsement of licence upon speeding conviction.

Mawson v Oxford [1987] Crim LR 131; [1987] RTR 398 HC QBD Under motorway regulations (1982) vehicles never entitled to stop on motorway carriageway: endorsement had to be ordered.

McCormick v Hitchins [1988] RTR 182 HC QBD Non-endorsement of licence (of owner of parked car — damaged by another — who refused breath test on basis that was not about to drive) was valid as were special reasons present.

Platten v Gowing (1982) 4 Cr App R (S) 386; [1983] Crim LR 184; [1982] TLR 615 CA Maximum penalties for exceeding general speed limit inapplicable to situation where exceed temporary (but not general) speed limit.

R v Ashford and Tenterden Magistrates' Court, ex parte Wood [1988] RTR 178 HC QBD Passenger-car owner guilty of failure to provide blood specimen could have licence endorsed.

R v Bromley Justices, ex parte Church [1970] Crim LR 655; [1970] RTR 182 HC QBD Successful action to have endorsement of licence (for carrying person on motor cycle in dangerous fashion) quashed as justices had no power to order such endorsement.

R v Cockermouth Justices and others, ex parte Patterson [1971] Crim LR 287; [1971] RTR 216 HC QBD Pre-Theft Act 1968 endorsement of licence/disqualification for taking and driving away car without owner's consent was improper.

R v Liskerrett Justices, ex parte Child [1972] RTR 141 HC QBD Endorsement quashed as accused when asked if would like to plead guilty by post was not given all facts on which prosecutor could later seek to rely.

R v Preston [1986] RTR 136 CA Cannot impose licence endorsement and discretionary disqualification; must be circumstances relating to facts making up offence to justify non-endorsement.

R v Reading Justices, ex parte Bendall [1982] RTR 30 HC QBD Disqualification/endorsement of licence quashed where person eligible for provisional licence drove unlicensed but in compliance with provisional licence conditions.

Robinson v Director of Public Prosecutions [1989] RTR 42 HC QBD No special reasons meriting non-endorsement of licence of solicitor speeding to arrive at place and deal with matter that could be (and was) adequately dealt with by his clerk.

Rumbles (Brian) v Poole (Anthony) (1980) 2 Cr App R (S) 50 CA Cannot endorse licence upon conviction for failure to stop when told to do so by traffic warden.

Walker v Rawlinson [1975] Crim LR 523; [1976] RTR 94 HC QBD Not special reason justifying non-endorsement of licence of peron guilty of speeding that did not appreciate that distance between street-lights on street meant had entered 30mph area.

SENTENCE (failure to provide specimen)

Daniels (Jeremy) v Director of Public Prosecutions (1992) 13 Cr App R (S) 482 CA May (unlike here) be special reasons meriting non-disqualification though failed to provide breath specimen absent reasonable excuse.

Hockin v Weston [1972] Crim LR 541t; (1971) 121 NLJ 690t; [1972] RTR 136; (1971) 115 SJ 675 HC QBD Could not be reasonable excuse for failure to provide specimen on part of person pleading guilty of said failure.

R v Gormley [1973] Crim LR 644; [1973] RTR 483 CA On appropriate sentencing of person who failed to provide specimen upon request and assaulted the arresting officer in the execution of his duty.

R v Huggans [1977] Crim LR 684 CA Could (but here did not) activate existing suspended sentence of person in control of car found guilty of failure to provide specimen.

R v Page [1981] RTR 132 CA On procedure when court seeking to draw inference adverse to accused; disparate sentencing proper for persons guilty of failure to provide specimen depending on whether proven/not proven to have been driving at relevant time.

R v Welsh [1974] RTR 478 CA £100 fine in total for throwing away of specimens provided and refusal to provide new specimen.

SENTENCE (fines)

R v Whelan (1969) 119 NLJ 676t CA Imprisonment unmerited by person for whom disqualification was not real punishment: fine ought to have been imposed.

SENTENCE (fixed penalties)

R v The Clerk to the Croydon Justices, ex parte Chief Constable of Kent [1989] Crim LR 910; (1990) 154 JP 118; [1991] RTR 257 HC QBD Under Transport Act 1982 fixed penalties are payable by unincorporated bodies.

SENTENCE (general)

R v Ling (1993) 157 JP 931 CA Sentences for various offences committed on same evening reduced as had been prompted by accused's diabetic attack after non-injection/eating chocolates/drinking alcohol.

R v McCausland; R v Quaile [1968] Crim LR 49; (1967) 111 SJ 998 CA On sentencing under twenty-one year olds: eighteen months' imprisonment for shopbreaking with intent and various road traffic offences.

R v McNeil (1993) 157 JP 469 CA Thirty months' detention appropriate for youth guilty of ram-raiding and crashing into police car in deliberate attempt to escape capture.

R v Ratcliffe (1977) 127 NLJ 1205 CA Fine reduced by one-third but costs order allowed stand in road traffic case where was some dispute as to means of offender.

SENTENCE (motor manslaughter)

R v Sherwood [1995] RTR 60 CA Seven years in young offender institution for motor manslaughter driver concurrent with eighteen months for dangerous driving merited by seriousness of manslaughter; maximum of two years' imprisonment for passenger guilty of theft where indictment had not spelt out facts meriting greater sentence.

R v Stubbs (Sydney) (1912-13) 8 Cr App R 238; (1912-13) XXIX TLR 421 CCA Twelve months' hard labour appropriate for motor car manslaughter; victim's negligence alleviated negligence of party convicted of manslaughter.

SENTENCE (parking)

Crossland v Chichester District Council [1984] RTR 181 CA Punitive amount payable if failed to display 'pay and display' parking ticket was a charge so local government measure complied with Road Traffic Regulation Act 1967, s 31(1) (as amended).

SENTENCE (penalty points)

Johnson v Finbow (1983) 5 Cr App R (S) 95; [1983] Crim LR 480; (1983) 147 JP 563; [1983] RTR 363; (1983) 127 SJ 411; [1983] TLR 203; [1983] 1 WLR 879 HC QBD Not stopping after accident and not reporting same though at different times were on same occasion for purposes of Transport Act 1981, s 19(1): on appropriate pentalty points for same.

Johnston v Over (1985) 149 JP 286; [1985] RTR 240; [1984] TLR 450 HC QBD On determining for purposes of Transport Act 1981 penalty points provisions whether offence committed on same occasion as another.

King (David John) v Luongo (Marcello) (1985) 149 JP 84; [1985] RTR 186; [1984] TLR 143 HC QBD On licence endorsement provisions of Transport Act 1981.

SENTENCE (probation)

R v Roughsedge (Attorney General's Reference (No 7 of 1993)) [1994] Crim LR 302; [1994] RTR 322 CA Imprisonment not probation appropriate where drove dangerously, then wounded police officer when resisting arrest but as probation seemed effective here appellate court did not substitute imprisonment.

SENTENCE (reckless driving)

Denny v Director of Public Prosecutions (1990) 154 JP 460; [1990] RTR 417 HC QBD Not improper for police officer to seek breath specimens on properly working breathalyser after failed attempts on defective breathalyser.

Director of Public Prosecutions v Warren (Frank John) [1992] Crim LR 200; (1992) 156 JP 753; [1992] RTR 129 HC QBD Failed prosecution for driving with excess alcohol as accused not given adequate opportunity to comment on what type of specimen would prefer to give (where breathalyser not functioning properly).

Director of Public Prosecutions v Welsh (1997) 161 JP 57; (1996) TLR 15/11/96 HC QBD Rounding down of readings/deduction of 6mg from final alcohol reading are common practices not immutable rules.

Myers v Director of Public Prosecutions (1993) 14 Cr App R (S) 247; [1993] Crim LR 84 HC QBD Fifteen year old reckless driver did merit period of detention at young offender institution.

R v Austin (Howard David) (1980) 2 Cr App R (S) 203; [1981] RTR 10 CA Consecutive sentences merited where in course of single occasion were guilty of reckless driving and causing bodily harm by wanton/furious driving.

R v Bannister (Stephen) (1990-91) 12 Cr App R (S) 314; [1991] Crim LR 71; [1991] RTR 1; [1990] TLR 491 CA Three months' imprisonment, two years' disqualification and driving test re-sit order merited by reckless driver.

R v Bilton (Frederick Stewart) (1985) 7 Cr App R (S) 103; [1985] Crim LR 611 CA Six months' imprisonment for reckless driving.

R v Blanchard (Leslie Norman) (1982) 4 Cr App R (S) 330; [1983] Crim LR 122 CA Six months' imprisonment for reckless driving/theft by good worker of good character.

R v Cambridge Magistrates' Court and the Crown Prosecution Service, ex parte Wong (1992) 156 JP 377; [1992] RTR 382 HC QBD Where taking prescribed cough medicine had unaware to person pushed them slightly over alcohol limit this could be special reason justifying non-disqualification — whether should be non-disqualification a matter for discretion of justices.

R v Carrier (Wayne Paul) (1985) 7 Cr App R (S) 57 CA Twelve months' imprisonment for reckless driving (repeated overtaking and slowing down until accident occurred).

R v Cartwright [1990] RTR 409 CA Two years' probation (with requirement that undergo medical treatment) substituted for twelve months' imprisonment imposed on manic depressive for inter alia reckless driving, criminal damage and drunk driving.

R v Charters (Thomas) (1990-91) 12 Cr App R (S) 446 CA Eighteen months' imprisonment for reckless driver.

R v Cox [1993] 2 All ER 19; (1993) 96 Cr App R 452; (1993) 157 JP 114; [1993] RTR 185; [1993] 1 WLR 188 CA Custodial sentence justifiable if to right-thinking members of the public justice would not otherwise be done; prevalence of offence/public concern relevant to seriousness.

R v Cunliffe (Anthony John) (1987) 9 Cr App R (S) 442 CA Roughly three and a half months' actual imprisonment/one years' disqualification for reckless driving.

R v Davies (Xavier Bartley) (1993) 14 Cr App R (S) 341 CA Fifteen months' imprisonment for driving recklessly (whilst being pursued by police).

R v Driver (Jabez) (1925-26) 19 Cr App R 86 CCA Four months in second division for bodily harm by wanton driving of trap conviction of middle-aged farmer of reasonably good character.

R v Elwood-Wade (Roger David) (1990-91) 12 Cr App R (S) 51 CA £750 fine/twelve month disqualification for reckless driving.

R v Emery [1985] RTR 415 CA Community service appropriate for young man of previously good character guilty of taking motor cycle without authority/gravely reckless driving.

R v Eynon (Graham) (1988) 10 Cr App R (S) 437 CA One hundred and eighty hour community service order imposed on reckless driver.

R v Ford [1982] RTR 5 CA Three year disqualification merited where overtook another on hill-crest.

R v Garraway (Desmond Turner) (1988) 10 Cr App R (S) 316 CA Three months' imprisonment for reckless (high speed) driving in city area while being chased by police.

R v Gay (David John) (1989) 11 Cr App R (S) 553 CA Six months' imprisonment for reckless driving on motorway after drinking (offender previously of good character).

R v Gibbon (Jason) (1992) 13 Cr App R (S) 479 CA Eighteen months' imprisonment for reckless driving.

R v Hanciles [1984] Crim LR 246 CA Four weeks' imprisonment (plus one year disqualification) for reckless driving.

R v Hourihane (Mark Dennis) (1993) 14 Cr App R (S) 357 CA Fifteen months' imprisonment for reckless driving (speed racing).

R v Hudson (Alan James) (1989) 89 Cr App R 51; (1989) 11 Cr App R (S) 65; [1989] RTR 206 CA Two years' imprisonment and five years' disqualification appropriate for person convicted of death by reckless driving who had previously good character/work record.

R v Jones (David George) (1991) 93 Cr App R 169; (1990-91) 12 Cr App R (S) 167 CA Nine months'/eighteen months' imprisonment plus disqualification for reckless driving/unlawful wounding justified by use of car to attack another.

R v Kerry [1992] RTR 232 CA On appropriate sentencing by CrCt for reckless driving of cars.

R v Khan (Nashad) (1990-91) 12 Cr App R (S) 352 CA Four months' imprisonment for reckless driver.

R v King (Daniel) [1991] RTR 62 CA Nine months' imprisonment/five year disqualification appropriate for previous offender guilty of grossly reckless driving (used car as battering ram/drove pedestrians from road).

R v Knowles (Christopher) (1993) 14 Cr App R (S) 224 CA Eighteen months' imprisonment for serious case of reckless driving (at high speed).

R v Moriarty (Terence Patrick) (1993) 14 Cr App R (S) 575 CA Reckless driving (of which police chase formed part) punishable only by custodial sentence.

R v Morton (Bernard Charles) (1992) 13 Cr App R (S) 315; [1992] Crim LR 70 CA Nine months' imprisonment for reckless driving.

R v Osborne (Michael Stuart) (1990-91) 12 Cr App R (S) 55; [1990] Crim LR 532 CA Three months' detention in young offender institution/eighteen months' disqualification for twenty year old guilty of reckless driving.

R v Phelan [1978] Crim LR 303/572 CA £250 fine plus three year disqualification for reckless driving.

R v Powell (James Thomas); R v Elliott (Jeffrey Terence); R v Daley (Frederick); R v Rafferty (Robert Andrew) (1984) 79 Cr App R 277 CA On sentencing for death by reckless driving; when disqualifying should have reference to accused's driving record.

R v Reay [1993] RTR 189 CA Guilty plea does not preclude possibility of maximum (her custodial) sentence; where caught 'in the act' credit for guilty plea nominal.

R v Redbridge Justices, ex parte Dent (1990) 154 JP 895 HC QBD Person who drafted erroneous Crown Prosecution Service summary of case did not have to be produced before court.

R v Sandhu (Bagicha Singh) (1987) 9 Cr App R (S) 540 CA Nine months' imprisonment/five years' disqualification appropriate for reckless drivers (road-racers).

R v Skinner (Roger Martin) (1986) 8 Cr App R (S) 166 CA Sentences for reckless driving/driving while disqualified convictions arising from same event to be concurrent.

R v Smith (Mark) (1993) 14 Cr App R (S) 617 CA Four months' imprisonment for reckless driving.

R v Staddon (1992) 13 Cr App R (S) 171; [1992] Crim LR 70; (1991) 141 NLJ 1004t; [1992] RTR 42; [1991] TLR 341 CA On distinction in sentencing for reckless driving/causing death by reckless driving.

R v Steel (Mark Jonathan) (1993) 96 Cr App R 121; (1993) 14 Cr App R (S) 218; [1992] Crim LR 904; (1994) 158 JP 439; [1993] RTR 415; [1992] TLR 351 CA On appropriate sentence for reckless driving: here twelve months' imprisonment substituted for fifteen.

R v Sutton (Peter) (1983) 5 Cr App R (S) 34; [1983] Crim LR 408 CA Three weeks'/months' imprisonment/disqualification for reckless driver (drove wrong way at speed on motorway).

R v Till (Geoffrey) (1982) 4 Cr App R (S) 158; [1982] Crim LR 540 CA Six weeks' imprisonment for reckless driver (who deliberately drove into another car on motorway).

R v Tout (Thomas) (1994) 15 Cr App R (S) 30 CA Twelve months' young offender detention for reckless driving (driving fast down road with parked cars from between which child ran out and was struck).

R v Williams (David Jude) (1990-91) 12 Cr App R (S) 767 CA Fifteen months' young offender detention for reckless driving during police chase.

R v Yapp (1993) 157 JP 312 CA Maximum sentence for reckless driving justified.

SENTENCE (speeding)

Attorney-General's References Nos 17 and 18 of 1996 (Mark Andrew Isteon and Lee Wardle) [1997] 1 Cr App R (S) 247 CA Twelve/nine months' imprisonment for speeding in built-up area allowed stand.

Platten v Gowing [1983] RTR 352 HC QBD On appropriate fine for breach of temporary speed limit (where did not breach normal speed limit).

R v Modeste (Floyd) (1984) 6 Cr App R (S) 221 CA Four months' imprisonment for speeding.

R v Robbins (Andrew) [1996] 1 Cr App R (S) 312 CA Three and a half years' imprisonment for driving at excessive speed in low maximum speed area.

R v Robson [1991] RTR 180 CA Three years' imprisonment appropriate for driver with alcohol guilty of prolonged speeding (despite passenger warnings) resulting in death of two passengers.

SENTENCE (suspended sentence)

R v Vanston [1972] Crim LR 57 CA Suspended sentence (for burglary) properly brought into effect upon later conviction for road traffic/road traffic-related cases.

SENTENCE (tachograph)

R v McCabe (James) (1989) 11 Cr App R (S) 154 CA Six months' suspended sentence/£750 fine for false tachograph entries.

R v Parkinson (Herbert) and others (1984) 6 Cr App R (S) 423 CA Seven days' imprisonment for falsification of tachograph records by drivers.

R v Raven (Cyril William) (1988) 10 Cr App R (S) 354 CA Nine months' imprisonment for eminence grise of haulage company guilty of unlawful interferences with tachograph machines on lorries it operated.

SENTENCE (theft of car)

R v Weisberg (Albert) and Walvisch (Henry) (1920-21) 15 Cr App R 103 CCA Nine months'/six months' imprisonment for theft of motor car by prisoners of previously good character.

SENTENCE (unauthorised taking of vehicle)

R v Brown (George Henry) (1931-32) 23 Cr App R 48 CCA Two year binding over for taking motor car for seventeen year old with one conviction but generally good character workwise.

R v Cornish (1965) 109 SJ 434 CCA Three years' probation order for young man (guilty of previous offences and here guilty of taking car without consent/driving while disqualified) on condition that attend Langley Trust house.

R v Dillon (Vincent James) (1983) 5 Cr App R (S) 439 CA Consecutve sentences merited by person guilty of driving with excess alcohol/taking vehicle without consent on same occasion.

R v Earle [1976] RTR 33 CA Disqualification quashed for person guilty of various offences (including taking car) who had been sentenced to three years.

R v Emery [1985] RTR 415 CA Community service appropriate for young man of previously good character guilty of taking motor cycle without authority/gravely reckless driving.

R v Giles (1991) 155 JP 1000 CA Unpermitted taking of motor vehicle carries maximum sentence of six months' imprisonment: suspended sentence supervision order therefore invalid.

R v Mulroy [1979] RTR 214 CA Disqualification merited where sentencing for taking conveyance without lawful authority/robbery but reduced to fifteen months given difficulties offender might otherwise face upon twelve months' imprisonment imposed for same offences.

SENTENCE (uninsured vehicle)

Johnston v Over (1984) 6 Cr App R (S) 420 CA Parking two uninsured vehicles on road (so as to remove parts from one to another) involved committing two offences on same occasion.

SENTENCE (vehicle taking)

R v Evans (Brandon) [1996] RTR 46 CA Person who pleaded guilty to aiding in ringing of cars sentenced to two years and three months' imprisonment.

R v Harrison (Michael Allen) (1967) 51 Cr App R 371 CA Six years' imprisonment for stealing of lorry loaded with valuable goods.

SENTENCE (violence)

R v Atkins (Jonathan Terence) (1993) 14 Cr App R (S) 146 CA Six weeks' imprisonment for motorist who struck another motorist once following road traffic incident.

R v Calvert (Martin) (1992) 13 Cr App R (S) 634 CA Eighteen months' imprisonment for stabbing man after motoring argument.

R v Ord (Daniel William); R v Ord (Stephen Frederick) (1990-91) 12 Cr App R (S) 12 CA Eight/six weeks' imprisonment for offences of causing grievous bodily harm/assault occasioning actual bodily harm occurring during dispute between drivers.

SENTENCE (wanton driving)

R v Lambert [1962] Crim LR 645 CCA Two years' imprisonment for wanton driving, five years' consecutive for receiving and a further three years' concurrent for more receiving.

SIGN (general)

Adams v Jenks (1937) 101 JP 393 HC KBD 'Stop, Road Traffic Officer' a proper traffic sign, failure to observe which an offence.

Boyd-Gibbins v Skinner [1951] 1 All ER 1049; [1951] 2 KB 379 HC KBD Speed limit signs evidence that area deemed a built up area under Road Traffic Act 1934, s 1(1)(b).

Brazier v Alabaster [1962] Crim LR 173t HC QBD Keep left traffic sign affects only those who have passed it.

Brooks v Jefferies [1936] 3 All ER 232; (1936) 80 SJ 856; (1936-37) LIII TLR 34 HC KBD Innocent failure to observe road traffic sign deemed offence.

Buffel v Cardox (Gt Britain), Ltd and another [1950] 2 All ER 878; (1950) 114 JP 564 CA On ignoring traffic signs.

Bursey v Barron [1972] Crim LR 486; [1971] RTR 273; (1971) 115 SJ 469 HC QBD Use of non-reflective material on reflective speed signs did not render signs invalid so could be charged for driving in excess of indicated speeds.

Cooper v Hall [1968] 1 All ER 185; [1968] Crim LR 116t; (1968) 132 JP 152; (1967) 117 NLJ 1243t; (1967) 111 SJ 928; [1968] 1 WLR 360 HC QBD Not defence to parking in restricted area that was no yellow line.

Cotterill v Chapman [1984] RTR 73 HC QBD Road markings were valid though for 30mm distance between lines 3mm less than prescribed minimum: de minimis principle applied.

Davies v Heatley [1971] Crim LR 244/371; [1971] RTR 145 HC QBD Conviction for non-coformity with prescribed road sign quashed as road sign did not conform with design regulations and so was not prescribed.

Evans v Cross [1938] 1 All ER 751; (1939-40) XXXI Cox CC 118; (1938) 102 JP 127; [1938] 1 KB 694; (1938) 85 LJ 225; [1938] 107 LJCL 304; (1938) 82 SJ 97; (1937-38) LIV TLR 354; (1938) WN (I) 42 HC KBD White lines on road not a traffic sign.

Hassan v Director of Public Prosecutions (1992) 156 JP 852; [1992] RTR 209 HC QBD Single yellow line ineffective, absent accompanying warning sign.

Hill v Baxter [1958] 1 All ER 193; (1958) 42 Cr App R 51; [1958] Crim LR 192; (1958) 122 JP 134; (1958) 108 LJ 138; [1958] 1 QB 277; (1958) 102 SJ 53; [1958] 2 WLR 76 HC QBD Dangerous driving/non-conformity with road sign are absolute offences; automatism defence rejected.

Hoy v Smith [1964] 3 All ER 670; [1965] Crim LR 49; (1965) 129 JP 33; (1964) 114 LJ 790; (1964) 108 SJ 841; [1964] 1 WLR 1377 HC QBD Must be able to see 'Stop' on school crossing sign for school crossing requirements to apply.

James v Cavey [1967] 1 All ER 1048; [1967] Crim LR 245t; (1967) 131 JP 306; [1967] 2 QB 676; (1967) 111 SJ 318; [1967] 2 WLR 1239 HC QBD Inadequate signing by local authority vitiated offence.

Kierman v Howard [1971] Crim LR 286; [1971] RTR 314; (1971) 115 SJ 350 HC QBD On what constitutes giving adequate information in or near parking zone of restrictions which apply.

Langley Cartage Company, Limited v Jenks; Adams v Same [1937] 2 All ER 525; (1934-39) XXX Cox CC 585; [1937] 2 KB 382; [1937] 106 LJCL 559; (1937) 156 LTR 529; (1937) 81 SJ 399; (1936-37) LIII TLR 654; (1937) WN (I) 194 HC KBD Board stating 'Stop, Road Traffic Officer, Bucks CC' a valid road sign.

Lord v Wolley [1956] Crim LR 493 Magistrates £2 fine for failure to observe 'Yield' sign placed on road on trial basis.

McKenzie v Director of Public Prosecutions [1997] Crim LR 232; [1997] RTR 175; (1996) TLR 14/5/96 HC QBD Successful appeal by taxi-driver against conviction for stopping to pick up fare on street with double white lines in centre.

O'Halloran v Director of Public Prosecutions (1990) 154 JP 837; [1990] RTR 62 HC QBD Driver must comply with arrow requiring that keep to left of double white line.

R v Blything (Suffolk) Justices, ex parte Knight [1970] Crim LR 471; [1970] RTR 218 HC QBD Was not defence where had been overtaking improperly that had been forced to continue in impropriety by virtue of how another car was being driven.

R v Warren (Geoffrey) [1972] Crim LR 117; [1971] RTR xxiv Sessions Successful prosecution for non-compliance with red traffic light diagonally across from driver (red light nearest him not working).

R v Watts [1955] Crim LR 787 Magistrates Person ought to have stopped car when signalled to do so by police constable even though traffic lights were green.

Ramoo son of Erulapan v Swee (Gan Soo) and another [1971] 3 All ER 320; [1971] RTR 401; (1971) 115 SJ 445; [1971] 1 WLR 1014 PC Liability of parties in collision at malfunctioning traffic lights.

Ryan v Smith [1967] 1 All ER 611; [1967] Crim LR 55; (1967) 131 JP 193; [1967] 2 QB 893; (1966) 110 SJ 854; [1967] 2 WLR 390 HC QBD Any passing of stop line after red light an offence.

Sharples v Blackmore [1973] Crim LR 248; [1973] RTR 249 HC QBD Speed limit sign where front complied with traffic signs regulations but rear did not.

Spencer v Silvester [1964] Crim LR 146; (1963) 107 SJ 1024 HC QBD Reversal of perverse finding by justices that was not careless driving to halt at stop sign.

Stubbs v Morgan [1972] Crim LR 443; [1972] RTR 459 HC QBD No requirement that traffic sign must be fit for illumination at night so as to be valid by day.

Swift v Barrett (1940) 104 JP 239; (1940) 163 LTR 154; (1940) 84 SJ 320; (1939-40) LVI TLR 650; (1940) WN (I) 155 HC KBD Where prosecuting driver for non-conformity with traffic sign prosecution must show sign was of prescribed proportions.

Tolhurst v Webster [1936] 3 All ER 1020; (1937) 101 JP 121; (1937) 83 LJ 10; (1937) 156 LTR 111; (1936) 80 SJ 1015; (1936-37) LIII TLR 174; (1937) WN (I) 37 HC KBD Word 'Halt' requires temporary standstill.

Walker v Dowswell [1977] RTR 215 HC QBD Was dangerous driving where failed to comply with traffic sign.

Walton v Hawkins [1973] Crim LR 187; [1973] RTR 366 HC QBD Failed appeal against conviction for being unlawfully positioned on road (appeal on basis that double white lines not of form prescribed in traffic signs regulations).

Wells v Woodward [1956] Crim LR 207 HC QBD Justices could (unless contrary proven) infer that one traffic light red where other traffic light green.

Woodriffe v Plowman [1962] Crim LR 326; (1962) 106 SJ 198 HC QBD Absent contrary evidence traffic signals are presumed to be lawfully placed.

SPECIAL VEHICLE (fire engine)

Coote v Winfield [1980] RTR 42 HC QBD Fire engine no longer such where no longer good for fire-fighting and had been re-painted for advertising purposes.

SPECIAL VEHICLE (general)

Anderson and Heeley Ltd v Paterson [1975] Crim LR 49; (1975) 139 JP 231; [1975] RTR 248; (1975) 119 SJ 115; [1975] 1 WLR 228 HC QBD On what constitutes a 'tower wagon'.

Phillips v Prosser [1976] Crim LR 262; [1976] RTR 300 HC QBD Private hire car not a vehicle servicing premises.

Plume v Suckling [1977] RTR 271 HC QBD Coach modified to carry passengers, kitchen items and stock car was a goods vehicle.

R v Department of Transport, ex parte Lakeland Plastics (Windermere) Ltd [1983] RTR 82; [1982] TLR 249 HC QBD Commercial exhibitioner's vehicle not a 'showman's goods vehicle'.

Robertson v Crew [1977] Crim LR 228; [1977] RTR 141 HC QBD For vehicle to be 'disabled' must have done more than simply taking out rotor arm.

Squires v Mitchell [1983] RTR 400 HC QBD On what constitutes a disabled vehicle.

Turberville v Wyer; Bryn Motor Co Ltd v Wyer [1977] RTR 29 HC QBD Where coil inexplicably fell from lorry was prima facie evidence that lorry insecurely loaded; that part of lorry carrying load possibly mistakenly described as trailer in information was irrelevant.

Wakeman v Catlow [1976] Crim LR 636; [1977] RTR 174 HC QBD On burden of proof on defendant seeking to prove that vehicle (here a Jeep) a land tractor.

SPECIAL VEHICLE (mobile crane)

Director of Public Prosecutions v Evans and another [1988] RTR 409 HC QBD Mobile crane in two parts could be one vehicle (not two as was charged) but could still nonetheless be in violation of motor vehicle construction and use regulations.

SPECIAL VEHICLE (road tanker)

West Cumberland By Products Ltd v Director of Public Prosecutions [1988] RTR 391 HC QBD On respective liabilities of road tanker operator and road tanker driver under Dangerous Substances (Conveyance by Road in Road Tankers and Tank Containers) Regulations 1981, reg 10 and Health and Safety at Work etc Act 1974, s 36.

SPECIMEN (admissibility)

R v Mitten [1965] 2 All ER 59; (1965) 49 Cr App R 216; [1965] Crim LR 313t; [1966] 1 QB 10; (1965) 109 SJ 554; [1965] 3 WLR 268 CCA Non-compliance with law in requesting specimen need not mean evidence inadmissible; police to offer to supply part of specimen when requesting it/shortly before.

R v Trump (Ronald Charles Henry) (1980) 70 Cr App R 300; [1980] Crim LR 379; [1980] RTR 274; (1980) 124 SJ 85 CA Blood specimen obtained without accused's consent admissible at trial judge's discretion.

SPECIMEN (blood)

Andrews v Director of Public Prosecutions [1992] RTR 1; [1991] TLR 223 HC QBD That medic mistakenly stated defendant's reason for refusing blood not a medical reason did not mean police officer had not performed duty to obtain medical opinion on excuse defendant offered.

Baldwin v Director of Public Prosecutions [1996] RTR 238; [1995] TLR 279 HC QBD Police officer requesting blood specimen and asking why any reason should not be given had discharged statutory requirements: taking of specimen valid.

Beauchamp-Thompson v Director of Public Prosecutions [1988] Crim LR 758; [1989] RTR 54 HC QBD Evidence showing that alcohol levels exceeded limit at moment of test but not while driving was inadmissible; that were mistaken as to strength of alcohol consumed not special reason meriting non-disqualification.

Beck v Watson [1979] Crim LR 533; [1980] RTR 90 HC QBD Could not rely on second blood specimen where obtained after first freely provided and only rendered incapable of analysis when dropped by police medic.

Bentley v Chief Constable of Northumbria (1984) 148 JP 266; [1984] RTR 276; [1983] TLR 642 HC QBD Judicial notice that 'Alcolyser' an approved device; absent certificate mere fact that test tube of blood bears name of person who gave blood on particular day inadequate evidence that blood is that of that person.

Chief Constable of Avon and Somerset v Kelliher [1986] Crim LR 635; [1987] RTR 305 HC QBD That trained operator could have been got from another police station did not mean that obtaining blood at police station with device but without operator was not impracticable.

Chief Constable of Kent v Berry [1986] Crim LR 748; [1986] RTR 321 HC QBD Where no adequate breathalyser at police station could remove defendant to other station (which had adequate breathalyser) for taking of blood specimen.

Cole v Director of Public Prosecutions [1988] RTR 224 HC QBD Doctor must specifically advise that driver's condition possibly resulting from drug-taking before blood specimen may be validly requested.

Davis v Director of Public Prosecutions [1988] Crim LR 249; [1988] RTR 156 HC QBD Test whether constable had reasonable cause to believe specimen cannot be provided/should not be required is an objective test.

Dear v Director of Public Prosecutions [1988] Crim LR 316; [1988] RTR 148 HC QBD Division of blood single blood sample (one portion for police, one for defendant) invalid where latter portion placed in jar containing few droplets of blood already taken from defendant's other arm.

Dhillon (Surinder Singh) v Director of Public Prosecutions (1993) 157 JP 420 HC QBD Person not properly given option to provide blood/urine specimen where police interpreted vague answer as refusal of same.

Director of Public Prosecutions v Berry [1995] TLR 571 HC QBD Breath test evidence available where could have given blood/urine sample instead but were incapable of understanding this in light of alcohol had taken.

Director of Public Prosecutions v Charles; Director of Public Prosecutions v Kukadia; Ruxton v Director of Public Prosecutions; Reaveley v Director of Public Prosecutions; Healy v Director of Public Prosecutions; McKean v Director of Public Prosecutions; Edge (James) v Director of Public Prosecutions [1996] RTR 247 HC QBD Slight procedural errors did not render specimens obtained inadmissible.

Director of Public Prosecutions v D; Director of Public Prosecutions v Rouse (1992) 94 Cr App R 185; [1991] Crim LR 911; (1991) 141 NLJ 104t; [1992] RTR 246; [1991] TLR 338 HC QBD Conversation between police officer and arrestee about blood/urine specimen option and request for breath specimens not an interview.

Director of Public Prosecutions v Dixon [1993] RTR 22 HC QBD Matter for subjective adjudication whether police officer has reasonable basis for belief that breathalyser malfunctioning (and so to require blood (as here) or urine specimen).

Director of Public Prosecutions v Elstob [1992] Crim LR 518; (1993) 157 JP 229; [1992] RTR 45; [1991] TLR 526 HC QBD Two-minute delay in obtaining and dividing specimen permissible; division did not have to take place before defendant.

Director of Public Prosecutions v Garrett (1995) 159 JP 561; [1995] RTR 302; [1995] TLR 55 HC QBD On procedure as regards taking of specimens: that was procedural error as regards taking of blood specimen does not mean urine specimen taken in accordance with procedure is inadmissible.

Director of Public Prosecutions v Gordon; Director of Public Prosecutions v Griggs [1990] RTR 71 HC QBD On need to inform defendant of alternative specimens that may be required and of consequences of not giving required specimen to police officer.

Director of Public Prosecutions v Hill-Brookes [1996] RTR 279 HC QBD Must tell person consequences of providing/not providing optional blood/urine specimen so that may properly exercise option.

Director of Public Prosecutions v Ward [1997] TLR 155 HC QBD No right to legal advice before decide whether to substitute blood/urine specimen for breath specimen.

Director of Public Prosecutions v Warren [1992] 4 All ER 865; [1993] AC 319; (1993) 96 Cr App R 312; (1993) 157 JP 297; (1992) 142 NLJ 1684; [1993] RTR 58; (1992) 136 SJ LB 316; [1992] TLR 520; [1992] 3 WLR 884 HL Procedure when obtaining blood/urine sample of accused.

Director of Public Prosecutions v Wythe [1996] RTR 137 HC QBD Absent medical doctor's opinion was unlawful to demand blood of person who had volunteered possibly plausible excuse as to why could not give said specimen.

Dixon v Director of Public Prosecutions (1994) 158 JP 430 HC QBD Police were wrong to obtain blood samples on basis that breathalyser defective where had no reasonable ground for believing breathalyser to be defective.

Doctorine v Watt [1970] Crim LR 648; [1970] RTR 305 HC QBD Person given adequate portion of specimen need not also be instructed how to store it properly.

Edge v Director of Public Prosecutions [1993] RTR 146 HC QBD Conviction quashed where motorist not asked whether any reason why doctor could/should not take specimen.

Ferriby v Sharman [1971] Crim LR 288; [1971] RTR 163 HC QBD Relevant time for determining blood-alcohol concentration is time at which blood specimen is taken.

Froggatt v Allcock [1975] Crim LR 461; [1975] RTR 372 HC QBD Justices could after hearing differing views on blood specimen analysis by two competent analysts elect to accept prosecution analyst's evidence and not that of defence analyst.

Gregson v Director of Public Prosecutions [1993] Crim LR 884 HC QBD Was adequate proof that blood sample taken from defendant was sample that was forensically examined.

Grix v Chief Constable of Kent [1987] RTR 103 HC QBD Whether refusal to provide blood specimen sprang from medical reasons not a matter policeman had to consider — that offered urine not relevant to sentencing for failure to provide blood.

Hawkins v Director of Public Prosecutions [1988] RTR 380 HC QBD Failed appeal where had sought to prove that had not received analysis evidence by mail at least seven days before hearing.

Hope v Director of Public Prosecutions [1992] RTR 305 HC QBD Person who provided breath specimens as required and who agreed to provide blood could then decide not to provide blood.

Horrocks v Binns [1986] RTR 202 HC QBD Failure to provide blood specimen charge dismissed where had not properly sought breath specimen.

Kemp v Chief Constable of Kent [1987] RTR 66 HC QBD Failure (where was physically and medically possible) to provide blood specimen on basis that already gave same to hospital for other reason not a reasonable excuse to failure to provide blood.

Langridge v Taylor [1972] RTR 157 HC QBD Container in which blood supplied to defendant clotted but remained capable of analysis was a 'suitable container'.

Macdonald v Skelt [1985] RTR 321 HC QBD On discretion of justices to allow prosecutor call additional evidence after has closed case.

McLellan v Director of Public Prosecutions [1993] RTR 401 HC QBD Conviction may only be based on blood specimen provided where police officer has required same after concluding that breathalyser not functioning properly.

Meade v Director of Public Prosecutions [1993] RTR 151 HC QBD Conviction quashed where motorist not asked whether any reason why doctor could/should not take specimen.

Nicholson v Watts [1973] Crim LR 246; [1973] RTR 208 CA On necessary quality of portion of blood sample given to defendant.

Ogburn v Director of Public Prosecutions; Williams (Gladstone) v Director of Public Prosecutions; Director of Public Prosecutions v Duffy; Director of Public Prosecutions v Nesbitt [1994] RTR 241 CA On procedure as regards requesting breath/blood/urine specimens.

Ogburn v Director of Public Prosecutions; Williams v Director of Public Prosecutions; Director of Public Prosecutions v Duffy; Director of Public Prosecutions v Nesbitt [1995] 1 Cr App R 383; (1994) 138 SJ LB 53; [1994] TLR 57; [1994] 1 WLR 1107 HC QBD Convictions quashed/ acquittals allowed stand where had been slight procedural irregularities in obtaining blood specimens from accused.

Paterson v Director of Public Prosecutions [1990] Crim LR 651; [1990] RTR 329 HC QBD Policeman's statement inadmissible (under Criminal Justice Act 1967, s 9) where stated to be of shorter length than was; defendant to be given real opportunity of advocating which of blood/urine specimens would prefer to give; inadequate proof that blood sample analysed was that of defendant.

Patterson v Charlton [1985] Crim LR 449; (1986) 150 JP 29; [1986] RTR 18 HC QBD Admission that were driving is enough proof of same; burden is on person over the blood-alcohol limit to prove had not driven while so.

Perry v McGovern [1986] RTR 240 HC QBD Drunk driving prosecution quashed where prosecutor led defendant to believe that latter's portion of blood specimen was incapable of analysis.

Poole v Lockwood [1980] Crim LR 730; (1980) 130 NLJ 909; [1981] RTR 285 HC QBD Blood specimen inadmissible where second urine specimen provided over an hour after requested discarded by police officer genuinely believing specimen inadequate.

Prince v Director of Public Prosecutions [1996] Crim LR 343 HC QBD Unnecessary to prove calibration of breathalyser correct where prosecution only seek to rely on blood specimen evidence.

Punshon v Rose (1968) 118 NLJ 1196t; (1969) 113 SJ 39 HC QBD That blood alcohol level in part attributable to drinking on night before apprehension (defendant also having drunk on day in question) not a special reason meriting non-disqualification.

R v Alyson [1976] RTR 15 CA Conviction quashed where accused alleged had consumed whisky between driving and providing specimen and jury led to believe that could convict nonetheless.

R v Byers [1972] Crim LR 718; [1971] RTR 383 CMAC Blood specimen analysis necessary where answer to policeman interpeted by court as indicating preparedness to supply specimen, not request for portion of same (which had not been given).

R v Cheshire Justices, ex parte Cunningham; Edwards v Director of Public Prosecutions; Glossop v Director of Public Prosecutions; R v Cheshire Justices, ex parte Burn; Hunt v Director of Public Prosecutions; R v Warrington Crown Court, ex parte Herbert; Director of Public Prosecutions v Sharpe; Director of Public Prosecutions v Swift; White v Director of Public Prosecutions; Wimalasundera v Director of Public Prosecutions [1995] RTR 287 HC QBD Driving with excess alcohol convictions quashed as procedure whereby blood specimens obtained procedurally flawed.

R v Cheshire Justices, ex parte Sinnott; R v South Sefton Justices, ex parte Chorley; R v Cheshire Justices, ex parte Burgess; R v Cheshire Justices, ex parte Vernon; R v Liverpool Justices, ex parte Cummins; R v Northwich Justices, ex parte McIlroy; R v Rhuddlan Justices, ex parte Rouski [1995] RTR 281 HC QBD Driving with excess alcohol convictions quashed as procedure whereby blood specimens obtained procedurally flawed.

R v Coates (Philip Charles) (1976) 63 Cr App R 71; [1977] RTR 77 CA Genuine but misplaced fear that providing blood specimen would endanger health not reasonable excuse for refusing to give same.

R v Elliot [1976] RTR 308 CA Jury could prefer conflicting evidence of one expert to that of another so long as believed evidence correct beyond reasonable doubt.

R v Forbes [1971] Crim LR 174; [1970] RTR 491 CA Valid request for blood specimen (though failed to mention urine option).

R v Graham (John) [1974] 2 All ER 561 CA Religious beliefs not valid ground for refusing blood specimen.

R v Green [1970] 1 All ER 408; [1970] Crim LR 289; (1970) 134 JP 208; [1970] RTR 193 Assizes On requiring blood/urine specimens of person in hospital.

R v Green (Hugh) [1980] RTR 415 CA Application for leave to appeal against conviction (on basis of impossible disparity between blood and breath tests) refused.

R v Harding (Anthony Raymond) (1974) 59 Cr App R 153; [1974] Crim LR 481; (1974) 124 NLJ 525t; [1974] RTR 325; (1974) 118 SJ 444 CA Fear of needles could be reasonable excuse for refusing blood specimen.

R v Harling [1970] 3 All ER 902; (1971) 55 Cr App R 8; [1970] Crim LR 536; (1971) 135 JP 29; [1970] RTR 441 CA Reasonable excuse to give blood does not mean have reasonable excuse to give urine.

R v Moore [1970] Crim LR 650; [1970] RTR 486 CA Constable to have reasonable suspicion regarding alcohol levels before requests breath specimen/may request blood or urine (neither preferable) for laboratory test; on appropriate warning to be given as to consequences of refusing blood and urine specimens.

R v Paduch (Jan) (1973) 57 Cr App R 676; [1973] Crim LR 533; [1973] RTR 493 CA Police could change request and ask for urine rather than blood specimen from driver.

R v Palfrey; R v Sadler [1970] 2 All ER 12; (1970) 54 Cr App R 217; [1970] Crim LR 231t; (1970) 134 JP 397; (1970) 120 NLJ 81t; [1970] RTR 127; (1970) 114 SJ 92; [1970] 1 WLR 416 CA Practice regarding blood tests.

R v Richards (Stanley) [1974] 3 All ER 696; (1974) 59 Cr App R 288; [1975] Crim LR 103; (1974) 138 JP 789; [1974] RTR 520; (1975) 119 SJ 66; [1975] 1 WLR 131 CA Forensic evidence qualifying blood analysis admissible on charge of driving while unfit to drink under Road Traffic Act 1972, s 5.

R v Tate [1977] RTR 17 CA Second analyst validly called after close of prosecution case where analysis conducted by two authorised analysts, the first of whom had relied on the second's notes when called.

R v Trump (Ronald Charles Henry) (1980) 70 Cr App R 300; [1980] Crim LR 379; [1980] RTR 274; (1980) 124 SJ 85 CA Blood specimen obtained without accused's consent admissible at trial judge's discretion.

R v Weil [1970] RTR 284 CA On what constiutes a proper part of blood specimen to be given to defendant.

R v Withecombe (David Ernest) (1969) 53 Cr App R 22; (1969) 133 JP 123; (1968) 112 SJ 949; [1969] 1 WLR 84 CA Blood/breath specimen not valid unless relevant procedural requirements of Road Safety Act 1967 complied with.

Ratledge v Oliver [1974] Crim LR 432; [1974] RTR 394 HC QBD Police officer could at same time notify doctor of intent to administer breath test/seek blood or urine sample of hospital patient.

Rawlins v Brown [1987] RTR 238 HC QBD Breath evidence admissible where accused (upon whom burden rested) failed to satisfy doctor that wanted to give blood sample.

Renshaw v Director of Public Prosecutions [1992] RTR 186 HC QBD On procedure as regards informing person of choice as to provision of sample (whether blood or urine).

Rowland v Thorpe [1970] 3 All ER 195; [1972] Crim LR 655; (1970) 134 JP 655; (1970) 120 NLJ 662t; [1970] RTR 406; (1970) 114 SJ 707 HC QBD Only severe mental condition/physical injuries justify refusing laboratory specimen; reasonable failure to give urine not per se reasonable excuse to refuse blood — must be excuse for latter.

Rowlands v Harper [1973] Crim LR 122; [1972] RTR 469 HC QBD Possible procedural error by police in dealing with blood specimen justified dismissal of information.

Sadler v Metropolitan Police Commissioner [1970] Crim LR 284t HL Refusal of leave to appeal question whether blood specimen given under Road Safety Act 1967 admissible in charge brought under the Road Traffic Act 1960.

Sivyer v Parker [1986] Crim LR 410; [1987] RTR HC QBD Not necessary to take blood specimen at hospital from person who had elected to substitute breath specimen with blood specimen.

Smith v Cole [1971] 1 All ER 200; [1970] Crim LR 701; (1971) 135 JP 97; [1970] RTR 459; (1970) 114 SJ 887 HC QBD On taking blood specimens.

Smith v Director of Public Prosecutions [1989] RTR 159 HC QBD Breath specimen evidence admissible where person had first refused alternative specimen, then provided blood specimen.

Stephenson v Clift [1988] RTR 171 HC QBD Preferability of one of competing blood analyses by competent analysts a question of fact for court.

Stokes v Sayers [1988] RTR 89 HC QBD On what prosecution must (and here failed to) prove for seeking of blood specimen in light of defectively operating breathalyser to be valid.

Sykes v Director of Public Prosecutions [1988] RTR 129 HC QBD Person could be convicted on basis of breath specimen evidence despite police officer's later invalid request for blood specimen.

Tee v Gough [1980] Crim LR 380; [1981] RTR 73 HC QBD Blood properly taken/analysed with driver's consent for excess alcohol charge admissible in driving while unfit through drink charge.

Turner v Director of Public Prosecutions [1978] Crim LR 754; [1996] RTR 274 HC QBD Must tell person what results refusal to provide blood/urine specimen entails.

Wade v Director of Public Prosecutions (1995) 159 JP 555; [1996] RTR 177; [1995] TLR 77 HC QBD Blood test evidence inadmissible where correct procedure not observed in obtaining specimen.

Wakeley v Hyams [1987] Crim LR 342; [1987] RTR 49 HC QBD Blood and breath specimens rightly deemed inadmissible evidence where statutory procedure as regarded obtaining same not complied with.

Walker v Hodgkins [1983] Crim LR 555; (1983) 147 JP 474; [1984] RTR 34; [1983] TLR 149 HC QBD Justices could take into account normal laboratory procedure when deciding whether to accept evidence regarding blood specimens.

Webb v Director of Public Prosecutions [1992] RTR 299 HC QBD Blood specimen validly requested where police officer decided that woman having difficulty in providing specimen was shocked and medically incapable of providing same.

White and another v Proudlock [1988] RTR 163 HC QBD Objective test as to whether breath specimen cannot be provided/should not be required for medical reasons: constable need not specifically consider this matter same before moving on to request blood specimen.

SPECIMEN (container)

Moore v Wilkinson [1969] Crim LR 493t; (1969) 119 NLJ 626t HC QBD Not necessary for police to place container containing blood sample inside a sealed envelope for it to be a 'suitable container' as required by the Road Traffic Act 1960.

SPECIMEN (failure to provide)

Baker v Foulkes [1975] 3 All ER 651; (1976) 62 Cr App R 64; [1976] Crim LR 67t; (1976) 140 JP 24; [1975] RTR 509; [1975] 1 WLR 1551 HL Police officer need not tell medical doctor of intention to caution of effects of failing to provide specimen.

Daniels v Director of Public Prosecutions (1992) 156 JP 543; (1991) 141 NLJ 1629t; [1992] RTR 140; [1991] TLR 535 HC QBD Though no special reasons justifying non-provision of specimen could exclude evidence as to seeking of specimens where was unfair (Police and Criminal Evidence Act 1984, s 78 applied).

Davies (Gordon Edward) v Director of Public Prosecutions [1990] Crim LR 60; (1990) 154 JP 336 HC QBD Failed appeal against conviction for refusing specimens by individual who had offered medical reasons for refusal (objective test applied as to credibility of reasons preferred).

Davies v Director of Public Prosecutions [1989] RTR 391 HC QBD Constable told by driver that was taking drug that would affect blood-alcohol levels could legitimately ask for specimens other than breath.

Director of Public Prosecutions v Ambrose (1992) 156 JP 493; [1992] RTR 285; [1991] TLR 496 HC QBD For there to be reasonable excuse not to give specimen doing so must be adverse to one's health or else one must be physically/mentally incapable of giving specimen.

Director of Public Prosecutions v Butterworth [1994] 3 All ER 289; [1995] 1 AC 381; [1995] 1 Cr App R 38; [1995] Crim LR 71; (1995) 159 JP 33; (1994) 144 NLJ 1043; [1994] RTR 330; [1994] TLR 418; [1994] 3 WLR 538 HL Unnecessary to specify offence for which breath sample sought in prosecution for failure to provide breath specimen.

Director of Public Prosecutions v Crofton [1994] RTR 279 HC QBD On deciding whether is reasonable medical excuse for failure to provide breath (here there was).

Director of Public Prosecutions v Daley [1992] RTR 155 HC QBD No evidence of stress/medical factor/reasonable excuse justifying failure to provide breath specimen.

Director of Public Prosecutions v Daley (No 2) [1994] RTR 107 HC QBD CrCt Finding that offender had done his best but failed to provide breath specimen not factually accurate and so not special reason justifying non-disqualification.

Director of Public Prosecutions v Eddowes [1990] Crim LR 428; [1991] RTR 35 HC QBD Person willing to and who tried but failed to give second breath specimen guilty of failure to provide same.

Director of Public Prosecutions v Radford [1995] RTR 86; [1994] TLR 195 HC QBD That suffered pain from accident in which involved not reasonable excuse for failure to provide second breath specimen.

Director of Public Prosecutions v Rose [1993] Crim LR 407; (1992) 156 JP 733 HC QBD Tiny distance which travelled/begrudgery over apprehension not special reasons justifying non-disqualification for failure to provide breath specimens.

Director of Public Prosecutions v Whalley [1991] Crim LR 211; (1992) 156 JP 661; [1991] RTR 161 HC QBD Refusal to provide breath specimen only reasonable if to give specimen would for mental/physical reasons be adverse to donor's health.

Director of Public Prosecutions v Wythe [1996] RTR 137 HC QBD Absent medical doctor's opinion was unlawful to demand blood of person who had volunteered possibly plausible excuse as to why could not give said specimen.

Edwards v Wood [1981] Crim LR 414 HC QBD Valid finding that there had not been refusal to provide specimen where specimen not provided but did not refuse to provide it.

Ely v Marle [1977] Crim LR 294; [1977] RTR 412 HC QBD Breathalyser must be made available to suspect in reasonable time but attempt to walk way from scene after ten minutes did render accused guilty of failure to provide specimen.

Glendinning v Bell [1973] Crim LR 57; [1973] RTR 52 HC QBD That drank alcohol after drove could be reasonable excuse for failure to provide breath specimen.

Grix v Chief Constable of Kent [1987] RTR 103 HC QBD Whether refusal to provide blood specimen sprang from medical reasons not a matter policeman had to consider — that offered urine not relevant to sentencing for failure to provide blood.

Horrocks v Binns [1986] RTR 202 HC QBD Failure to provide blood specimen charge dismissed where had not properly sought breath specimen.

Hoyle v Walsh (1969) 53 Cr App R 61 CA Not guilty of failing to provide breath specimen where failure arose through defect in breathalyser — arrest without warrant for same therefore invalid.

Jarvis v Director of Public Prosecutions [1996] RTR 192 HC QBD Not allowing doctor to take blood after breathalyser failed was failure to provide specimen; no judicial notice as to number of breathalysers in particular police station.

Kemp v Chief Constable of Kent [1987] RTR 66 HC QBD Failure (where was physically and medically possible) to provide blood specimen on basis that already gave same to hospital for other reason not a reasonable excuse to failure to provide blood.

Mallows v Harris [1979] Crim LR 320; [1979] RTR 404 HC QBD Where defence does not raise matter of reasonable excuse prosecution have discharged burden in respect of same; excuse not pertaining to capacity to provide specimen could not be reasonable excuse.

Metropolitan Police Commissioner v Curran [1976] 1 All ER 162; (1976) 62 Cr App R 131; [1976] Crim LR 198; (1976) 140 JP 77; (1976) 126 NLJ 90t; [1976] RTR 61; (1976) 120 SJ 49; [1976] 1 WLR 87 HL Can be guilty of not giving specimen even if not guilty of drunk driving.

Murray v Director of Public Prosecutions [1993] Crim LR 968; (1994) 158 JP 261; [1993] RTR 209; [1993] TLR 56 HC QBD Warning that failure to give specimen renders one open to prosecution is absolutely mandatory.

Parker v Smith [1974] RTR 500 HC QBD On respective burdens of proof where is sought to raise (here failed) reasonable excuse to charge of failure to provide specimen.

R v Brush [1968] 3 All ER 467; (1968) 52 Cr App R 717; [1968] Crim LR 619t; (1968) 132 JP 579; (1968) 118 NLJ 1053; (1968) 112 SJ 806; [1968] 1 WLR 1740 CA Specimen may be valid though no warning given about effects of failing to supply same; if warning an issue must be trial within trial to decide on matter; issue not a matter for jury.

R v Chapman (Paul) [1977] RTR 190 CA Jury are judges of fact: to direct that they 'must' rather than 'may' find breathalyser to have been available a misdirection.

R v Chippendale [1973] Crim LR 314; [1973] RTR 236 CA On when warning to be given as regards consequences of failure to provide requested specimen.

R v Clarke [1969] 2 All ER 1008; (1969) 53 Cr App R 438; [1969] Crim LR 441; (1969) 113 SJ 109; (1969) 133 JP 546; (1969) 113 SJ 428; [1969] 1 WLR 1109 CA Any clearly worded request suffices in asking to take breath test; any form of declining is a refusal; any reasonable excuse a defence.

R v Ferguson (Frank) (1970) 54 Cr App R 415; [1970] Crim LR 652; [1970] RTR 395; (1970) 114 SJ 621 CA Direction to convict in prosecution for drunk driving where had been refusal to provide breath specimen.

R v Godden [1972] Crim LR 656; [1971] RTR 462 CA Was refusal to give specimen where would only give same if was taken by own doctor.

R v Haslam [1972] RTR 297 CA Person charged with failure to provide breath/laboratory specimen could be acquitted of former charge but convicted of latter charge.

R v Holah [1973] 1 All ER 106; (1973) 57 Cr App R 186; [1973] Crim LR 59; (1973) 137 JP 106; (1972) 122 NLJ 1012t [1973] RTR 74; (1972) 116 SJ 845; [1973] 1 WLR 127 CA Person not failing to provide specimen if gives enough breath for reading; police must tell motorist why being arrested.

R v John [1974] 2 All ER 561; (1974) 59 Cr App R 75; [1974] Crim LR 670t; (1974) 138 JP 492; (1974) 124 NLJ 293t; [1974] RTR 332; (1974) 118 SJ 348; [1974] 1 WLR 624 CA Religious-philosphocial beliefs not reasonable excuse for refusing to give specimen — excuse must relate to physical capacity.

R v Mackey [1977] RTR 146 CA On what constitutes failure to provide breath specimen (where do not expressly decline to give same).

R v McAllister (William Joseph) (1974) 59 Cr App R 7; [1974] Crim LR 716; [1974] RTR 408 CA On what constitutes refusal to give blood/urine specimen.

R v Reid (Philip) [1973] 3 All ER 1020; (1973) 57 Cr App R 807; [1973] Crim LR 760; (1974) 138 JP 51; [1973] RTR 536; (1972) 116 SJ 565; (1973) 117 SJ 681; [1973] 1 WLR 1283 CA That get out of car need not mean not driving/attempting to drive; mistaken belief that not obliged to give specimen not failure for not doing so; whether was refusal a factual question for jury.

R v Rey (Frederick Brian) (1978) 67 Cr App R 244; [1978] RTR 413 CA Valid arrest for failure to provide specimen where had provided same but in inadequate amount.

R v Rothery (Henry Michael) (1976) 63 Cr App R 231; [1976] Crim LR 691t; (1976) 126 NLJ 790t; [1976] RTR 550 CA If steal portion of own specimen retained by police are guilty of theft but not failure to provide specimen.

R v Wagner [1970] Crim LR 535; [1970] RTR 422; (1970) 114 SJ 669 CA Was refusal to provide breath specimen where requesting constable did not have necessary equipment and accused refused to wait.

Salter v Director of Public Prosecutions [1992] RTR 386 HC QBD No evidence that accused had been incapable of understanding warning as to consequences of not providing specimen so no reasonable excuse for non-provision.

Smith (Nicholas) v Director of Public Prosecutions [1992] RTR 413 HC QBD That accused did what could to provide specimen not reasonable excuse justifying non-provision.

Solesbury v Pugh [1969] 2 All ER 1171; (1969) 53 Cr App R 326; [1969] Crim LR 381t; (1969) 133 JP 544; (1969) 119 NLJ 438t; (1969) 113 SJ 429; [1969] 1 WLR 1114 HC QBD Failing to provide specimen without reasonable excuse an offence.

Stepniewski v Commissioner of Police of the Metropolis [1985] Crim LR 675; [1985] RTR 330 HC QBD Was failure to give (breath) specimen where gave one specimen and then refused to provide second.

Walker v Lovell [1975] 3 All ER 107; (1978) 67 Cr App R 249; [1975] Crim LR 720; (1975) 139 JP 708; (1975) 125 NLJ 820t; [1975] RTR 377; (1975) 119 SJ 544; [1975] 1 WLR 1141 HL Where enough breath to provide specimen arrest for failure to provide specimen unlawful and this being so there can be no conviction for drunk driving.

Webb v Director of Public Prosecutions [1992] RTR 299 HC QBD Blood specimen validly requested where police officer decided that woman having difficulty in providing specimen was shocked and medically incapable of providing same.

Williams v Osborne (1975) 61 Cr App R 1; [1975] Crim LR 166; [1975] RTR 181 HC QBD Failure to provide specimen must spring from offender's personal situation (mental and physical condition relevant).

Wilson v Cummings [1983] RTR 347 HC QBD Constable not required to inspect breathalyser crystals where despite offender's placing device to lips evidence showed was no air in bag.

Young (Paula Anne) v Director of Public Prosecutions [1992] Crim LR 893; (1993) 157 JP 606; [1992] RTR 328; [1992] TLR 154 HC QBD Drunkenness could be medical reason for refusing breath specimen/police officer's seeking alternative specimen.

SPECIMEN (general)

Anderton v Lythgoe [1985] Crim LR 158; [1985] RTR 395; (1984) 128 SJ 856; [1984] TLR 596; [1985] 1 WLR 222 HC QBD Police could not rely on breath specimen as definitive evidence of accused's guilt where failed to inform accused when arrested him of his right to give blood/urine sample.

Baker v Foulkes [1975] 3 All ER 651; (1976) 62 Cr App R 64; [1976] Crim LR 67t; (1976) 140 JP 24; [1975] RTR 509; [1975] 1 WLR 1551 HL Police officer need not tell medical doctor of intention to caution of effects of failing to provide specimen.

Beck v Sager [1979] Crim LR 257; [1979] RTR 475 HC QBD Was not offence of failing to provide person where (foreigner) did not understand what were told by police officer.

Bourlet v Porter [1973] 2 All ER 800; (1974) 58 Cr App R 1; [1974] Crim LR 53; (1973) 137 JP 649; [1973] RTR 293; (1973) 117 SJ 489; [1973] 1 WLR 866 HL Need not be given opportunity of second test at police station; specimen requested from hospital patient need not be provided while patient.

Braddock v Whittaker [1970] Crim LR 112; [1970] RTR 288; (1969) 113 SJ 942 HC QBD Adequate evidence of non-contamination of sample/that sample analysed was that provided by defendant.

Brooks v Ellis [1972] 2 All ER 1204; [1972] Crim LR 439; (1972) 136 JP 627; [1972] RTR 361; (1972) 116 SJ 509 HC QBD Request for specimen must be within immediate time of driving.

Buller v Easton (1969) 119 NLJ 996t HC QBD Giving of second breath specimen and then of blood specimen could legitimately be done at two different stations.

Butler v Easton [1970] Crim LR 45t; [1970] RTR 109; (1969) 113 SJ 906 HC QBD Breath test/specimen request/specimen provision to occur at same police station: otherwise laboratory certificate inadmissible.

Campbell v Tormey [1969] 1 All ER 961; (1969) 53 Cr App R 99; [1969] Crim LR 150t; (1969) 133 JP 267; (1968) 118 NLJ 1196t; (1968) 112 SJ 1023; [1969] 1 WLR 189 HC QBD Cannot ask person who has fully completed drive to take breath test; must arrest person before can request breath test/laboratory specimen.

Clark v Stenlake [1972] RTR 276 HC QBD On determining suitability of container in which part of specimen supplied to defendant.

Cronkshaw v Rydeheard [1969] Crim LR 492t; (1969) 113 SJ 673 HC QBD Successful appeal against drunk driving conviction where specimen handed to appellant capable of analysis only by way of gas chromotography.

Davies v Director of Public Prosecutions [1989] RTR 391 HC QBD Constable told by driver that was taking drug that would affect blood-alcohol levels could legitimately ask for specimens other than breath.

Dickson v Atkins [1972] Crim LR 185; [1972] RTR 209 HC QBD Certificate of analysis admissible despite spelling errors where independently established that specimen analysed that of defendant.

Director of Public Prosecutions v Byrne (1991) 155 JP 601; [1991] RTR 119; [1990] TLR 649 HC QBD On giving option to detainee as to which type of specimen would prefer to provide.

Director of Public Prosecutions v Carey [1969] 3 All ER 1662; [1970] AC 1072; (1970) 54 Cr App R 119; [1970] Crim LR 107t; (1970) 134 JP 59; [1970] RTR 14; (1969) 113 SJ 962; [1969] 3 WLR 1169 HL On breathalyser and specimen testing.

Director of Public Prosecutions v Dear (1988) 87 Cr App R 181 HC QBD Medical procedure for extracting analysis specimen must be rigorously observed.

Director of Public Prosecutions v Poole (1992) 156 JP 571; [1992] RTR 177; [1991] TLR 163 HC QBD Person must have option of giving alternative specimens presented to them but if impede officer from relaying option are not later entitled to acquittal on basis that option not given.

Director of Public Prosecutions v Warren (Frank John) [1992] Crim LR 200; (1992) 156 JP 753; [1992] RTR 129 HC QBD Failed prosecution for driving with excess alcohol as accused not given adequate opportunity to comment on what type of specimen would prefer to give (where breathalyser not functioning properly).

Doyle v Leroux [1981] Crim LR 631; [1981] RTR 438 HC QBD On when abuse of process merits halting of prosecution; person's destroying own specimen after informed that would not be prosecuted (and then was) did not have special reason meriting non-disqualification.

Earl v Roy [1969] 2 All ER 684; [1969] Crim LR 151; (1969) 133 JP 427; (1968) 112 SJ 1023; [1969] 1 WLR 1050 HC QBD Specimen to be proper must be capable of examination.

Gumbley v Cunningham [1989] 1 All ER 5; [1989] AC 281; (1989) 88 Cr App R 273; [1989] Crim LR 297; (1990) 154 JP 686; (1988) 138 NLJ 356; [1989] RTR 49; (1989) 133 SJ 84; [1989] 2 WLR 1 HL Back-calculation of alcohol levels permissible if understandable/shows excess alcohol.

Gumbley v Cunningham; Gould v Castle [1987] 3 All ER 733; (1988) 86 Cr App R 282; [1987] Crim LR 776; (1987) 137 NLJ 788; [1988] QB 170; [1988] RTR 57; (1987) 131 SJ 1064; [1987] 3 WLR 1072 HC QBD Back-calculation of alcohol levels admissible if understandable and indicates excess alcohol.

Hawkins v Ebbutt [1975] Crim LR 465; [1975] RTR 363 HC QBD Prosecution blood analysis inadmissible where portion supplied to defendant had been in inadequate container that resulted in blood being incapable of analysis.

Hirst v Wilson [1969] 3 All ER 1566; [1970] Crim LR 106; [1970] RTR 67; (1969) 113 SJ 906; [1970] 1 WLR 47 HC QBD Arrest for failing to provide proper specimen — though medical reason — lawful: drunk driving conviction sustained.

Hobbs v Clark [1988] RTR 36 HC QBD On need to specifically inform defendant that can substitute blood/urine specimen for that of breath specimen which proves to have less than 50mg of alcohol to 100ml of breath.

Holling v Director of Public Prosecutions (1991) 155 JP 609; [1992] RTR 192 HC QBD On giving option to detainee as to which type of specimen would prefer to provide.

Hudson v Hornsby [1972] Crim LR 505; [1973] RTR 4 HC QBD Where defence pleading without notice to prosecution that part specimen supplied could not be analysed prosecution ought to have been allowed call rebutting evidence.

Jones v Brazil [1970] RTR 449; [1971] Crim LR 47 HC QBD On who may qualify as medical practititoner to be notified when wish to obtain specimen from hospital patient: here casualty officer sufficed.

Jones v Roberts [1973] Crim LR 123; [1973] RTR 26 HC QBD Person refusing specimen need not be given an hour within which to change mind.

Kelly v Dolbey [1984] RTR 67 HC QBD Arrest of driver after failed to inflate bag but crystals afforded negative reading was valid (though crystals later found at station to give positive reading): appellant properly convicted inter alia of failure to give specimen.

Kidd v Kidd; Ley v Donegani [1968] 3 All ER 226; (1968) 52 Cr App R 659; [1968] Crim LR 509t; (1968) 132 JP 536; (1968) 118 NLJ 958t; [1969] 1 QB 320; (1968) 112 SJ 602; [1968] 3 WLR 734 HC QBD Contents of syringe may be divided into more than two parts yet be valid specimens.

Kierman v Willcock [1972] Crim LR 248; [1972] RTR 270 HC QBD Specimen supplied to accused must be capable of analysis at time handed over; that random part specimen selected by prosecution was capable of analysis meant could infer same of defendant's specimen.

Millard v Director of Public Prosecutions (1990) 91 Cr App R 108; [1990] Crim LR 600; (1990) 154 JP 627; [1990] RTR 201; [1990] TLR 119 HC QBD Is irrebuttable presumption that proportion of alcohol in breath/blood/urine at time of offence not lower than that in specimen.

Nugent v Hobday [1972] Crim LR 569; [1973] RTR 41 HC QBD Part specimen supplied must be capable of analysis using ordinary equipment/skill.

Oxford v Lowton [1978] Crim LR 295; [1978] RTR 237 HC QBD Not necessary to be outside patient's presence/hearing when informing doctor of intention to require specimen of patient.

Pascoe v Nicholson [1981] 2 All ER 769; [1981] Crim LR 839; (1981) 145 JP 386; (1981) 131 NLJ 732; [1981] RTR 421; (1981) 125 SJ 499; [1981] 1 WLR 1061 HL Blood/urine specimen need not be provided in police station at which requested.

Piggott v Sims [1973] RTR 15 HC QBD Analyst's certificate admissible on behalf of prosecution after prosecution closed case.

Pine v Collacott (1985) 149 JP 8; [1985] RTR 282; [1984] TLR 496 HC QBD Need not show that constable (at station where no breathalyser) specifically considered alternative specimen could seek and plumped for one type of specimen rather than another.

R v Bolton Justices, ex parte Scally; R v Bolton Justices, ex parte Greenfield; R v Eccles Justices, ex parte Meredith; R v Trafford Justices, ex parte Durran-Jorda [1991] Crim LR 550; (1991) 155 JP 501; [1991] 1 QB 537; [1991] RTR 84; (1990) 134 SJ 1403; [1990] TLR 639; [1991] 2 WLR 239 HC QBD Excessive blood-alcohol concentration convictions quashed where very likely that blood samples taken had been contaminated.

R v Bowell [1974] Crim LR 369t; (1974) 124 NLJ 271t; [1974] RTR 273; (1974) 118 SJ 367 CA On blood-alcohol concentration prosecution had to prove under Road Traffic Act 1961, s 6(2), to secure drunk driving conviction.

R v Carpenter (1979) 129 (2) NLJ 810; [1980] RTR 65 CA Evidence of specimen analysis admissible though defendant had sought to challenge breath tests.

R v Clwd Justices, ex parte Charles (1990) 154 JP 486 HC QBD Drunk driving conviction quashed where accused had not been given option of providing blood/urine specimens.

R v Coomaraswamy (Subramaniam) (1976) 62 Cr App R 80; [1976] Crim LR 260t; (1975) 125 NLJ 1167t; [1976] RTR 21 CA Prosecution must show blood-alcohol concentration exceeded prescribed limit not that figure arrived at precisely correct.

R v Dolan [1969] 3 All ER 683; (1969) 53 Cr App R 556; [1970] Crim LR 43; (1969) 133 JP 696; (1969) 119 NLJ 996t; [1970] RTR 43; (1969) 113 SJ 818; [1969] 1 WLR 1479 CA Issue for jury if person alleges was not warned what failure to comply with specimen request entailed.

R v Durrant [1969] 3 All ER 1357; (1970) 54 Cr App R 24; [1970] Crim LR 39t; (1970) 134 JP 57; [1970] RTR 420; (1969) 113 SJ 905; [1970] 1 WLR 29 CA Invalid specimen test if accused drank after ceased driving.

R v Epping Justices, ex parte Quy [1993] Crim LR 970 HC QBD Fear of needles could be medical reason for refusing blood sample.

R v Hamilton [1970] 3 All ER 284; [1970] Crim LR 651t; (1970) 120 NLJ 968; [1970] RTR 417 CA Analyst's blood certificate when alcohol obtained after driving is not certificate of blood whilst driving.

R v Haslam [1972] RTR 297 CA Person charged with failure to provide breath/laboratory specimen could be acquitted of former charge but convicted of latter charge.

R v Hegarty [1977] RTR 337 CA Drunk driving conviction quashed where prosecution failed to explain discrepancy between doctor's clinical examination of accused and later specimen results.

R v Hyams [1972] 3 All ER 651; (1973) 57 Cr App R 103; [1972] Crim LR 645; (1972) 136 JP 842; [1973] RTR 68; (1972) 116 SJ 886; [1973] 1 WLR 13 CA Cannot request specimen after two specimens given.

R v Jones (Colin) [1974] RTR 117 CA Part specimen placed with defendant's personal items had been 'supplied' to defendant.

R v Jones (EJM) [1970] 1 All ER 209; (1970) 54 Cr App R 148; (1970) 134 JP 215; (1969) 119 NLJ 1141t; [1970] RTR 56; [1970] 1 WLR 211 CA Matter for jury whether was driving; specimen request may be made immediately after driving.

R v Lennard [1973] 2 All ER 831; (1973) 57 Cr App R 542; [1973] Crim LR 312t; (1973) 137 JP 585; (1973) 123 NLJ 298t; [1973] RTR 252; (1973) 117 SJ 284; [1973] 1 WLR 483 CA Mental/physical inability or threat to health only valid grounds for refusing specimen.

R v Lewis [1965] Crim LR 50t; (1965) 49 Cr App R 26; (1964) 108 SJ 863 CCA Doctor, unlike police constable, not required to offer defendant a portion of the urine specimen which the doctor requests from the defendant.

121

R v Marr [1977] RTR 168 CA Person validly convicted of driving with excess alcohol despite massive discrepancy between prosecution/defence analyses of specimen provided.

R v Moore (Richard) [1975] Crim LR 722; [1975] RTR 285 CA Onus on defendant to establish defence that arrest invalid; jury could reasonably decide that was adequte connection between specimen analysed and driving behaviour three hours prior to giving specimen.

R v Nicholls [1972] 2 All ER 186; (1972) 56 Cr App R 382; [1972] Crim LR 380; (1972) 136 JP 481; (1972) 122 NLJ 196t; [1972] RTR 308; (1972) 116 SJ 298; [1972] 1 WLR 502 CA Requirement to provide specimen not heard/understood although constable believed was understood was valid requirement of specimen.

R v Nixon [1969] 2 All ER 688; (1969) 53 Cr App R 432; [1969] Crim LR 378; (1969) 133 JP 520; (1969) 119 NLJ 578; (1969) 113 SJ 388; [1969] 1 WLR 1055 CA Prosecution must show sample was adequate and capable of examination.

R v Price [1963] 3 All ER 938; (1964) 48 Cr App R 65; [1964] Crim LR 60; (1964) 128 JP 92; (1964) 114 LJ 41; [1964] 2 QB 76; (1963) 107 SJ 933; [1963] 3 WLR 1027 CCA Must offer to supply part of specimen to suspect when request it/very shortly before.

R v Pursehouse [1970] 3 All ER 218; (1970) 54 Cr App R 478; [1970] Crim LR 651; (1970) 134 JP 682; (1970) 120 NLJ 920; [1970] RTR 494 CA Police need not specify statutory requirements after two refusals to give any sample.

R v Richardson (John) [1975] 1 All ER 905; (1975) 60 Cr App R 136; [1975] Crim LR 163t; (1975) 139 JP 362; (1975) 125 NLJ 17t; [1975] RTR 173; (1975) 119 SJ 152; [1975] 1 WLR 321 CA Crown must show accused drove/attempted to drive prior to specimen being requested to sustain conviction for refusing latter.

R v Roberts (Gwylim Ian) (1964) 48 Cr App R 300; [1964] Crim LR 531g CCA On police officer's duty to offer portion of suspect's specimen to suspect.

R v Rutter [1977] RTR 105 CA On proving that sample analysed not that of defendant; no need for laboratory analysis to be physically conducted by authorised analyst/s.

R v Sharp [1968] 3 All ER 182; (1968) 52 Cr App R 607; [1968] Crim LR 452; (1968) 132 JP 491; (1968) 118 NLJ 734; [1968] 2 QB 564; (1968) 112 SJ 602; [1968] 3 WLR 333 CA Element of specimen to be provided to accused may be provided within reasonable time of leaving police station.

R v Shaw [1974] RTR 458 CA Absent evidence as to police/laboratory error jury could infer that specimen before court was that of accused.

R v Somers [1963] 3 All ER 808; (1964) 48 Cr App R 11; [1963] Crim LR 845; (1964) 128 JP 20; (1963) 113 LJ 821; (1963) 107 SJ 813; [1963] 1 WLR 1306 CCA Doctor's evidence based not on own tests but expert findings is admissible.

R v Wright (John) (1975) 60 Cr App R 114; [1975] RTR 193 CA Portion of specimen supplied to accused to be capable of analysis at that moment and for reasonable time thereafter.

Regan v Director of Public Prosecutions [1989] Crim LR 832; [1990] RTR 102 HC QBD Person properly informed of right to give either blood or urine where this choice was apparent from information given by police officer.

Roney v Matthews (1975) 61 Cr App R 195; [1975] Crim LR 394; [1975] RTR 273; (1975) 119 SJ 613 HC QBD That second specimen provided one minute over maximum period prescribed by statute did not render it inadmissible as specimen.

Rowlands v Hamilton [1971] 1 All ER 1089; (1971) 55 Cr App R 347; [1971] Crim LR 366t; (1971) 135 JP 241; [1971] RTR 153; (1971) 115 SJ 268; [1971] 1 WLR 647 HL Specimen test inadmissible if drank after stopped driving nor evidence to show would have been over limit anyway.

Rush v Director of Public Prosecutions [1994] RTR 268; [1994] TLR 182 HC QBD Acquittal necessary where seemed was procedural error in seeking specimens — in course of unrecorded talk accused was allegedly dissuaded by police from offering alternative specimen.

Sasson v Taverner [1970] 1 All ER 215; (1970) 134 JP 244; (1970) 120 NLJ 12t; [1970] RTR 63; (1970) 114 SJ 75; [1970] 1 WLR 338 HC QBD Specimen request off-road following pursuit after request on road a valid request.

Scott v Baker [1968] 2 All ER 993; (1968) 52 Cr App R 566; [1968] Crim LR 393t; (1968) 132 JP 422; (1968) 118 NLJ 493t; [1969] 1 QB 659; (1968) 112 SJ 425; [1968] 3 WLR 796 HC QBD Specimen not valid unless relevant procedural requirements of Road Safety Act 1967 complied with.

Sharp v Spencer [1987] Crim LR 420 HC QBD Bona fide comment by police officer which led road traffic offender not to give blood specimen in lieu of breath specimens did not invalidate process.

Smith (Dennis Edward) v Director of Public Prosecutions (1990) 154 JP 205; (1989) 133 SJ 389 HC QBD Prosecution could rely on breath specimen given by accused where latter waived option had to give blood specimen (though had subsequently changed mind and given blood specimen).

Spicer v Holt [1976] 3 All ER 71; [1977] AC 987; (1976) 63 Cr App R 270; [1977] Crim LR 364; (1976) 140 JP 545; (1976) 126 NLJ 937t; [1976] RTR 389; (1976) 120 SJ 572; [1976] 3 WLR 398 HL Unlawfully arrested person cannot be compelled to give specimen.

Spicer v Holt [1976] Crim LR 139; (1976) 126 NLJ 44t; [1976] RTR 1 HC QBD Analysis of specimen given by unlawfully arrested person was not admissible in evidence.

Thomas v Henderson [1983] RTR 293 HC QBD Wrong for prosecution/defence to accept defence scientist's written statement without having him called to testify/to challenge evidence of analyst by way of inference without calling analyst or relevant witnesses.

Ward v Keene [1970] RTR 177 HC QBD On what constitutes an adequate part of a specimen to give to a defendant.

Williams v Mohamed [1976] Crim LR 577; [1977] RTR 12 HC QBD Analysis evidence admissible where possible failure to offer defendant part of specimen did not result in unfairness.

Young (Paula Anne) v Director of Public Prosecutions [1992] Crim LR 893; (1993) 157 JP 606; [1992] RTR 328; [1992] TLR 154 HC QBD Drunkenness could be medical reason for refusing breath specimen/police officer's seeking alternative specimen.

SPECIMEN (hospital)

Bosley v Long [1970] 3 All ER 286; [1970] Crim LR 591; (1970) 134 JP 652; (1970) 120 NLJ 968; [1970] RTR 432; (1970) 114 SJ 571; [1970] 1 WLR 1410 HC QBD Specimen reuested at hospital to be provided at hospital; medic may object anytime after notified of proposal to take specimen.

Burke v Jobson [1972] Crim LR 187 HC QBD Police officer in evidence-in-chief must give full account of notification to doctor that intended to seek specimen from patient in doctor's care.

Burn v Kernohan [1973] Crim LR 122; [1973] RTR 82 HC QBD Police officer's evidence as to what told medic and what medic said to constable was admissible.

Edwards v Davies [1982] RTR 279 HC QBD Requirement of specimen valid where properly requested same of hospital patient though received it after patient discharged self.

Foulkes v Baker [1975] RTR 50; (1975) 119 SJ 777 HC QBD Prosecution must prove medical practitioner in charge of patient informed of intention to request latter for specimen.

Goodley v Kelly [1973] RTR 125; [1973] Crim LR 125 HC QBD Hospital procedure followed irregular course but contained essential steps so was valid.

Griffiths v Willett [1979] Crim LR 320; [1979] RTR 195 HC QBD Conviction quashed where evidence did not establish that constable had reasonable suspicion of hospital patient having alcohol in body so justifying requirement of specimen.

R v Burton-upon-Trent Justices, ex parte Woolley [1996] Crim LR 340; (1995) 159 JP 165; [1995] RTR 139; [1994] TLR 589 HC QBD On procedure as regards taking specimen from person in hospital.

R v Crowley (1977) 64 Cr App R 225; [1977] Crim LR 426; [1977] RTR 153 CA Cannot request specimen at roadside and obtain same without further ado when person removed to hospital; on requirement that medical officer in charge of suspect (normally duty casualty officer) be notified that specimen sought.

SPECIMEN (portion)

R v Green [1970] 1 All ER 408; [1970] Crim LR 289; (1970) 134 JP 208; [1970] RTR 193 Assizes On requiring blood/urine specimens of person in hospital.

R v Knightley [1971] 2 All ER 1041; (1971) 55 Cr App R 390; [1971] Crim LR 426t; (1971) 121 NLJ 409t; [1971] RTR 409; (1971) 115 SJ 448; [1971] 1 WLR 1073 CA Must tell medic of intention to give statutory warning when notifying of intent to seek specimen.

R v Porter (Edward Charles Thomas) (1973) 57 Cr App R 290; [1973] Crim LR 124t; (1972) 122 NLJ 1134t; (1973) 123 NLJ 612t; [1973] RTR 116; (1973) 117 SJ 36 CA On when 'hospital procedure'/'police station' procedure appropriate for taking of blood specimen.

Ratledge v Oliver [1974] Crim LR 432; [1974] RTR 394 HC QBD Police officer could at same time notify doctor of intent to administer breath test/seek blood or urine sample of hospital patient.

SPECIMEN (portion)

R v Mitten [1965] 2 All ER 59; (1965) 49 Cr App R 216; [1965] Crim LR 313t; [1966] 1 QB 10; (1965) 109 SJ 554; [1965] 3 WLR 268 CCA Non-compliance with law in requesting specimen need not mean evidence inadmissible; police to offer to supply part of specimen when requesting it/shortly before.

R v Rothery (Henry Michael) (1976) 63 Cr App R 231; [1976] Crim LR 691t; (1976) 126 NLJ 790t; [1976] RTR 550 CA If steal portion of own specimen retained by police are guilty of theft but not failure to provide specimen.

SPECIMEN (urine)

Dhillon (Surinder Singh) v Director of Public Prosecutions (1993) 157 JP 420 HC QBD Person not properly given option to provide blood/urine specimen where police interpreted vague answer as refusal of same.

Director of Public Prosecutions v Berry [1995] TLR 571 HC QBD Breath test evidence available where could have given blood/urine sample instead but were incapable of understanding this in light of alcohol had taken.

Director of Public Prosecutions v D; Director of Public Prosecutions v Rouse (1992) 94 Cr App R 185; [1991] Crim LR 911; (1991) 141 NLJ 1004t; [1992] RTR 246; [1991] TLR 338 HC QBD Conversation between police officer and arrestee about blood/urine specimen option and request for breath specimens not an interview.

Director of Public Prosecutions v Dixon [1993] RTR 22 HC QBD Matter for subjective adjudication whether police officer has reasonable basis for belief that breathalyser malfunctioning (and so to require blood (as here) or urine specimen).

Director of Public Prosecutions v Garrett (1995) 159 JP 561; [1995] RTR 302; [1995] TLR 55 HC QBD On procedure as regards taking of specimens: that was procedural error as regards taking of blood specimen does not mean urine specimen taken in accordance with procedure is inadmissible.

Director of Public Prosecutions v Hill-Brookes [1996] RTR 279 HC QBD Must tell person consequences of providing/not providing optional blood/urine specimen so that may properly exercise option.

Director of Public Prosecutions v Ward [1997] TLR 155 HC QBD No right to legal advice before decide whether to substitute blood/urine specimen for breath specimen.

Director of Public Prosecutions v Warren [1992] 4 All ER 865; [1993] AC 319; (1993) 96 Cr App R 312; (1993) 157 JP 297; (1992) 142 NLJ 1684; [1993] RTR 58; (1992) 136 SJ LB 316; [1992] TLR 520; [1992] 3 WLR 884 HL Procedure when obtaining blood/urine sample of accused.

Gabrielson v Richards [1975] Crim LR 722; [1976] RTR 223 HC QBD Third urine specimen ought not to have been relied upon by prosecution where second had been given within hour of request.

Hayes v Director of Public Prosecutions [1993] Crim LR 966; [1994] RTR 163 HC QBD Police officer not required to caution suspect of possible consequences of not providing specimen where latter not being required to give specimen.

Hope v Director of Public Prosecutions [1992] RTR 305 HC QBD Person who provided breath specimens as required and who agreed to provide blood could then decide not to provide blood.

Nugent v Ridley [1987] Crim LR 640; [1987] RTR 412 HC QBD Third specimen could be sent for laboratory examination (so long as procedures in acquiring sample complied with).

Over v Musker (1984) 148 JP 759; [1985] RTR 84; [1984] TLR 120 HC QBD One minute delay between giving two urine specimens did not mean specimens were not two separate specimens.

Poole v Lockwood [1980] Crim LR 730; (1980) 130 NLJ 909; [1981] RTR 285 HC QBD Blood specimen inadmissible where second urine specimen provided over an hour after requested discarded by police officer genuinely believing specimen inadequate.

Prosser v Dickeson [1982] RTR 96 HC QBD Apparent provision of two urine specimens inside two minutes was actualy provision of one specimen.

R v Beckett [1976] Crim LR 140 CrCt Urinating into toilet and not container was provision of specimen of urine.

R v Dawson [1976] RTR 533 CA Jury reasonably found that correct specimen correctly analysed/ properly convicted despite disparity in readings.

R v Forbes [1971] Crim LR 174; [1970] RTR 491 CA Valid request for blood specimen (though failed to mention urine option).

R v Green [1970] 1 All ER 408; [1970] Crim LR 289; (1970) 134 JP 208; [1970] RTR 193 Assizes On requiring blood/urine specimens of person in hospital.

R v Harling [1970] 3 All ER 902; (1971) 55 Cr App R 8; [1970] Crim LR 536; (1971) 135 JP 29; [1970] RTR 441 CA Reasonable excuse to give blood does not mean have reasonable excuse to give urine.

R v Kershberg [1976] RTR 526 CA Analyst signing certificate must control but need not implement every stage of analysis; conviction quashed in light of lingering doubts about validity of analysis set against defendant's early release from custody.

R v Lewis [1965] Crim LR 50t; (1965) 49 Cr App R 26; (1964) 108 SJ 863 CCA Doctor, unlike police constable, not required to offer defendant a portion of the urine specimen which the doctor requests from the defendant.

R v Moore [1970] Crim LR 650; [1970] RTR 486 CA Constable to have reasonable suspicion regarding alcohol levels before requests breath specimen/may request blood or urine (neither preferable) for laboratory test; on appropriate warning to be given as to consequences of refusing blood and urine specimens.

R v Moore (John) [1978] RTR 384 CA Analysis of third urine specimen admissible in evidence.

R v Orrell [1972] Crim LR 313; [1972] RTR 14 CA On proving that bottle of urine placed in evidence was specimen obtained from defendant.

R v Paduch (Jan) (1973) 57 Cr App R 676; [1973] Crim LR 533; [1973] RTR 493 CA Police could change request and ask for urine rather than blood specimen from driver.

R v Radcliffe [1977] RTR 99 CA Urine specimen obtained in compliance with statutory require- ments admissible.

R v Welsby [1972] Crim LR 512; [1972] RTR 301 CA First urine specimen provided can be disposed of.

Ratledge v Oliver [1974] Crim LR 432; [1974] RTR 394 HC QBD Police officer could at same time notify doctor of intent to administer breath test/seek blood or urine sample of hospital patient.

Renshaw v Director of Public Prosecutions [1992] RTR 186 HC QBD On procedure as regards informing person of choice as to provision of sample (whether blood or urine).

Rowland v Thorpe [1970] 3 All ER 195; [1972] Crim LR 655; (1970) 134 JP 655; (1970) 120 NLJ 662t; [1970] RTR 406; (1970) 114 SJ 707 HC QBD Only severe mental condition/physical injuries justify refusing laboratory specimen; reasc;*ahle failure to give urine not per se reasonable excuse to refuse blood — must be excuse for latter.

Standen v Robertson [1975] Crim LR 395; [1975] RTR 329 HC QBD Drunk driving prosecution could rely on second urine specimen provided over an hour after requested.

Turner v Director of Public Prosecutions [1978] Crim LR 754; [1996] RTR 274 HC QBD Must tell person what results refusal to provide blood/urine specimen entails.

Wareing (Alan) v Director of Public Prosecutions (1990) 154 JP 443 [1990] TLR 41 HC QBD Dismissal of drunk driving charge on basis that defendant not given option by police of providing urine specimen was not dismissal on basis of technicality.

SPEEDING (general)

Aitken v Yarwood [1964] Crim LR 602; (1964) 128 JP 470; [1965] 1 QB 327; (1964) 108 SJ 381; [1964] 3 WLR 64 HC QBD Vehicle being driven at high speed to convey policeman to court to give evidence was being used for 'police purposes' (Road Traffic Act 1960, s 25).

Baker v Sweet [1966] Crim LR 51g HC QBD Burden f proving that 50 mph (Temporary Speed Limit) (England and Wales) Order 1964 inapplicable to instant case rested on defence.

Barna v Hudes Merchandising Corporation and another [1962] Crim LR 321 CA Surpassing speed limit does not per se mean are negligent/in breach of statutory duty.

Beresford v Justices of St Albans (1905-06) XXII TLR 1 HC KBD Valid speeding conviction where evidence as to identity was rebuttable but convicted person had not sought to rebut it.

Blenkin v Bell [1952] 1 All ER 1258; (1952) 116 JP 317; (1952) 102 LJ 276; [1952] 2 QB 620; (1952) 96 SJ 345; [1952] 1 TLR 1216; (1952) WN (I) 247 HC QBD Special speed limits do not apply to unloaded passenger/goods vehicle.

Boyd-Gibbins v Skinner [1951] 1 All ER 1049; (1951) 115 JP 360; [1951] 2 KB 379; [1951] 1 TLR 1159; (1951) WN (I) 267 HC KBD Speed limit signs evidence that area deemed a built up area under Road Traffic Act 1934, s 1(1)(b).

Bracegirdle v Oxley; Bracegirdle v Cobley [1947] 1 All ER 126; [1947] KB 349; [1947] 116 LJR 815; (1947) 176 LTR 187; (1947) 91 SJ 27; (1947) LXIII TLR 98; (1947) WN (I) 54 HC KBD Speed under speed limit can be excessive.

Briere v Hailstone [1969] Crim LR 36; (1968) 112 SJ 767 HC QBD Road lamps that were 1½ feet over 200 yards apart (prescribed distance) were nonetheless valid indicators that were driving in area of restricted speed.

Brighty v Pearson [1938] 4 All ER 127; (1939-40) XXXI Cox CC 177; (1938) 102 JP 522; (1938) 82 SJ 910 HC KBD Evidence of two police officers on speed at different times not enough to secure speeding conviction.

Bryson v Rogers [1956] 2 All ER 826; [1956] Crim LR 570t; (1956) 120 JP 454; (1956) 106 LJ 507; [1956] 2 QB 404; (1956) 100 SJ 569; [1956] 3 WLR 495 HC QBD Dual-purpose vehicle covered by special speed limits whether or not conveying goods.

Bursey v Barron [1972] Crim LR 486; [1971] RTR 273; (1971) 115 SJ 469 HC QBD Use of non-reflective material on reflective speed signs did not render signs invalid so could be charged for driving in excess of indicated speeds.

Burton v Gilbert (1983) 147 JP 441; [1984] RTR 162; [1983] TLR 319 HC QBD Evidence obtained from properly functioning radar device was corroborative of evidence of policeman that accused had been speeding.

Cooper v Hawkins [1904] 2 KB 164; (1902-03) 47 SJ 691; (1903-04) 52 WR 233 HC KBD Local authority restrictions on locomotive speed under Locomotives Act 1865 could not apply to Crown locomotive driven by Crown servant on Crown business.

Crossland v Director of Public Prosecutions [1988] 3 All ER 712; [1988] Crim LR 756; (1989) 153 JP 63; [1988] RTR 417 HC QBD Police officer's expert opinion together with facts on which opinion based sufficient to ground conviction for excessive speed.

Darby v Director of Public Prosecutions (1995) 159 JP 533; [1995] RTR 294; [1994] TLR 555 HC QBD 'GR Speedman' speedometer evidence admissible without certificate: trained person accustomed to using same could testify to its proper working order.

Director of Public Prosecutions v Holtham [1990] Crim LR 600; (1990) 154 JP 647; [1991] RTR 5; [1990] TLR 132 HC QBD On what constitutes a 'motor tractor'.

Du Cros v Lambourne (1907-09) XXI Cox CC 311; (1906) 70 JP 525; [1907] 1 KB 40; (1906) 41 LJ 701; [1907] 76 LJKB 50; (1906-07) XCV LTR 782; (1906-07) XXIII TLR 3 HC KBD Owner convicted as principal for excessively fast driving by driver over/with whom had control/ whom was in car.

Durnell v Scott [1939] 1 All ER 183; (1939) 87 LJ 86; (1939) 83 SJ 196 HC KBD Guilty of speeding offence where public if present would be endangered.

Elwes v Hopkins (1907-09) XXI Cox CC 133; (1906) 70 JP 262; [1906] 2 KB 1; [1906] 75 LJKB 450; (1906) XCIV LTR 547 HC KBD Evidence of customary traffic on highway admissible in prosecution for excessive speed 'in all the circumstances' (Motor Car Act 1903, s 1).

Ex parte JEB Stone (1909) 73 JP 444; (1908-09) XXV TLR 787 HC KBD Driving car at twenty-three miles per hour through village justified speeding conviction under Motor Car Act 1903, s 1.

Hood v Lewis [1976] Crim LR 74; [1976] RTR 99 HC QBD Not defence to speeding charge that did not see speed signs as should have appreciated from distance between street-lights that had entered 30mph area.

Jessop v Clarke (1908) 72 JP 358; (1908-09) XCIX LTR 28; (1907-08) XXIV TLR 672 HC KBD Person stopped and told officer thought he was speeding and would prosecute if another officer agreed was warning of intended prosecution.

Kent v Stamps [1982] RTR 273 HC QBD Justices could rely on own local knowledge when deciding accuracy of speedometer evidence.

Lyon v Oxford [1983] RTR 257 HC QBD Conviction quashed as speeding legislation vaguely worded so defendant given benefit of higher of two speeding limits (in excess of which had not driven).

Maddox v Storer [1962] 1 All ER 831; [1962] Crim LR 328; (1962) 126 JP 263; (1962) 112 LJ 306; [1963] 1 QB 451; (1962) 106 SJ 372 HC QBD 'Adapted' does not mean 'altered so as to be apt' but 'fit and apt for'.

Manning v Hammond [1951] 2 All ER 815; (1951) 115 JP 600; (1951) 101 LJ 611; (1951) WN (I) 525 HC KBD Goods vehicles to which thirty mile maximum speed applied.

Melhuish v Morris [1938] 4 All ER 98; (1938) 86 LJ 311; (1938) 82 SJ 854 HC KBD Evidence of two police officers based on one speedometer not enough to secure speeding conviction.

Musgrove v Kennison (1901-07) XX Cox CC 874; (1905) 69 JP 341; (1905) XCII LTR 865; (1904-05) 49 SJ 567; (1904-05) XXI TLR 600 HC KBD Royal Park notice pertaining to maximum driving speeds therein but not laid before Parliament was valid.

Nicholas v Penny [1950] 2 All ER 89; (1950) 114 JP 335; [1950] 2 KB 467; (1950) 100 LJ 275; (1950) 94 SJ 437; [1950] 66 (1) TLR 1122; (1950) WN (I) 302 HC KBD Can convict for excessive speed on evidence of police officer whose evidence is grounded on speedometer reading.

Parkes v Cole (1922) 86 JP 122; (1922) 127 LTR 152; (1922) WN (I) 123 HC KBD Speeding conviction quashed where had been no warning of intended prosecution as required under Motor Car Act 1903.

Plancq v Marks (1907-09) XXI Cox CC 157; (1906) 70 JP 216; (1906) XCIV LTR 577; (1905-06) 50 SJ 377; (1905-06) XXII TLR 432 HC KBD Could be convicted of excessive speed on evidence of constable using stop watch.

Police v Dormer [1955] Crim LR 252 Magistrates Electrician carrying testing tools and equipment in his van was not carrying 'goods'.

Popperwell v Cockerton [1968] 1 All ER 1038; [1968] Crim LR 336; (1968) 132 JP 231; (1968) 112 SJ 175; [1968] 1 WLR 438 HC QBD On excessive speed of dual-purpose vehicle.

Pritchard v Dyke (1934-39) XXX Cox CC 1; (1933) 97 JP 179; (1933) 149 LTR 493; (1932-33) XLIX TLR 473 HC KBD No evidence on speed necessary to sustain speeding conviction where weight of vehicle over twelve tons.

R v Dailloz (Max) (1908) 1 Cr App R 258 CCA Can be found guilty of driving furiously where were driving too fast shortly prior to/after accident.

R v Marsham, ex parte Chamberlain (1907-09) XXI Cox CC 510; (1907) 71 JP 445; [1907] 2 KB 638; (1907) 42 LJ 429; [1907] 76 LJKB 1036; (1907-08) XCVII LTR 396; (1906-07) 51 SJ 592; (1906-07) XXIII TLR 629; (1907) WN (I) 163 HC KBD Need not endorse licence upon first conviction where is for speeding through Royal Park.

R v Petersfield Justices, ex parte Levy [1981] RTR 204 HC QBD Was arguable point that Road Traffic Regulation Act 1967, s 78A not an offence-creating provision.

R v Plowden, ex parte Braithwaite (1911-13) XXII Cox CC 114; (1909) 73 JP 266; [1909] 2 KB 269; [1909] 78 LJKB 733; (1909) 100 LTR 856; (1908-09) XXV TLR 430; (1909) WN (I) 87 HC KBD Royal Park speeding regulations valid.

Roberts v Director of Public Prosecutions [1994] Crim LR 926; [1994] RTR 31; [1993] TLR 307 HC QBD Could not take judicial notice of approval by Secretary of State of 'Kustom Falcon' radar gun just because police officer had used same on many occasions.

Russell v Beesley [1937] 1 All ER 527; (1937) 81 SJ 99; (1936-37) LIII TLR 298 HC KBD No rule that evidence of police officer on speed of accused's car by reference to own speedometer requires corroboration.

Saunders v Johns [1965] Crim LR 49t HC QBD Was no case to answer where did not establish that person guilty of speeding offence was in fact the defendant.

Sellwood v Butt [1962] Crim LR 841; (1962) 106 SJ 835 HC QBD Car possessing device for measuring engine speed but not speedometer did not comply with requirement that every car have a speed measuring device.

Smith v Boon (1899-1901) XIX Cox CC 698; (1901) LXXXIV LTR 593; (1900-01) 45 SJ 485; (1900-01) XVII TLR 472; (1900-01) 49 WR 480 HC KBD Excessive speed determined by reference to traffic on road not immediately around motor vehicle concerned.

Spittle v Kent County Constabulary [1985] Crim LR 744; [1986] RTR 142 HC QBD Properly convicted of speeding though some of lamps on road (distance between which enabled person to gauge speed limit) were not lit.

Staunton v Coates (1924) 88 JP 193; (1924) 59 LJ 681; [1925] 94 LJCL 95; (1925) 132 LTR 199; (1924-25) 69 SJ 126; (1924-25) XLI TLR 33 HC KBD Driver of heavy motor-car to be prosecuted for speeding need not be given warning of intention to prosecute (Locomotives on Highways Act 1896, s 6, as amended).

Swain v Gillett [1974] Crim LR 433; [1974] RTR 446 HC QBD Police speedometer reading could corroborate evidence of officer as to speed of car without need to prove accuracy of speedometer.

Tribe v Jones [1961] Crim LR 835t; (1961) 105 SJ 931 HC QBD Driving in excess of permitted speed limit did not per se mean were driving dangerously.

Welton v Taneborne (1907-09) XXI Cox CC 702; (1908) 72 JP 419; (1908-09) XCIX LTR 668; (1907-08) XXIV TLR 873 HC KBD Could not be convicted for excessive speed where speed taken into account in dangerous driving conviction arising from same facts.

Woolley v Moore [1952] 2 All ER 797; (1952) 116 JP 601; (1952) 102 LJ 639 [1953] 1 QB 43; (1952) 96 SJ 749; [1952] 2 TLR 673; (1952) WN (I) 480 HC QBD Unloaded goods vehicle not subject to special conditions of carrier's licence.

STAGE CARRIAGE (general)

Chapman v Kirke [1948] 2 All ER 556; (1948) 112 JP 399; [1948] 2 KB 450; [1949] 118 LJR 255; (1948) LXIV TLR 519; (1948) WN (I) 357 HC KBD Negligent injury by tram opens driver to conviction under Stage Carriages Act 1832, s 48.

Covington v Wright [1963] 2 All ER 212; (1963) 113 LJ 366; (1963) 107 SJ 477; [1963] 2 WLR 1232 HC QBD Dishonesty not necessary for fare avoidance charge; full fare is that less any amount paid.

Middlemas v McAleer [1979] RTR 345 HC QBD Unlicensed use of public service vehicle (coach) as contract carriage.

Osborne v Richards [1932] All ER 833; (1931-34) XXIX Cox CC 524; (1932) 96 JP 377; [1933] 1 KB 283; (1933) 102 LJCL 44; (1932) 147 LTR 419; (1931-32) XLVIII TLR 622; (1932) WN (I) 189 HC KBD 'Contract carriages' offence contrary to Road Traffic Act 1930, s 72(10).

Phesse v Fisher (1915) 79 JP 174; [1915] 1 KB 572; [1915] 84 LJKB 277; (1914) WN (I) 438 HC KBD Overcrowding of tramcar: what was 'inside' and what was 'outside' for purpose of London County Council (Tramways and Improvements) Act 1913, s 27.

R v Farnborough Urban District Council, ex parte Aldershot District Traction Co (1919) WN (I) 271 HC KBD Improper exercise by local authority of its discretion whether or not to renew omnibus licence.

R v Traffic Commissioners for the Metropolitan Traffic Area and another Ex parte Licensed Taxi Drivers' Association Ltd [1984] RTR 197 HC QBD Application for road service licence authorising stage carriage service properly made.

Railway Executive v Henson (1949) 113 JP 333; (1949) WN (I) 242 HC KBD Company operating omnibus service at Railway Executive's behest did not require road service licence.

Victoria Motors (Scarborough), Ltd and another v Wurzal [1951] 1 All ER 1016; (1951) 115 JP 333; [1951] 2 KB 520; (1951) 101 LJ 247; (1951) 95 SJ 382; [1951] 1 TLR 837; (1951) WN (I) 233 HC KBD Special occasion must be special in itself, not just so to parties being carried.

Westminster Coaching Services, Ltd v Piddlesden; Hackney Wick Stadium, Ltd v Same [1933] All ER 379; (1931-34) XXIX Cox CC 660; (1933) 97 JP 185; (1933) 149 LTR 449; (1932-33) XLIX TLR 475 HC KBD Unlawful use of vehicle as stage carriage, not contract carriage.

White v Cubitt (1931-34) XXIX Cox CC 80; (1930) 94 JP 60; [1930] 1 KB 443; (1930) 69 LJ 10; (1930) 99 LJCL 129; (1930) 142 LTR 427; (1929) 73 SJ 863; (1929-30) XLVI TLR 99; (1929) WN (I) 266 HC KBD Was plying for hire though did not advertise or seek business, merely stood around.

TAXI (advertisement)

Atkins v Green [1970] Crim LR 653; [1970] RTR 332 HC QBD Failed prosecution for advertising cabs without making clear were unlicensed (advertisement contained term 'non-hackney').

Kingcabs (Southall) Ltd v Gordon [1970] RTR 115 HC QBD Successful prosecution for advertising 'cabs' without making it clear that vehicles concerned were unlicensed cabs.

TAXI (general)

Abrahams v Bartlett [1970] RTR 276 HC QBD Could not on same facts be guilty both of not stopping meter after fare finished and of starting meter before new fare began (here found guilty of former).

Adamson v Waveney District Council [1997] 2 All ER 898; [1997] TLR 94 HC QBD On procedure as regards taking into account spent convictions of applicant for hackney carriage licence.

Adur District Council v Fry [1997] RTR 257 HC QBD Licensed private hire firm operating in one district who received call from another district to carry out private hire journey in that district not guilty of 'operation' of private hire business in district where unlicensed.

Alker v Woodward [1962] Crim LR 313t HC QBD On what it means for unlicensed hackney cab to be 'plying for hire'.

Armitage v Walton [1976] Crim LR 70t; (1975) 125 NLJ 1022t; [1976] RTR 160 HC QBD Failed prosecution for unlawful advertisement of mini-cab service as advertisement made clear that vehicles hired not licensed cabs.

Attorney-General v Sharp (1930) 70 LJ 62; (1928-29) XLV TLR 628 HC ChD Picking-up passengers who had bought tickets in garage outside area where bus company allowed ply for hire was plying for hire in that area by bus company.

Balsdon v Heard [1965] Crim LR 232g HC QBD Failed prosecution for unlicensed plying for hire (noting to set car at issue apart from an ordinary private car).

Bassam v Green [1981] Crim LR 626; [1981] RTR 362 HC QBD Hackney carriage licensed and used as such could not be divested of that attribute; was offence for hackney carriage driver to seek/take booking fee where ride arranged via radio service.

Benjamin v Cooper [1951] 2 All ER 907; (1951) 115 JP 632; [1951] 2 TLR 906; (1951) WN (I) 553 HC KBD Plying for hire in forecourt of tube station not offence as is private place.

Benson v Boyce [1997] RTR 226 HC QBD Prosecution do not have to prove (unlicensed) private hire of vehicle at particular time in issue in order to secure conviction for same (that vehicle used (unlicensed) as private hire vehicle the defining feature of offence).

Blackpool and Fleetwood Tramroad Co v Bailey [1920] 89 LJCL 784 HC KBD Valid conviction for unlicensed use of tramcar as hackney carriage.

Bowers v Gloucester Corporation [1963] 1 All ER 437; [1963] Crim LR 276; (1963) 127 JP 214; (1963) 113 LJ 268; [1963] 1 QB 881; [1963] 2 WLR 386 HC QBD Can be convicted of any two offences for hackney carriage licence to be revoked.

Braintree District Council v Howard [1993] RTR 193 HC QBD Was offence for private vehicle hire operator licensed in one area to take telephone bookings at number in area where unlicensed and to drive in area where unlicensed.

Breame v Anderson and another [1971] RTR 31; (1971) 115 SJ 36 CA Private hire vehicle did not by advertising telephone number to call for hire indicate that was for immediate hire.

Britain v ABC Cabs (Camberley) Ltd [1981] RTR 395 HC QBD Not offence where hackney carriage owner in one licensing area collected passenger from (without parking or plying for hire in) other licensing area.

Bygraves v Dicker [1923] All ER 473; [1923] 2 KB 585; [1923] 92 LJCL 1021; (1923) 129 LTR 688; (1923) WN (I) 151 HC KBD Registered proprietor of hackney carriage may be sued for negligence of driver.

Cinnamond and others v British Airports Authority [1980] RTR 220; (1980) 124 SJ 221; [1980] 1 WLR 582 CA Ban on certain car-hire drivers (who persistently hassled arriving passengers) entering Heathrow Airport save as bona fide travellers was valid.

Cinnamond and others v British Airports Authority [1979] RTR 331 HC QBD Ban on certain car-hire drivers (who persistently hassled arriving passengers) entering Heathrow Airport save as bona fide travellers was valid.

Cogley v Sherwood; Car Hire Group (Skyport) Ltd v Sherwood; Howe v Kavanaugh; Howe v Kavanaugh; Car Hire Group (Skyport) Ltd v Kavanaugh [1959] 2 All ER 313; [1959] Crim LR 464t; (1959) 123 JP 377; (1959) 109 LJ 330; [1959] 2 QB 311; (1959) 103 SJ 433 HC QBD Hackney carriage not plying for hire unless shown for hire.

Cook v Southend Borough Council (1990) 154 JP 145; [1990] 2 QB 1; [1990] 2 WLR 61 CA Council could appeal as aggrieved person against justices' order to pay damages to/costs of hackney driver whose licence it had withdrawn.

Cook v Southend Borough Council (1987) 151 JP 641 HC QBD Local authority could be aggrieved person and could therefore appeal against decision of justices to allow appeal against authority's revocation of hackney carriage licence.

Crack v Holt (1926-30) XXVIII Cox CC 319; (1927) 91 JP 36; (1927) 136 LTR 511; (1926-27) XLIII TLR 231 HC KBD Licence required when plying for hire wherever did so.

Crawley Borough Council v Crabb [1996] RTR 201 HC QBD Intervening period of disqualification between period licenced to drive and period applied for licence not counted when calculating duration held licence upon private hire licence application.

Crawley Borough Council v Ovenden (1992) 156 JP 877; [1992] RTR 60 HC QBD Long-term hire exemption from private vehicle hire licensing requirements inapplicable to long-time contract which was immediately terminable and of no fixed duration.

D'Arcy v Evesham Motors Ltd, Burgin and Heward [1971] RTR 35 Magistrates Sign on private hire vehicle contravened legislative requirement that such signs not indicate vehicle could presently be hired.

Darlington Borough Council v Wakefield (Paul) (1989) 153 JP 481 HC QBD Appeal to magistrates from refusal by local council of hackney carriage licence ought to have been by way of full rehearing.

Director of Public Prosecutions v Computer Cab Co Ltd and others [1996] RTR 130 HC QBD Persons prohibited from plying for hire in central London did not do so where accepted passengers within that area who had arranged all details of prospective trip with radio booking service.

Dittah v Birmingham City Council [1993] Crim LR 610; [1993] RTR 356 HC QBD Was offence for private vehicle hire operators licensed in one area to use vehicles/drivers licensed in different area.

Docherty v South Tyneside Borough [1982] TLR 358 CA Council decision to increase number of Hackney carriage licences was valid.

Dunning v Maher (1912) 76 JP 255 HC KBD Liverpool Corporation bye-law requiring lighting of taxi-meter was valid.

Eccles v Kirke [1949] 1 All ER 428; (1949) 113 JP 175; (1949) LXV TLR 133 HC KBD Non-compliance with requirement of sign showing taxi and how many persons carries.

Eldridge v British Airports Authority [1970] 2 All ER 92; [1970] Crim LR 284; (1970) 134 JP 414; [1970] 2 QB 387; [1970] RTR 270; (1970) 114 SJ 247; [1970] 2 WLR 968 HC QBD Taxi can 'stand' in such a way that is not obliged to take passengers.

Ely v Godfrey (1921-25) XXVII Cox CC 191; (1922) 86 JP 82; (1922) 126 LTR 664 HC KBD Conviction for taking greater fare than permitted quashed as relevant by-law deemed inapplicable.

Ghafoor and others v Wakefield Metropolitan District Council [1990] RTR 389; [1990] TLR 597 HC QBD Failed appeal against refusal of hackney carriage licence: council not required to issue licences beyond number of carriages required in its area of authority.

Gilbert v McKay [1946] 1 All ER 458; (1946) 110 JP 186; (1946) 174 LTR 196; (1946) 90 SJ 201; (1945-46) LXII TLR 226 HC KBD Though no contract there was an unlawful plying for hire.

Goodman v Serle [1947] 2 All ER 318; (1947) 111 JP 492; [1947] KB 808; (1947) 97 LJ 389; [1948] 117 LJR 381; (1947) 177 LTR 521; (1947) 91 SJ 518; (1947) LXIII TLR 395; (1947) WN (I) 225 HC KBD Successful appeal against conviction for agreeing on bargained fare within area of City of London where stanadardised fares applicable.

Green v Turkington; Green v Cater; Craig v Cater [1975] Crim LR 242; [1975] RTR 322; (1975) 119 SJ 356 HC QBD Private hire car with sticker-signs on it reading 'Speedicars Ltd' did suggest car was used for carrying of passengers for hire/reward contrary to London Cab (No 2) Order 1973.

Gross v O'Toole [1983] RTR 376 HC QBD Heathrow Airport byelaw prohibiting unpermitted offers of services includes only offers relating to carrying on of trade/business.

Hawkins v Edwards (1899-1901) XIX Cox CC 692; [1901] 70 LJK/QB 597; (1901) LXXXIV LTR 532; (1900-01) 45 SJ 447; (1901) WN (I) 88; (1900-01) 49 WR 487 HC KBD Any carriage occasionally used to stand/ply for hire is a hackney carriage under Town Police Clauses Act 1847, s 38.

House and others v Reynolds [1977] 1 All ER 689; [1977] Crim LR 42; (1977) 141 JP 202; (1976) 126 NLJ 1219t; [1977] RTR 135; (1976) 120 SJ 820; [1977] 1 WLR 88 HC QBD Booking fee 'more than the fee' allowed under byelaws.

Hulin v Cook and another (1977) 127 NLJ 791t; [1977] RTR 345; (1977) 121 SJ 493 HC QBD Person plying for hire on railway premises required local authority licence and nor special permission to do so on rail premises.

Hunt v Morgan [1948] 2 All ER 1065; (1949) 113 JP 67; [1949] 1 KB 233; (1948) 98 LJ 687; (1949) LXV TLR 15; (1948) WN (I) 506 HC KBD Taxi must be at rank/in special stationary position before obliged to accept fare.

Jones v Short (1900) 35 LJ 99; [1900] 69 LJCL 473; (1899-1900) 44 SJ 211; (1899-1900) 48 WR 251 HC QBD Unlicensed hackney driver plying for hire at stand provided and owned by rail company not guilty of unlicensed plying for hire on street.

Kaye v Hougham [1964] Crim LR 544g; (1964) 108 SJ 358 HC QBD Was not 'alighting' from taxi where driver/passenger left it to get change from nearby restaurant so driver could be prosecuted for parking violation.

Kemp v Lubbock (1919) 83 JP 270; [1920] 1 KB 253; [1920] 89 LJCL 239; (1920) 122 LTR 220; (1919) WN (I) 269 HC KBD Infant in arms need not be paid for as extra person under London Hackney Carriage Act 1853.

Khan v Evans [1985] RTR 33 HC QBD On what constitutes 'plying for hire': successful prosecution for unauthorised plying for hire on railway property.

Leeds City Council v Azam and another [1989] RTR 66 HC QBD Defendant must prove on balance of probabilities that falls within Local Government (Miscellaneous Provisions) Act 1976, s 75(1).

Levinson v Powell [1967] 3 All ER 796; [1967] Crim LR 542t; (1968) 132 JP 10; (1967) 117 NLJ 784t; (1967) 111 SJ 871; [1967] 1 WLR 1472 HC QBD Taxi not a goods vehicle: waiting restrictions apply; taxi driving to place requested to observe law.

London County Council v Fairbank (1911) WN (I) 96 HC KBD Are only guilty of unlicensed keeping of carriage if possess and use it (mere possession not enough).

Nottingham City Council v Woodings [1994] RTR 72; [1993] TLR 115 HC QBD Driver of obvious minicab not plying for hire when parked and got out of same to enter toilet nor when was coming back but was plying for hire when sat into car and told enquirers was free for hire.

Pettigrew v Barry [1984] TLR 447 HC QBD Unlicensed private hire driver ought to have been found to be plying for hire where waited alongside hackney carriage stand.

Pitts v Lewis (1989) 153 JP 220; [1989] RTR 71 HC QBD On defence under Local Government (Miscellaneous Provisions) Act 1976, s 75, that vehicle used under contract of hire for not less than seven days where are charged operating unlicensed private hire vehicle.

R v Assistant Commissioner of Police of the Metropolis, ex parte Howell [1985] RTR 181 HC QBD Asisstant police commissioner entitled to refuse hackney carriage licence to driver on basis of medical evidence provided inter alia by medic of driver's choice.

R v Assistant Commissioner of Police of the Metropolis, ex parte Howell [1986] RTR 52 CA Refusal to renew hackney carriage licence quashed where applicant not afforded opportunity to plead case.

R v Blackpool Borough Council, ex parte Red Cab Taxis Limited and others (1994) 158 JP 1069; [1994] RTR 402; [1994] TLR 270 HC QBD Insertion by council of new term in private hire licences improper as had been done without hearing interested parties.

R v Bradford Corporation, ex parte Minister of Transport (1926) 135 LTR 227 HC KBD Mandamus rule absolute requiring Bradford Corporation to grant certain hackney carriage licenses authorised by Minister for Transport under Roads Act 1920, s 14(3).

R v Brighton Corporation, ex parte Thomas Tilling, Limited (1916) 80 JP 219 HC KBD Mandamus ordered/refused in respect of properly/improperly refused omnibus/char-a-banc licence.

R v British Airports Authority, ex parte Wheatley [1983] RTR 147; [1982] TLR 278 HC QBD Licensed taxi driver picking up passenger not affected by Gatwick Airport byelaw requiring permission of airport authority before could offer services.

R v British Airports Authority, ex parte Wheatley [1983] RTR 466; [1983] TLR 320 CA Licensed taxi driver picking up passenger was covered by Gatwick Airport byelaw requiring permission of airport authority before could offer services.

R v Commissioner of Police of the Metropolis, ex parte Humphreys; R v Commissioner of Police of the Metropolis, ex parte Randall (1911) 75 JP 486; (1910-11) 55 SJ 726 HC KBD Metropolitan Police Commissioner cannot refuse hackney-cab licences out of hand on basis of precept that are to be unavailable to those who enjoy cab on hire-purchase terms.

R v Great Yarmouth Borough Council, ex parte Sawyer [1989] RTR 297 CA Failed action against decision of council de-regulating local taxi provision.

R v Manchester City Justices, ex parte McHugh; Manchester City Council (Intervening); R v Manchester City Council, ex parte Reid [1989] RTR 285 HC QBD Council could require new hackney carriage licence holders to make provision for wheelchair passengers whilst not imposing same requirement on old licence holders.

R v Prestwich Corporation, ex parte Gandz and others (1945) 109 JP 157 HC KBD Non-interference by HC in local authority's good faith application of hackney carriage licensing system.

R v Reading Borough Council, ex parte Egan; R v Reading Borough Council, ex parte Sullman [1990] RTR 399 HC QBD Council required to grant licences where not satisified was no significant demand for taxis unmet.

R v Solomons (1911-13) XXII Cox CC 178; (1909) 2 Cr App R 288; (1909) 73 JP 467; [1909] 2 KB 980; (1910) 79 LJKB 8; (1909-10) 101 LTR 496; (1908-09) XXV TLR 747 CCA Taxicab driver of cab owned by company received fares for/on behalf of company for purposes of Larceny Act 1901.

R v Weymouth Corporation, ex parte Teletax (Weymouth), Ltd [1947] 1 All ER 779; (1947) 111 JP 303; [1947] KB 583; (1947) 97 LJ 221; [1947] 116 LJR 899; (1947) 177 LTR 210; (1947) WN (I) 134 HC KBD Hackney carriage licence attaches to vehicle not owner.

Robertson v Bannister [1973] Crim LR 46; [1973] RTR 109 HC QBD Onus of proof on taxi-driver offering services at Heathrow to prove authorised to do so by British Airports Authority.

Rodgers v Taylor [1987] RTR 86 HC QBD Exception to restricted waiting provisions applied to hackney carriages awaiting hire not to taxis left parked and unattended on street.

Rose v Welbeck Motors, Ltd and another [1962] 2 All ER 801; [1962] Crim LR 566t; (1962) 126 JP 413; (1962) 112 LJ 570; (1962) 106 SJ 470 HC QBD That colourfully advertised mini-cab hanging around bus stop created prima facie case that was plying for hire.

Rosenbloom v McDonnell [1957] Crim LR 809t HC QBD Absolute discharge for cab-driver guilty of unlawfully demanding more than the proper fare.

Royal Borough of Windsor and Maidenhead v Khan (Mahoob) (1994) 158 JP 500; [1994] RTR 87; [1993] TLR 255 HC QBD Not guilty of unlicensed operation of hire vehicle in area as was licensed in and worked from another area to which person hiring had telephoned to make arrangements.

Shepherd v Hack (1917) 81 JP 210; [1917] 86 LJCL 1480; (1917-18) 117 LTR 154 HC KBD Affirmation of conviction for refusal to convey passenger.

Solihull Metropolitan Borough Council v Cars (Silverline) (1989) 153 JP 209; [1989] RTR 142 HC QBD Court may vary condition under which private hire car disc/licence to be displayed but cannot obviate condition.

St Albans District Council v Taylor [1991] Crim LR 852; (1992) 156 JP 120; [1991] RTR 400 HC QBD That allowed unlicensed person to carry out complementary trips in unlicensed vehicle when short-staffed did not mean operator of private vehicle hire business had not contravened licensing legislation.

Stevenage Borough Council v Younas [1990] RTR 405 HC QBD Council could refuse licences despite evidence that at one particular spot near particular nightclub was significant unmet demand for taxis.

Vant v Cripps [1964] Crim LR 594 HC QBD Failed appeal against conviction for unlawful plying for hire.

Willingale v Norris (1907-09) XXI Cox CC 737; (1908) 72 JP 495; [1909] 1 KB 57; [1909] 78 LJKB 69; (1908-09) XCIX LTR 830 HC KBD Breach of regulations under London Hackney Carriage Act 1850, s 4 punishable in way prescribed in London Hackney Carriage Act 1853, s 19.

Wurzal v Addison [1965] 1 All ER 20; [1965] Crim LR 116g; (1965) 129 JP 86; (1965) 115 LJ 42; [1965] 2 QB 131; (1964) 108 SJ 1046; [1965] 2 WLR 131 HC QBD Method of payment immaterial: if payment is hire for reward; 'adapted' in Road Traffic Act refers to capability.

Yakhya v Tee [1984] RTR 122; [1983] TLR 286 HC QBD Mere display of telephone number on car-roof did not contravene requirement that non-taxi not have sign on roof indicating is taxi.

Yates v Gates [1970] 1 All ER 754; [1970] Crim LR 233t; (1970) 134 JP 274; (1970) 120 NLJ 105t; [1970] 2 QB 27; [1970] RTR 135; (1970) 114 SJ 93; [1970] 2 WLR 593 HC QBD Driver cannot carry other person absent express consent of existing passenger; person driving taxi must have taxi licence.

Yorkshire (Woollen District) Electric Tramways Limited v Ellis (1901-07) XX Cox CC 795; (1904-05) 53 WR 303 HC KBD Light railway carriage not an omnibus nor a hackney carriage.

Young and Allen v Solihull Metropolitan Borough Council (1989) 153 JP 321 HC QBD On what consttitues 'hackney carriage'/'street' for purposes of Town Police Clauses Act 1847.

Young and another v Scampion [1989] RTR 95 HC QBD On ingredients of offence of hackney carriage licensed in one area undertaking to stand/ply for hire/drive in another area in which unlicensed.

TRACTOR (general)

Dennis v Leonard (1926-30) XXVIII Cox CC 621; (1929) 141 LTR 94 HC KBD Petrol-driven tractor was motor car for purposes of Motor Cars (Use and Construction) Order 1904.

Director of Public Prosecutions v Holtham [1990] Crim LR 600; (1990) 154 JP 647; [1991] RTR 5; [1990] TLR 132 HC QBD On what constitutes a 'motor tractor'.

Keyse v Sainsbury and another [1971] Crim LR 291; [1971] RTR 218 HC QBD Failed prosecution for improper construction and use of tractor where parts complained of were held on but not part of tractor.

Millard v Turvey [1968] 2 All ER 7; [1968] Crim LR 277; (1968) 132 JP 286; (1968) 118 NLJ 397; [1968] 2 QB 390; (1968) 112 SJ 235; [1968] 2 WLR 1192 HC QBD Chassis not a 'motor tractor'.

William Gwennap (Agricultural), Ltd and another v Amphlett [1957] 2 All ER 605; (1957) 121 JP 487; (1957) 107 LJ 507; (1957) 101 SJ 592 HC QBD Motor tractor towing combine harvester two vehicles.

TRAFFIC EXAMINER (general)

Swan v The Vehicle Inspectorate (1997) 161 JP 293; [1997] RTR 187; (1996) 140 SJ LB 263 HC QBD Vehicle Inspectorate traffic examiner who is not authorised to prosecute is not a prosecutor within meaning of the Road Traffic Offenders Act 1988, s 6(1).

TRAFFIC REGULATION (general)

Brownsea Haven Properties, Ltd v Poole Corporation (1957) 121 JP 571; (1957) 107 LJ 554 HC ChD Improper to use powers under Town Police Clauses Act 1847 to test prospective one-way street system under Road Traffic Act 1930: order doing so was void.

Chorlton v Liggett (1910) 74 JP 458 HC KBD Conviction for not driving horse and cart as close to kerb as possible quashed.

Corfe Transport Ltd v Gwynedd County Council [1984] RTR 79; [1983] TLR 126 HC QBD Construction of Road Traffic Regulation Act 1967, ss 1(5) and 84B(1).

Etherington v Carter [1937] 2 All ER 528 HC KBD Order under Town Police Clauses Act 1847, s 21, governing summer trading was valid.

Gouldie v Pringle [1981] RTR 525 HC QBD Justices' finding that person did not breach prohibition on right-hand turns when did U-turn unassailable on facts.

Hoffman v Thomas [1974] 2 All ER 233; [1974] Crim LR 122; (1974) 138 JP 414; (1974) 124 NLJ 36t; [1974] RTR 182; (1974) 118 SJ 186; [1974] 1 WLR 374 HC QBD Police may regulate traffic to protect life and property not for traffic census purposes.

Kentesber v Waumsley [1980] Crim LR 383; [1980] RTR 462 HC QBD Was offence not to stop for traffic warden regulating traffic but not offence to then continue without being waved on by warden.

Pontin v Price (1934-39) XXX Cox CC 44; (1933) 97 JP 315; (1934) 150 LTR 177 HC KBD Conviction for failing to keep to line of traffic into which directed by police officer acting in course of duty.

Post Office and Mr A Harris v London Borough of Richmond; Post Office v London Borough of Richmond [1994] Crim LR 940; (1994) 158 JP 919; [1995] RTR 28; [1994] TLR 276 HC QBD Breach of permit condition (permit allowing certain lorries travel in certain areas at restricted hours) did not give rise to criminal liability.

R v Warwickshire County Council, ex parte Boyden; Boyden v Warwickshire County Council (1992) 156 JP 1 HC QBD Failed attempt to have road traffic regulation orders found invalid.

Wright v Howard [1972] Crim LR 710t; (1972) 122 NLJ 610t; [1973] RTR 12 HC QBD On meaning of phrase 'right-hand turn into' in City of Oxford traffic order.

TRAFFIC WARDEN (general)

R v Saunders [1978] Crim LR 98 CrCt Person properly convicted for non-compliance with signal of traffic warden (a non-endorseable offence).

UNAUTHORISED TAKING OF VEHICLE (attempt)

R v Cook [1964] Crim LR 56t; (1964) 114 LJ 59 CCA Judge to decide whether facts can constitute attempt, jury to decide whether (as here) facts did constitute attempt.

R v Marchant (Stephen); R v McCallister (Stephen) (1985) 80 Cr App R 361 CA Was attempt to take conveyance without authority even though did not use conveyance.

UNAUTHORISED TAKING OF VEHICLE (general)

A (an infant) v Bundy (1961) 125 JP 89; (1961) 105 SJ 40 HC QBD That were passenger in car knowing it to be uninsured/have been taken without owner's consent inadequate per se to establish guilt of driving while uninsured/taking and driving away car.

Chief Constable of Avon and Somerset Constabulary v Jest [1986] RTR 372 HC QBD Defendant's thumb print on internal mirror did not prove had taken car without consent for own (uninsured) use.

D (an infant) v Parsons [1960] 2 All ER 493; [1960] Crim LR 711; (1960) 124 JP 375; (1960) 110 LJ 493; (1960) 104 SJ 605; [1960] 1 WLR 797 HC QBD Not guilty of taking and driving away where although involved in driving away is no proof were involved in taking.

Director of Public Prosecutions v Spriggs [1993] Crim LR 622; (1993) 157 JP 1143; [1994] RTR 1; [1993] TLR 36 HC QBD One person's taking without consent an abandoned car which another had previously taken without consent was taking car without consent.

McKnight v Davies [1974] Crim LR 62; (1973) 123 NLJ 1066t; [1974] RTR 4; (1973) 117 SJ 940 HC QBD Lorry driver who after work took employer's lorry to public house, then home, returning it next morning guilty of taking vehicle without lawful authority.

Mowe v Perraton [1952] 1 All ER 423; (1951-52) 35 Cr App R 194; (1952) 116 JP 139; (1952) 96 SJ 182; (1952) WN (I) 96 HC QBD Not taking and driving away if vehicle at outset in one's lawful possession.

Neal v Gribble and others (1979) 68 Cr App R 9; [1978] Crim LR 500; [1978] RTR 409 HC QBD Horse/bridled horse not conveyance/not adapted for carriage of person within meaning of Theft Act 1968, s 12.

R v Ambler [1979] RTR 217 CA Owner's post facto statement that would (had he been asked) have loaned vehicle to person pleading guilty to taking vehicle without authority cannot be relied upon as defence.

R v Bogacki and others [1973] 2 All ER 864; (1973) 57 Cr App R 593; [1973] Crim LR 385t; (1973) 137 JP 676; [1973] QB 832; [1973] RTR 384; (1973) 117 SJ 355; [1973] 2 WLR 937 CA Must be movement of vehicle after possession assumed for there to be taking without authority.

R v Bow (Dennis Arthur) (1977) 64 Cr App R 54; [1977] Crim LR 176t; (1977) 127 NLJ 43t; [1977] RTR 6 CA Was taking conveyance without authority to freewheel another's vehicle downhill.

R v Diggin (Richard Joseph) (1981) 72 Cr App R 204; [1980] Crim LR 656; [1981] RTR 83 CA Defendant ought to have known from start of journey that was being conveyed in vehicle taken without authority to be guilty of same.

R v Gannon (Kevin) (1988) 87 Cr App R 254; [1988] RTR 49 CA Drunkenness precluded defence on basis of belief had lawful authority for unauthorised taking of conveyance.

R v Hogdon [1962] Crim LR 563t CCA Conviction for taking/driving away motor car quashed where did not prove absence of consent to same.

R v McGill [1970] Crim LR 290 CA Was taking and driving without owner's consent to borrow vehicle for one purpose, not return it and then use for different purpose.

R v Pearce [1961] Crim LR 122 CCA On whether taking and driving away motor vehicle without owner's consent a continuing offence.

R v Peart [1970] 2 All ER 823; (1970) 54 Cr App R 374; [1970] Crim LR 479; (1970) 134 JP 547; (1970) 120 NLJ 481t; [1970] 2 QB 673; [1970] RTR 376; (1970) 114 SJ 418; [1970] 3 WLR 63 CA Consent to use motor vehicle exists even if obtained by false pretence about proposed trip.

R v Phipps (Owen Roger Charles); R v McGill (John Peter) (1970) 54 Cr App R 300; [1970] RTR 209 CA Is taking and driving away for person who has borrowed car for certain reason to use it afterwards in way to which owner would not consent — passenger consenting to situation guilty of similar offence.

R v Richardson [1958] Crim LR 480 CCA Irrelevant to guilt of persons who knowingly drove vehicle without consent of owner whether they took the car themselves or from some other.

R v Roberts (1964) 48 Cr App R 296; [1964] Crim LR 472; (1964) 128 JP 395; (1964) 114 LJ 489; [1965] 1 QB 85; (1964) 108 SJ 383; [1964] 3 WLR 180 CCA Not taking and driving away to release handbrake, set lorry going, then jump free and allow freewheel downhill.

R v Stokes [1982] Crim LR 695; [1983] RTR 59 CA Person to be conveyed within vehicle before can be convicted of taking same without authority.

R v Tolhurst; R v Woodhead [1962] Crim LR 489; (1962) 112 LJ 138; (1962) 106 SJ 16 CCA On who constitutes owner of car in context of taking and driving away car.

R v Wibberley [1965] 3 All ER 718; (1966) 50 Cr App R 51; [1966] Crim LR 53; (1966) 130 JP 58; (1965-66) 116 NLJ 189; [1966] 2 QB 214; (1965) 109 SJ 877; [1966] 2 WLR 1 CCA Was taking without owner's consent where took with consent but later used without consent.

R v Yates (Nicholas Peter) (1986) 82 Cr App R 232; [1986] RTR 68 CA Imposition of penalty points and disqualification for driving without owner's consent/insurance improper; disqualification periods to be concurrent.

Ross v Rivenall [1959] 2 All ER 376; [1959] Crim LR 589; (1959) 123 JP 352; (1959) 103 SJ 491 HC QBD On taking and driving away motor vehicle.

Shimmell v Fisher and others [1951] 2 All ER 672; (1951-52) 35 Cr App R 100; (1951) 115 JP 526; (1951) 101 LJ 483; (1951) 95 SJ 625; [1951] 2 TLR 753; (1951) WN (I) 484 HC KBD Need not drive car to commit offence of taking and driving away.

Tolley v Giddings and another [1964] 1 All ER 201; (1964) 48 Cr App R 105; [1964] Crim LR 231t; (1964) 128 JP 182; (1964) 114 LJ 174; [1964] 2 QB 354; (1964) 108 SJ 36; [1964] 2 WLR 471 HC QBD Was offence of taking and driving away that did not amount to theft of car (Road Traffic Act 1960, s 217).

Whittaker and another v Campbell [1983] 3 All ER 582; (1983) 77 Cr App R 267; [1983] Crim LR 812; [1984] QB 318; [1984] RTR 220; (1983) 127 SJ 714; [1983] TLR 398; [1983] 3 WLR 676 HC QBD That consent to use vehicle fraudulently obtained did not mean no consent thereto.

UNAUTHORISED TAKING OF VEHICLE (mens rea)

C (A) (a minor) v Hume [1979] Crim LR 328; [1979] RTR 424 HC QBD Ingredients of taking vehicle without authority (mens rea particularly important as accused had just reached age of doli capax).

R v MacPherson [1973] Crim LR 457; [1973] RTR 157 CA Prosecution need only prove that did not have authority to take conveyance: no need to prove specific intent.

UNDER-AGE DRIVING (general)

R v Saddleworth Justices, ex parte Staples [1968] 1 All ER 1189; (1968) 132 JP 275; (1968) 118 NLJ 373; (1968) 112 SJ 336; [1968] 1 WLR 556 HC QBD Prosecution could choose between under-age driving charge or driving while disqualified charge.

UNINSURED DRIVING (general)

Allen v John [1955] Crim LR 383 HC QBD Failed prosecution for uninsured driving.

Biddle v Johnston [1965] Crim LR 494; (1965) 109 SJ 395 HC QBD Certificate of insurance that was not a contract of no aid in deciding whether was valid insurance contract for purposes of the Road Traffic Act 1960.

Blows v Chapman [1947] 2 All ER 576 HC KBD Non-disqualification of labourer unknowingly driving uninsured vehicle over highway under employer's instructions.

Boldizsar v Knight [1980] Crim LR 653; [1981] RTR 136 HC QBD Person accepting lift in uninsured car which discovered was being driven without owner's consent guilty/not guilty of allowing self to be carried in same/using same without insurance.

Bryan v Forrow [1950] 1 All ER 294 HC KBD 'Paid driver' included person employed by another in turn reimbursed by owner: insured.

Carnill v Rowland [1953] 1 All ER 486; [1953] 1 WLR 380 HC QBD Not guilty of driving uninsured vehicle where insurance contract unclear but insurer states that views itself 'on risk'.

Chief Constable of Norfolk v Fisher (1992) 156 JP 93 HC QBD Permission that car could be used by insured driver only ought to have been communicated by owner to prosepective driver.

Corfield v Groves and another [1950] 1 All ER 488; (1950) 94 SJ 225 HC KBD Payment by Motor Insurers Bureau did not mean no loss sustained through inability to recover from defendant; car owner liable for damages owed by unisured driver whom had permitted to drive car.

Durrant v Maclaren [1956] Crim LR 632t HC QBD Failed prosecution for uninsured driving by person who obtained insurance by falsely completing policy application form.

Edwards v Griffiths [1953] 2 All ER 874 HC QBD '[D]isqualified' means court disqualification not mere refusal of licence.

Elliott v Grey [1960] Crim LR 63; (1960) 124 JP 58; (1959) 109 LJ 720; [1960] 1 QB 367; (1959) 103 SJ 921; [1959] 3 WLR 956 HC QBD Guilty of use of uninsured motor vehicle where car could be moved though not driven.

Evans v Walkden and another [1956] 1 WLR 1019 HC QBD Father (giving son a driving lesson) and (unlicensed) son guilty of uninsured driving.

Evens v Lewis [1964] Crim LR 472t HC QBD Publican not insured where hired car and completed insurance proposal form under pretence that was printer from garage that did not rent cars to publicans.

Goodbarne v Buck and another (1939-40) XXXI Cox CC 338; (1939) 103 JP 393 HC KBD Not offence to drive with voidable insurance (even if knew was voidable).

Goodbarne v Buck and another (1939-40) XXXI Cox CC 380; [1940] 109 LJCL 837; (1940) 162 LTR 259; (1939-40) LVI TLR 433 CA Not causing/permitting another to drive uninsured where provide money for voidable insurance policy.

John T Ellis, Ltd v Hinds [1947] 1 All ER 337; [1947] KB 475 HC KBD Motor vehicle deemed insured though driven by unlicensed driver.

Langman v Valentine and another [1952] 2 All ER 803 HC QBD Learner driver and instructors both drivers: learner driver unlicenced but covered by instructor's insurance policy as driver.

Leggate v Brown [1950] 2 All ER 564 HC KBD Insurance policy covering negligent (not unlawful) use valid.

Lloyd v Singleton [1953] 1 All ER 291; [1953] 1 QB 357; [1953] 2 WLR 278 HC QBD Employee could be properly charged with driving unlicenced vehicle owned by employer.

Lyons v May [1948] 2 All ER 1062 HC KBD Permitting vehicle to be used whilst uninsured.

Milstead v Sexton [1964] Crim LR 474g HC QBD Steering of uninsured car being towed from one place to another was causing unroadworthy/uninsured use of towed car.

Philcox v Carberry [1960] Crim LR 563 HC QBD Onus on person charged with uninsured driving to prove he had valid policy of insurance.

Police v Wright [1955] Crim LR 714 Magistrates Person riding motorised pedal cycle on which motor did not work acquitted of driving an uninsured vehicle.

Reay v Young and another [1949] 1 All ER 1102 HC KBD That openly committed offence in lonely area at slow speed for short distance justified non-disqualification for permitting uninsured driving.

Richards v Port of Manchester Insurance Company Limited and another (1935) 152 LTR 261 HC KBD Owner of car but not insurance company guilty of permitting use of car by uninsured person contrary to Road Traffic Act 1930, s 35(2).

Rogerson v Stephens [1950] 2 All ER 144 HC KBD No offence of uninsured driving of motor vehicle and trailer.

Starkey v Hall (1936) 80 SJ 347 HC KBD Were not complying with law where had valid insurance policy in respect of car but did not have insurance certificate.

Williamson v O'Keefe [1947] 1 All ER 307 HC KBD Any person driving uninsured car — whether or not are owner — commits offence.

UNLAWFUL USE OF VEHICLE (general)

Everall v Barnwell [1944] KB 333; (1944) WN (I) 64 HC KBD Motor fuel used for purposes specified in application for motor fuel coupons so no offence.

James and Son, Ltd v Smee; Green v Burnett and another [1954] 3 All ER 273; (1954) 104 LJ 729; [1955] 1 QB 78; [1954] 3 WLR 631 HC QBD Employer not responsible for causing vehicle to be used unlawfully where permitted (not committed) act.

UNLICENSED DRIVING (general)

Richardson v Baker [1976] Crim LR 76 HC QBD On when employer may be said to 'use' vehicle driven by employee.

Smith v Jenner [1968] Crim LR 99t; (1967) 117 NLJ 1296t; (1968) 112 SJ 52 HC QBD Successful appeal by driving instructor against conviction for aiding and abetting learner driver in unlicensed driving of car.

WANTON DRIVING (general)

R v Crowden (Harry) (1910-11) 6 Cr App R 190 CCA Meaning of 'wanton' in charge of wanton driving contrary to Offences Against the Person Act 1861, s 35.

ALPHABETICAL INDEX

Anderton v Waring [1986] RTR 74 HC QBD Was not reasonable excuse to non-provision of breath specimen charge that had tried one's best but machine did not record sample as being provided.

Anderton, McQuaid v [1980] 3 All ER 540; (1980) 144 JP 456; [1980] RTR 371; (1981) 125 SJ 101; [1981] 1 WLR 154 HC QBD Steering/braking a car being towed is 'driving'.

Anderton, O'Brien v [1979] RTR 388 HC QBD 'Italjet' a motor vehicle (had two wheels/22cc engine/seat/handle bars).

Anderton, Oldfield v [1986] Crim LR 189; (1986) 150 JP 40; [1986] RTR 314 HC QBD First breath sample could be admitted in evidence where failed to give second sample and could convict of drunk driving on basis of that sample.

Anderton, Regan v [1980] Crim LR 245; (1980) 144 JP 82; [1980] RTR 126 HC QBD Person could still be deemed to be driving car (for purposes of taking breath test) though car not actually moving.

Andrews v Director of Public Prosecutions [1937] 2 All ER 552; (1936-38) 26 Cr App R 34; (1937) 101 JP 386; (1937) 83 LJ 304; [1937] 106 LJCL 370; (1937) 81 SJ 497; (1936-37) LIII TLR 663; (1937) WN (I) 188 HL Can be convicted of reckless driving where negligence insufficient to sustain manslaughter charge in case where victim dies.

Andrews v Director of Public Prosecutions [1992] RTR 1; [1991] TLR 223 HC QBD That medic mistakenly stated defendant's reason for refusing blood not a medical reason did not mean police officer had not performed duty to obtain medical opinion on excuse defendant offered.

Andrews v HE Kershaw, Ltd and another [1951] 2 All ER 764; (1951) 115 JP 568; [1952] 1 KB 70; (1951) 101 LJ 581; (1951) 95 SJ 698; [1951] 2 TLR 867; (1951) WN (I) 510 HC KBD Overhang measured when tailboard up.

Andrews, Maynard v [1973] RTR 398 HC QBD No 'totting up' where person only convicted of single offence inside three years preceding commission of offence for which being sentenced here.

Andrews, R v (1934-39) XXX Cox CC 576; (1937) 156 LTR 464 HL Death through negligent driving to be treated as death through any form of negligence: law of manslaughter remains the same.

Anelay, Browne v [1997] TLR 302 HC QBD On when driver deemed to have taken over vehicle for purposes of tachograph legislation.

Angell, R v [1977] Crim LR 682 CA £250 fine/five years' disqualification for causing death by dangerous driving, plus £50 fine/one years' disqualification for driving with excess alcohol fully merited.

Anker and another, Arthur and another v [1996] 3 All ER 783; (1996) 146 NLJ 86; [1997] QB 564; [1996] RTR 308; [1995] TLR 632; [1996] 2 WLR 602 CA Unauthorised parking on land (consequences of which were posted) meant took risk of clamping/having to pay for removal of clamp.

Another v Probert [1968] Crim LR 564 HC QBD Giving misleading indications when driving was careless driving.

Anthony v Jenkins [1972] Crim LR 596; [1971] RTR 19 HC QBD Person who had stopped car and run away from it was not driving when constable formed opinion that had been drinking alcohol.

Appleby (RW), Ltd and another v Vehicle Inspectorate [1994] RTR 380 HC QBD Non-liability of parent company for non-provision of tachograph records by subsidiary to employee; tachograph necessary when driving from depot to docks to collect passengers (no notion of 'positioning' journey under the tachograph legislation).

Appleyard v Bingham [1914] 1 KB 258; (1913) 48 LJ 660; [1914] 83 LJKB 193; (1913) WN (I) 300 HC KBD Breach of petrol storage regulations.

Archer v Blacker [1965] Crim LR 165t; (1965) 109 SJ 113 HC QBD On onus of proof as regards proving that notice of road traffic prosecution served.

Archer v Woodard [1959] Crim LR 461 HC QBD That drank whisky on empty stomach not special reason justifying reduced disqualification period.

Archer, Lovell v; Lovell v Ducket [1971] Crim LR 240; (1971) 121 NLJ 128t; [1971] RTR 237; (1971) 115 SJ 157 HC QBD Provisional licence to be endorsed where holder of same does not comply with conditions of licence.

Argyle Motors (Birkenhead), Ltd, Jones v [1967] Crim LR 244t; (1967) 111 SJ 279 HC QBD On whether carriage of rubbish in motor dealers' vehicle was breach of terms of general trade licence.

Arif (Mohammed), R v (1985) 7 Cr App R (S) 92; [1985] Crim LR 523 CA Can under Powers of Criminal Courts Act 1973, s 44, disqualify person from driving for period exceeding that for which are imprisoned.

Arlidge, Siddle C Cook, Ltd v [1962] Crim LR 319; (1962) 106 SJ 154 HC QBD On what constitutes abnormal indivisible load.

Arlidge, Sunter Brothers Ltd and another v [1962] 1 All ER 510; [1962] Crim LR 320; (1962) 126 JP 159; (1962) 112 LJ 240; (1962) 106 SJ 154 HC QBD Load an 'abnormal indivisible load' if would be undue risk/damage in dividing.

Armitage v Walton [1976] Crim LR 70t; (1975) 125 NLJ 1022t; [1976] RTR 160 HC QBD Failed prosecution for unlawful advertisement of mini-cab service as advertisement made clear that vehicles hired not licensed cabs.

Armstrong v Clark [1957] 1 All ER 433; (1957) 41 Cr App R 56; [1957] Crim LR 256; (1957) 121 JP 193; (1957) 107 LJ 138; [1957] 2 QB 391; (1957) 101 SJ 208; [1957] 2 WLR 400 HC QBD Diabetic having injected insulin can be charged of driving under influence of drug.

Armstrong v Ogle (1926-30) XXVIII Cox CC 253; (1926) 90 JP 146; [1926] 2 KB 438; (1926) 61 LJ 474; [1926] 95 LJCL 908; (1926) 135 LTR 118; (1925-26) XLII TLR 553; (1926) WN (I) 156 HC KBD Conviction of omnibus driver for unlicensed plying for hire.

Armstrong, Kerr v (1973) 123 NLJ 638t; [1974] RTR 139; [1973] Crim LR 532 HC QBD That person involved in accident (which did not involve another) telephoned police to inform them of same was not special reason justifying non-disqualification.

Arnold Transport (Rochester), Ltd v British Transport Commission and others [1963] 1 QB 457 CA Transport Tribunal could not refuse licence on basis that was not in applicant's best interests.

Arnold v Chief Constable of Kingston-upon-Hull [1969] 3 All ER 646; [1969] Crim LR 442; (1969) 133 JP 694; (1969) 113 SJ 409; [1969] 1 WLR 1499 HC QBD On taking of tests 'there or nearby'.

Arnold, Carrimore Six Wheelers, Ltd v [1949] 2 All ER 416; (1949) 113 JP 456; (1949) LXV TLR 506; (1949) WN (I) 349 HC KBD Tractor covered by limited trade licence carrying goods in attached trailer not so licenced an offence: trailer not part of tractor.

Arnold, Grays Haulage Co, Ltd v [1966] 1 All ER 896; [1966] Crim LR 224; (1966) 130 JP 196; (1965-66) 116 NLJ 445t; (1966) 110 SJ 112; [1966] 1 WLR 534 HC QBD Must be actual knowledge on employer's part to sustain charge of permitting driver to drive for longer than law allows.

Arnold, R v [1964] Crim LR 664 Sessions Was not 'driving' to be steering one motor vehicle being pushed along by another.

Arnott, Wilson v [1977] 2 All ER 5; [1977] Crim LR 43; (1977) 141 JP 278; (1976) 126 NLJ 1220t; [1977] RTR 308; (1976) 120 SJ 820; [1977] 1 WLR 331 HC QBD Permissible to wait in parking bay whose use suspended.

Arrowsmith v Jenkins [1963] Crim LR 353; (1963) 127 JP 289; (1963) 113 LJ 350; (1963) 107 SJ 215; [1963] 2 WLR 856 HC QBD Person who does act that leads to obstruction of highway which did not intend is nonetheless guilty of obstruction of highway.

Arthur and another v Anker and another [1996] 3 All ER 783; (1996) 146 NLJ 86; [1997] QB 564; [1996] RTR 308; [1995] TLR 632; [1996] 2 WLR 602 CA Unauthorised parking on land (consequences of which were posted) meant took risk of clamping/having to pay for removal of clamp.

Arthur Sanderson (Great Broughton), Ltd v Vickers [1964] Crim LR 474g; (1964) 108 SJ 425 HC QBD Licensing authority when issuing 'A' carrier's licence could specify which trailers might be used.

Baines, R v (1970) 54 Cr App R 481; [1970] Crim LR 590; (1970) 120 NLJ 733t; [1970] RTR 455; (1970) 114 SJ 669 CA That person who had been drinking only drove to lend assistance to person stranded without petrol in remote area after latter telephoned for help not special reason justifying non-disqualification.

Baker v Cole [1971] 3 All ER 680; [1966] Crim LR 453; (1971) 135 JP 592; [1972] RTR 43; [1971] 1 WLR 1788 HC QBD Factors to be considered as mitigating when sentencing for third traffic conviction.

Baker v Esau [1972] Crim LR 559; [1973] RTR 49 HC QBD Ambulance trailer and car together constituted single four-wheeled trailer.

Baker v Foulkes [1975] 3 All ER 651; (1976) 62 Cr App R 64; [1976] Crim LR 67t; (1976) 140 JP 24; [1975] RTR 509; [1975] 1 WLR 1551 HL Police officer need not tell medical doctor of intention to caution of effects of failing to provide specimen.

Baker v Sweet [1966] Crim LR 51g HC QBD Burden of proving that 50 mph (Temporary Speed Limit) (England and Wales) Order 1964 inapplicable to instant case rested on defence.

Baker v Williams [1956] Crim LR 204t HC QBD Justices found to have misdirected themselves on facts (which showed defendant must have been driving dangerously).

Baker's Transport (Southampton), Ltd, British Transport Commission v (1960) 110 LJ 286 TrTb On form of evidence required when objecting to variation of ('B') licence.

Baker, Foulkes v [1975] RTR 50; (1975) 119 SJ 777 HC QBD Prosecution must prove medical practitioner in charge of patient informed of intention to request latter for specimen.

Baker, Richardson v [1976] Crim LR 76 HC QBD On when employer may be said to 'use' vehicle driven by employee.

Baker, Scott v [1968] 2 All ER 993; (1968) 52 Cr App R 566; [1968] Crim LR 393t; (1968) 132 JP 422; (1968) 118 NLJ 493t; [1969] 1 QB 659; (1968) 112 SJ 425; [1968] 3 WLR 796 HC QBD Specimen not valid unless relevant procedural requirements of Road Safety Act 1967 complied with.

Balch v Beeton [1970] Crim LR 285; [1970] RTR 138 HC QBD Scrap trade different from car dealership/repair trade so was offence to use vehicle licensed for dealership/repair trade to transport scrap.

Baldessare (Cyril), R v (1930-31) Cr App R 70; (1931) 144 LTR 185; (1930) WN (I) 193 CCA Passenger in recklessly driven car may be guilty of criminal negligence by driving.

Baldwin v Director of Public Prosecutions [1996] RTR 238; [1995] TLR 279 HC QBD Police officer requesting blood specimen and asking why any reason should not be given had discharged statutory requirements: taking of specimen valid.

Baldwin v Worsman [1963] 2 All ER 8; [1963] Crim LR 364; (1963) 127 JP 287; (1963) 113 LJ 349; (1963) 107 SJ 215; [1963] 1 WLR 326 HC QBD Vehicle capable of reversing does not cease to be such because means of reverse shut off.

Baldwin, Hargreaves v (1905) 69 JP 397; (1905-06) XCIII LTR 311; (1904-05) XXI TLR 715 HC KBD Speed relevant to reckless driving charge even though speeding a different offence (Motor Car Act 1903, s 1(1)).

Baldwin, Taylor v [1976] Crim LR 137; [1976] RTR 265 HC QBD Police officer need not be in uniform when forms suspicion that driver was drinking but must be (and here was so) when administers breath test.

Balfour Beatty and Co Ltd v Grindey [1974] Crim LR 120; [1975] RTR 156 HC QBD Was wrong to charge company with 'use' of vehicle with defective indicator and not 'causing or permitting' same to be on road.

Ball, Percival v (1937) WN (I) 106 HC KBD Notice of intended prosecution for careless driving deemed adequate.

Ball, R v [1974] RTR 296 CA Four-month disqualification merited in case where sentencing court acted under mistaken view that twelve month disqualification mandatory for dangerous driving.

Ball, R v; R v Loughlin (1966) 50 Cr App R 266; [1966] Crim LR 451; (1965-66) 116 NLJ 978t; (1966) 110 SJ 510 CCA Offence of causing death by dangerous driving an absolute offence so driver liable though acting under instructions of another.

Barrett, Wiltshire v [1965] 2 All ER 271; (1965) 129 JP 348; [1966] 1 QB 312; (1965) 109 SJ 274; [1965] 2 WLR 1195 CA Valid arrest if police officer has reasonable suspicion are committing road traffic offence/are released once discover innocent or no case.

Barron, Bursey v [1972] Crim LR 486; [1971] RTR 273; (1971) 115 SJ 469 HC QBD Use of non-reflective material on reflective speed signs did not render signs invalid so could be charged for driving in excess of indicated speeds.

Barry, Pettigrew v [1984] TLR 447 HC QBD Unlicensed private hire driver ought to have been found to be plying for hire where waited alongside hackney carriage stand.

Bartholomew, R v (1907-09) XXI Cox CC 556; (1908) 72 JP 79; [1908] 1 KB 554; [1908] 77 LJCL 275; (1907-08) XXIV TLR 238 CCR Coffee stall on highway not common nuisance.

Bartlett, Abrahams v [1970] RTR 276 HC QBD Could not on same facts be guilty both of not stopping meter after fare finished and of starting meter before new fare began (here found guilty of former).

Barton, Fraser v (1974) 59 Cr App R 15 HC QBD On special circumstances justifying non-disqualification.

Bason v Vipond; Same v Robson [1962] 1 All ER 520; [1962] Crim LR 320; (1962) 126 JP 178; (1962) 112 LJ 241; (1962) 106 SJ 221 HC QBD Test whether vehicle exceeds length restrictions is in its ordinary position as constructed.

Bassam v Green [1981] Crim LR 626; [1981] RTR 362 HC QBD Hackney carriage licensed and used as such could not be divested of that attribute; was offence for hackney carriage driver to seek/take booking fee where ride arranged via radio service.

Bastick, Stone v [1965] 3 All ER 713; (1966) 130 JP 54; (1965-66) 116 NLJ 189; [1967] 1 QB 74; (1965) 109 SJ 877; [1965] 3 WLR 1233 HC QBD Format of disqualification certificates: only offence for which disqualified to appear.

Bates, Mutton v (1983) 147 JP 459; [1984] RTR 256; [1983] TLR 490 HC QBD On duty under Road Traffic Act 1972, s 25 to stop as soon as practicable after accident and report it to the police.

Bates, R v [1973] 2 All ER 509; (1973) 57 Cr App R 757; [1973] Crim LR 449t; (1973) 137 JP 547; (1973) 123 NLJ 493t; [1973] RTR 264; (1973) 117 SJ 395; [1973] 1 WLR 718 CA Whether driving/attempting to drive ultimately a matter for jury.

Batty, Glendinning v [1973] Crim LR 763; (1973) 123 NLJ 927t; [1973] RTR 405 HC QBD Inappropriate to stretch the law so as to find 'special reasons' justifying non-disqualification.

Baxter v Matthews [1958] Crim LR 263 Magistrates Valid conviction under Highway Act 1835, s 78, for negligent interruption of free passage of carriage on the highway.

Baxter, Hill v [1958] 1 All ER 193; (1958) 42 Cr App R 51; [1958] Crim LR 192; (1958) 122 JP 134; (1958) 108 LJ 138; [1958] 1 QB 277; (1958) 102 SJ 53; [1958] 2 WLR 76 HC QBD Dangerous driving/non-conformity with road sign are absolute offences; automatism defence rejected.

Baxter, Knight v [1971] Crim LR 368; [1971] RTR 270; (1971) 115 SJ 350 HC QBD That had drunk alcohol on empty stomach not special reason justifying non-disqualification.

BCF Transport Co Ltd v Townend (1960) 110 LJ 10 TrTb Failed appeal against addition to 'A' licence of trailer.

Beames v Director of Public Prosecutions [1989] Crim LR 659; [1990] RTR 362 HC QBD Driver could not use unexpired time of another driver and (once expired) add coins to meter up to maximum time (Parking Places and Controlled Zone (Manchester) Order 1971).

Beard v Wood [1980] Crim LR 384; [1980] RTR 454 HC QBD Road Traffic Act 1972, s 159 empowered uniformed police constable to stop traffic independent of any common law powers.

Beard, Parsley v [1978] RTR 263 HC QBD On mechanics of breath test.

Beardmore, Morris v [1980] 2 All ER 753; [1981] AC 446; (1980) 71 Cr App R 256 (also HC QBD); [1979] Crim LR 394; (1980) 144 JP 331; (1980) 130 NLJ 707; [1980] RTR 321; (1980) 124 SJ 512; [1980] 3 WLR 283 HL Request to take breath test invalid where police officer trespassing on property of person requested at time of request.

Beardmore, Morris v [1979] 3 All ER 290; (1980) 71 Cr App R 256 (also HL); (1980) 144 JP 30; [1980] QB 105; [1979] RTR 393; (1979) 123 SJ 300; [1979] 3 WLR 93 HC QBD Request to take breath test valid though police officer trespassing on property of person requested at time of request.

Beardsley, R v [1979] RTR 472 CA Eighteen months' imprisonment, fine, order to pay prosecution costs in relation to offence for which convicted (to maximum of £100) for person convicted of driving with excess alcohol.

Beauchamp-Thompson v Director of Public Prosecutions [1988] Crim LR 758; [1989] RTR 54 HC QBD Evidence showing that alcohol levels exceeded limit at moment of test but not while driving was inadmissible; that were mistaken as to strength of alcohol consumed not special reason meriting non-disqualification.

Beaumont, Police v [1958] Crim LR 620 Magistrates Successful plea of automatism (consequent upon pneumonia) in careless driving prosecution.

Beaumont, R v [1964] Crim LR 665; (1964) 114 LJ 739 CCA Occupation road deemed not to road on which could be guilty of offence of drunk driving.

Beck v Sager [1979] Crim LR 257; [1979] RTR 475 HC QBD Was not offence of failing to provide person where (foreigner) did not understand what were told by police officer.

Beck v Scammell [1985] Crim LR 794; [1986] RTR 162 HC QBD Constable's written amendments to breathalyser printout/printout copy (that should read BST not GMT) did not render printout inadmissible.

Beck v Watson [1979] Crim LR 533; [1980] RTR 90 HC QBD Could not rely on second blood specimen where obtained after first freely provided and only rendered incapable of analysis when dropped by police medic.

Beck, Steff v [1987] RTR 61 HC QBD Driver failed to take all reasonable precautions regarding safety of passengers where pulled off from stop before pensioner had seated herself.

Beckett, R v [1976] Crim LR 140 CrCt Urinating into toilet and not container was provision of specimen of urine.

Beckford, R v [1996] 1 Cr App R 94; [1995] RTR 251; [1995] TLR 26 CA That car unavailable did not preclude prosecution continuing where charged with causing death by careless driving after drinking.

Bedford, Neal v [1965] 3 All ER 250; [1965] Crim LR 614; (1965) 129 JP 534; (1965) 115 LJ 677; [1966] 1 QB 505; (1965) 109 SJ 477; [1965] 3 WLR 1008 HC QBD Must yield to pedestrians at pedestrian crossing.

Beeby (Leslie Albert), R v (1983) 5 Cr App R (S) 56 CA Twelve month disqualification/£750 fine for person who caused death by reckless driving when fell asleep at the wheel.

Beech, Director of Public Prosecutions v [1992] Crim LR 64; (1992) 156 JP 31; (1991) 141 NLJ 1004t; [1992] RTR 239; [1991] TLR 342 HC QBD Not reasonable excuse for non-provision of breath specimen that had made self so drunk could not understand what were instructed to do.

Beer v Clench (WH) (1930) Limited (1934-39) XXX Cox CC 364; (1936) 80 SJ 266 HC KBD Employer's working hour records admissible in prosecution for permitting worker-driver to drive above permitted work limit.

Beer v Davies [1958] 2 All ER 255; (1958) 42 Cr App R 198; [1958] Crim LR 398t; (1958) 122 JP 344; (1958) 108 LJ 345; [1958] 2 QB 187; (1958) 102 SJ 383 HC QBD Information rightly dismissed where non-delivery of notice not accused's fault.

Beer v TM Fairclough and Sons Limited (1934-39) XXX Cox CC 551; (1937) 101 JP 157; (1937) 156 LTR 238; (1937) 81 SJ 180; (1936-37) LIII TLR 345 HC KBD Employer to ensure worker-driver has rest periods, not to police how he spends them.

Beesley, Russell v [1937] 1 All ER 527; (1937) 81 SJ 99; (1936-37) LIII TLR 298 HC KBD No rule that evidence of police officer on speed of accused's car by reference to own speedometer requires corroboration.

Beeton, Balch v [1970] Crim LR 285; [1970] RTR 138 HC QBD Scrap trade different from car dealership/repair trade so was offence to use vehicle licensed for dealership/repair trade to transport scrap.

Bolton Justices, ex parte Scally, R v; R v Bolton Justices, ex parte Greenfield; R v Eccles Justices, ex parte Meredith; R v Trafford Justices, ex parte Durran-Jorda [1991] Crim LR 550; (1991) 155 JP 501; [1991] 1 QB 537; [1991] RTR 84; (1990) 134 SJ 1403; [1990] TLR 639; [1991] 2 WLR 239 HC QBD Excessive blood-alcohol concentration convictions quashed where very likely that blood samples taken had been contaminated.

Bond and others, Williams v [1986] Crim LR 564; [1986] RTR HC QBD Was breach of excess hour regulations where four drivers drove four vehicles in which an extra driver sat at different times.

Bond, Amalgamated Roadstone Corporation, Ltd v [1963] 1 All ER 682; [1963] Crim LR 290; (1963) 127 JP 254; (1963) 113 LJ 234; (1963) 107 SJ 316; [1963] 1 WLR 618 HC QBD Spreader used in conjunction with tractor a 'land implement'.

Bond, R v [1968] 2 All ER 1040; (1968) 52 Cr App R 505; [1968] Crim LR 622; (1968) 118 NLJ 734t; (1968) 112 SJ 908; [1968] 1 WLR 1885 CA Life disqualification unmerited for driving while disqualified.

Bond, Ross Hillman Ltd v [1974] 2 All ER 287; (1974) 59 Cr App R 42; [1974] Crim LR 261; (1974) 138 JP 428; [1974] QB 435; [1974] RTR 279; (1974) 118 SJ 243; [1974] 2 WLR 436 HC QBD Person to have actual knowledge of facts that are unlawful use to be convicted of causing/permitting same.

Bool, Saycell v [1948] 2 All ER 83; (1948) 112 JP 341; (1948) 98 LJ 315; (1948) LXIV TLR 421; (1948) WN (I) 232 HC KBD Freewheeling is 'driving'.

Boon, Smith v (1899-1901) XIX Cox CC 698; (1901) LXXXIV LTR 593; (1900-01) 45 SJ 485; (1900-01) XVII TLR 472; (1900-01) 49 WR 480 HC KBD Excessive speed determined by reference to traffic on road not immediately around motor vehicle concerned.

Booth v Director of Public Prosecutions [1993] RTR 379 HC QBD Approval of finding that drawing of empty trailer to test same was drawing of goods vehicle requiring operator's licence.

Boothby, Slender v (1985) 149 JP 405 HC QBD Breathalyser incapable of printing correct date on printout not a reliable device.

Borthwick v Vickers [1973] Crim LR 317; [1973] RTR 390 HC QBD Justices could rely on own knowledge of local area when deciding if certain trip must have involved using public road.

Bosley v Long [1970] 3 All ER 286; [1970] Crim LR 591; (1970) 134 JP 652; (1970) 120 NLJ 968; [1970] RTR 432; (1970) 114 SJ 571; [1970] 1 WLR 1410 HC QBD Specimen reuested at hospital to be provided at hospital; medic may object anytime after notified of proposal to take specimen.

Bosomworth, Farrow v [1969] Crim LR 320t; (1969) 119 NLJ 390; (1969) 113 SJ 368 HC QBD Home Office circular could not be relied on to establish that breathalyser an approved device.

Boss v Measures [1989] Crim LR 582; [1990] RTR 26 HC QBD Local authority could validly require that information as to identity of driver be given in writing.

Bostock v Ramsey Urban District Council (1899-1900) XVI TLR 18 HC QBD Not malicious prosecution to bring proceedings for obstruction of highway by employee against employer.

Boswell, R v; R v Elliott (Jeffrey); R v Daley (Frederick); R v Rafferty (1984) 6 Cr App R (S) 257; [1984] Crim LR 502; [1984] RTR 315; [1984] TLR 395; [1984] 1 WLR 1047 CA On sentencing (disqualification) for death by reckless driving.

Botwood v Phillips [1976] Crim LR 68; [1976] RTR 260 HC QBD On what constitutes 'dock road'.

Boulton v Pilkington [1981] RTR 87 HC QBD Authority to wait in restricted area when loading/unloading meant for trade purposes not while collecting take-away.

Bourlet v Porter [1973] 2 All ER 800; (1974) 58 Cr App R 1; [1974] Crim LR 53; (1973) 137 JP 649; [1973] RTR 293; (1973) 117 SJ 489; [1973] 1 WLR 866 HL Need not be given opportunity of second test at police station; specimen requested from hospital patient need not be provided while patient.

Bove, R v [1970] 2 All ER 20; [1970] Crim LR 353; (1970) 134 JP 418 HC QBD Person not driving when refused breath test was unlawfully arrested.

Bove, R v (1970) 54 Cr App R 316; [1970] Crim LR 471; [1970] RTR 261; (1970) 114 SJ 418; [1970] 1 WLR 949 CA Person not driving when refused breath test was unlawfully arrested.

Bow (Dennis Arthur), R v (1977) 64 Cr App R 54; [1977] Crim LR 176t; (1977) 127 NLJ 43t; [1977] RTR 6 CA Was taking conveyance without authority to freewheel another's vehicle downhill.

Bowell, R v [1974] Crim LR 369t; (1974) 124 NLJ 271t; [1974] RTR 273; (1974) 118 SJ 367 CA On blood-alcohol concentration prosecution had to prove under Road Traffic Act 1961, s 6(2), to secure drunk driving conviction.

Bowen v Wilson (1926-30) XXVIII Cox CC 298; (1927) 91 JP 3; [1927] 1 KB 507; (1926) 62 LJ 376; [1927] 96 LJCL 183; (1927) 136 LTR 310; (1925-26) 70 SJ 1161; (1926-27) XLIII TLR 77; (1926) WN (I) 308 HC KBD Flaw in brake drum not flaw in brake/s.

Bowers v Gloucester Corporation [1963] 1 All ER 437; [1963] Crim LR 276; (1963) 127 JP 214; (1963) 113 LJ 268; [1963] 1 QB 881; [1963] 2 WLR 386 HC QBD Can be convicted of any two offences for hackney carriage licence to be revoked.

Bowers v Worthington [1982] RTR 400 HC QBD Interpretation and application of Road Vehicles (Registration and Licensing) Regulations 1971, reg 35(4).

Bowler, Lomas v [1984] Crim LR 178 HC QBD Way in which defendant drove considered in tandem with conflicting analyses of urine samples meant drunk driving conviction valid.

Bowman v Director of Public Prosecutions [1990] Crim LR 600; (1990) 154 JP 524; [1991] RTR 263; [1990] TLR 52 HC QBD Justices can rely on local knowledge in determining whether particular car park a public place.

Bowman, Vickers v [1976] Crim LR 77; [1976] RTR 165 HC QBD Vehicle was used as express carriage carrying passengers at separate fares even though money collected weekly by one passenger from all and then handed to driver.

Bowman-Shaw, Dilks v [1981] RTR 4 HC QBD Decision that overtaking in left-hand lane of motorway not inconsiderate driving was valid.

Bowra v Dann Catering Ltd [1982] RTR 120 HC QBD Catering firm providing portable equipment (including toilets) at entertainment venues were not travelling showmen; portable toilets here not meant for circus/fun-fair so firm/its drivers not exempt from driver-records legislation.

Bowsher, R v [1973] Crim LR 373; [1973] RTR 202 CA Driving while disqualified an absolute offence; disqualifications ordered under Road Traffic Act 1962, s 5(3) to be consecutive.

Boxell (Michael John), R v (1989) 11 Cr App R (S) 269 CA Four years' imprisonment for causing three deaths by reckless driving (speeding after drinking).

Boxer v Snelling [1972] Crim LR 441; [1972] RTR 472; (1972) 116 SJ 564 HC QBD On what is a 'vehicle' (here for purposes of Road Traffic Regulation Act 1967, s 31/Folkestone Traffic Regulation Order 1970, Article 15).

Boyce v Absalom [1974] Crim LR 192; [1974] RTR 248 HC QBD Constable cannot under Road Traffic Acts demand driving licence/other driving documents of person not driving.

Boyce, Benson v [1997] RTR 226 HC QBD Prosecution do not have to prove (unlicensed) private hire of vehicle at particular time in issue in order to secure conviction for same (that vehicle used (unlicensed) as private hire vehicle the defining feature of offence).

Boyd-Gibbins v Skinner [1951] 1 All ER 1049; (1951) 115 JP 360; [1951] 2 KB 379; [1951] 1 TLR 1159; (1951) WN (I) 267 HC KBD Speed limit signs evidence that area deemed a built up area under Road Traffic Act 1934, s 1(1)(b).

Boyes, Pearson v [1953] 1 All ER 492; (1953) 117 JP 131; (1953) 97 SJ 134; [1953] 1 WLR 384 HC QBD Not using goods vehicle as goods vehicle when simply using to tow caravan.

Bracegirdle v Oxley (1947) 111 JP 131 HC KBD Speed on occasion can of itself be indicative of dangerous driving.

Bracegirdle v Oxley; Bracegirdle v Cobley [1947] 1 All ER 126; [1947] KB 349; [1947] 116 LJR 815; (1947) 176 LTR 187; (1947) 91 SJ 27; (1947) LXIII TLR 98; (1947) WN (I) 54 HC KBD Speed under speed limit can be excessive.

Bradburn v Richards (1975) 125 NLJ 998t; [1976] RTR 275 HC QBD Justices could not extend period for variation of sentence allowed under Road Traffic Act 1972, s 101.

Braddock v Whittaker [1970] Crim LR 112; [1970] RTR 288; (1969) 113 SJ 942 HC QBD Adequate evidence of non-contamination of sample/that sample analysed was that provided by defendant.

Bradfield and Sonning Justices, ex parte Holdsworth, R v [1971] 3 All ER 755; [1972] Crim LR 542t; (1971) 135 JP 612; (1971) 121 NLJ 664t; [1972] RTR 108; (1971) 115 SJ 608 HC QBD Lifting of disqualification.

Bradford Corporation, ex parte Minister of Transport, R v (1926) 135 LTR 227 HC KBD Mandamus rule absolute requiring Bradford Corporation to grant certain hackney carriage licenses authorised by Minister for Transport under Roads Act 1920, s 14(3).

Bradford v Wilson (1984) 78 Cr App R 77; [1983] Crim LR 482; (1983) 147 JP 573; [1984] RTR 116 HC QBD Driving after inhalation of toluene (element of glue) was driving under influence of drug.

Bradley, Adams v [1975] Crim LR 168; [1975] RTR 233 HC QBD That drank alcoholic beverage in mistaken belief that was beverage of lower alcohol content not special reason justifying non-disqualification.

Bradley, Sparrow v [1985] RTR 122 HC QBD Evidence obtained via second breath test at police station using different breathalyser was admissible.

Bradley, Steetway Sanitary Cleansers, Ltd v (1961) 105 SJ 444 HC QBD Vehicle being used to transport effluent was a goods vehicle in which goods being transported for hire/reward.

Bradley, Sweetway Sanitary Cleansers, Ltd v [1961] 2 All ER 821; [1961] Crim LR 553; (1961) 125 JP 470; (1961) 111 LJ 582; [1962] 2 QB 108; [1961] 3 WLR 196 HC QBD Vehicle transporting effluent a goods vehicle; because bringing effluent to farm (without charge) part of removing it from tank (for which charge) was for reward.

Bradshaw and another, R v [1994] TLR 693 CA Extended driving test after disqualification period expired unmerited by person found guilty of aggravated vehicle-taking but who was said to have been passenger in relevant vehicle.

Brady, North West Traffic Area Licensing Authority v [1981] Crim LR 407; [1981] RTR 265 CA Licence application 'made' where completed/mailed in December 1976 though not received by licensing authority until January 1977.

Brady, North West Traffic Area Licensing Authority v [1979] Crim LR 397; (1979) 129 (2) NLJ 712; [1979] RTR 500 HC QBD Licence application not 'made' when completed/mailed in December 1976 but when received by licensing authority in January 1977.

Braham v Director of Public Prosecutions (1995) 159 JP 527; [1996] RTR 30; [1994] TLR 684 HC QBD Breath specimen obtained when police after tip-off went to house of person and arrested her when she refused to take breath test (later taken at police station) was admissible.

Braintree District Council v Howard [1993] RTR 193 HC QBD Was offence for private vehicle hire operator licensed in one area to take telephone bookings at number in area where unlicensed and to drive in area where unlicensed.

Bray, Evans v [1976] Crim LR 454; [1977] RTR 24 HC QBD Disqualification merited where driver with excess alcohol bringing tablets to wife in emergency had not duly considered possible alternatives such as calling emergency services.

Brazier v Alabaster [1962] Crim LR 173t HC QBD Keep left traffic sign affects only those who have passed it.

Brazier, Pilbury v [1950] 2 All ER 835; (1950) 114 JP 548; [1950] 66 (2) TLR 763 HC KBD That insurers considered themselves bound a special reason justifying non-disqualification.

Brazil, Jones v [1970] RTR 449; [1971] Crim LR 47 HC QBD On who may qualify as medical practititoner to be notified when wish to obtain specimen from hospital patient: here casualty officer sufficed.

Breame v Anderson and another [1971] RTR 31; (1971) 115 SJ 36 CA Private hire vehicle did not by advertising telephone number to call for hire indicate that was for immediate hire.

Brentford Magistrates' Court, ex parte (Clarke) Robert Anthony, R v; Clarke v Hegarty (1986) 150 JP 495; [1987] RTR 205 HC QBD Where two breath specimen readings were the same they could both be admitted in evidence.

Brentwood Magistrates' Court, ex parte Richardson, R v (1992) 95 Cr App R 187; (1992) 156 JP 839; [1993] RTR 374; [1992] TLR 1 HC QBD Construction of 'totting-up' provisions of Road Traffic Act 1988 so as to determine whether mandatory disqualification of defendant required.

Brewer v Metropolitan Police Commissioner [1969] Crim LR 149t; (1969) 133 JP 185; (1968) 112 SJ 1022; [1969] 1 WLR 267 HC QBD Worker who unknowingly inhaled alcohol fumes at factory (and had no constructive knowledge of same) had special reasons justifying non-disqualification.

Brewer v Metropolitan Police Commissioner (1969) 53 Cr App R 157 CA Lacing of drinks/ inadvertent ingestion of alcoholic fumes could be special reason justifying non-disqualification.

Briddon, Cook and another v [1975] RTR 505; (1975) 119 SJ 462 HC QBD Was improper use of articulated vehicle of exceptional length to carry load that could be carried on standard or shorter than standard articulated vehicle.

Briere v Hailstone [1969] Crim LR 36; (1968) 112 SJ 767 HC QBD Road lamps that were 1½ feet over 200 yards apart (prescribed distance) were nonetheless valid indicators that were driving in area of restricted speed.

Bright, Guyll v [1987] RTR 104 HC QBD Burden of proving any exception/exemption/proviso/ excuse/qualification to excise licensing requirements rests on person seeking to establish same.

Brighton Corporation, ex parte Thomas Tilling, Limited, R v (1916) 80 JP 219 HC KBD Mandamus ordered/refused in respect of properly/improperly refused omnibus/char-a-banc licence.

Brighty v Pearson [1938] 4 All ER 127; (1939-40) XXXI Cox CC 177; (1938) 102 JP 522; (1938) 82 SJ 910 HC KBD Evidence of two police officers on speed at different times not enough to secure speeding conviction.

Bristol Crown Court, ex parte Jones, R v; Jones v Chief Constable of Avon and Somerset Constabulary (1986) 83 Cr App R 109; (1986) 150 JP 93; [1986] RTR 259 HC QBD Was not careless driving where facts giving rise to alleged offence had been immediate reaction to sudden emergency.

Bristow, Director of Public Prosecutions v (1997) 161 JP 35; (1996) TLR 28/10/96 HC QBD Objective test as to whether emergency which led person to drive after drinking excess alcohol a special reason jurifying non-disqualification.

Britain v ABC Cabs (Camberley) Ltd [1981] RTR 395 HC QBD Not offence where hackney carriage owner in one licensing area collected passenger from (without parking or plying for hire in) other licensing area.

Britannia Hygienic Laundry Company, Limited, Phillips v [1923] 1 KB 539; (1923) CC Rep XII 28; [1923] 92 LJCL 389; (1923) 128 LTR 690; (1922-23) 67 SJ 365; (1922-23) XXXIX TLR 207; (1923) WN (I) 47 HC KBD Apportionment of liability for motor car with hidden defect.

Britannia Hygienic Laundry Company, Limited, Phillips v [1924] 93 LJCL 5; (1923) 129 LTR 777; (1922-23) XXXIX TLR 530 CA Regulations concerning use and construction of motor cars did not create personal remedy against repairers who returned car with latent defect which meant car did not comply with regulations.

British Airports Authority, Cinnamond and others v [1980] RTR 220; (1980) 124 SJ 221; [1980] 1 WLR 582 CA Ban on certain car-hire drivers (who persistently hassled arriving passengers) entering Heathrow Airport save as bona fide travellers was valid.

British Airports Authority, Cinnamond and others v [1979] RTR 331 HC QBD Ban on certain car-hire drivers (who persistently hassled arriving passengers) entering Heathrow Airport save as bona fide travellers was valid.

British Airports Authority, Eldridge v [1970] 2 All ER 92; [1970] Crim LR 284; (1970) 134 JP 414; [1970] 2 QB 387; [1970] RTR 270; (1970) 114 SJ 247; [1970] 2 WLR 968 HC QBD Taxi can 'stand' in such a way that is not obliged to take passengers.

British Airports Authority, ex parte Wheatley, R v [1983] R⁻R 147; [1982] TLR 278 HC QBD Licensed taxi driver picking up passenger not affected by Gatwick Airport byelaw requiring permission of airport authority before could offer services.

British Airports Authority, ex parte Wheatley, R v [1983] RTR 466; [1983] TLR 320 CA Licensed taxi driver picking up passenger was covered by Gatwick Airport byelaw requiring permission of airport authority before could offer services.

British Airports Authority, Kennett v [1975] Crim LR 106; [1975] RTR 164; (1975) 119 SJ 137 HC QBD On application of regulations pertaining to car brakes (Motor Vehicles (Construction and Use) Regulations 1973).

British Car Auctions Ltd v Wright [1972] 3 All ER 462; [1972] Crim LR 562t; (1972) 122 NLJ 680t; [1972] RTR 540; [1972] 1 WLR 1519 HC QBD Car auctioneer seeking bids not offering to sell car — conviction for selling unroadworthy vehicle quashed.

British Engineering Contractors, Ltd v British Transport Commission and others (1960) 110 LJ 176 TrTb 'B' licence restricted to fifteen mile area granted in respect of low-loading vehicle.

British Fuel Company Limited, R v [1983] Crim LR 747 CrCt Conviction for not requiring driver-employee to keep record book of driving times.

British Gypsum Ltd v Corner [1982] RTR 308 HC QBD Bowser which carried goods subject to driver's records requirements and required operator's licence.

British Railways Board and another, Carman Transport, Ltd v (1963) 113 LJ 770 TrTb On seeking of licence on foot of traffic coming from outside area of deciding licensing authority.

British Railways Board and others, Munson v [1965] 3 All ER 41; (1965) 115 LJ 709; [1966] 1 QB 813; (1965) 109 SJ 597; [1965] 3 WLR 781 CA Answer in previous licence application on normal user relevant in late licence application.

British Railways Board v Leinster Ferry Transport, Ltd (1963) 113 LJ 770 TrTb Reference to 'other kinds of transport' in Road Traffic Act 1960, s 173(4), meant other kinds of land transport.

British Railways Board, Warwickshire County Council and others v [1969] 1 WLR 1117 CA Construction of clause 'until . . . time for . . . appeal . . . has expired' in Road Traffic Commissioners granting of licence (essentially meant licences granted not operative until appeal procedure exhausted).

British Road Services Ltd and another v Owen [1971] 2 All ER 999; [1971] Crim LR 290; (1971) 135 JP 399; [1971] RTR 372; (1971) 115 SJ 267 HC QBD Regard to entire route to see if unsuitable at any point.

British Road Services Ltd and another v Wurzal [1971] 3 All ER 480; [1972] Crim LR 537; (1971) 135 JP 557; [1972] RTR 45; (1971) 115 SJ 724; [1971] 1 WLR 1508 HC QBD Trailer used to transport goods from England to Continent and vice versa is more than 'temporarily in Great Britain'.

British Telecommunications plc, Director of Public Prosecutions v [1991] Crim LR 532; (1991) 155 JP 869; [1990] TLR 742 HC QBD Prosecution evidence as to defectiveness of trailer brakes admissible even though nature of its examiner's scrutiny of same had rendered proper examination by defence impossible.

British Transport Commission and others v BH Cecil and Sons (1959) 109 LJ 524 TrTb Reduction in number of vehicles made subject to 'A' licence where had been illegal user of same.

British Transport Commission and others v McKelvie and Co, Ltd (1960) 110 LJ 733 TrTb On use of trailers by A-licence holders.

British Transport Commission and others v McKelvie and Co, Ltd (1959) 109 LJ 394 TrTb Reduction in number of artculated vehicles licensed where did not come to Tribunal with clean hands.

British Transport Commission and others v McKelvie and Co, Ltd (1960) 110 LJ 255 TrTb Are 'providing' transport facility if are eager/prepared to do so.

British Transport Commission and others, Arnold Transport (Rochester), Ltd v [1963] 1 QB 457 CA Transport Tribunal could not refuse licence on basis that was not in applicant's best interests.

Bryan v Forrow [1950] 1 All ER 294 HC KBD 'Paid driver' included person employed by another in turn reimbursed by owner: insured.

Bryant v Marx (1931-34) XXIX Cox CC 545; (1932) 96 JP 383; (1932) 147 LTR 499; (1932) 76 SJ 577; (1931-32) XLVIII TLR 624 HC KBD Obstructing road under Motor Vehicles (Construction and Use) Regulations 1931, r 74(1) includes obstructing footpath.

Bryson v Rogers [1956] 2 All ER 826; [1956] Crim LR 570t; (1956) 120 JP 454; (1956) 106 LJ 507; [1956] 2 QB 404; (1956) 100 SJ 569; [1956] 3 WLR 495 HC QBD Dual-purpose vehicle covered by special speed limits whether or not conveying goods.

Buchanan v Motor Insurers' Bureau [1955] 1 All ER 607; [1955] 1 WLR 488 HC QBD 'Road' not road under Road Traffic Act 1930, s 121(1) as no public access by right/tolerance.

Buck and another, Goodbarne v (1939-40) XXXI Cox CC 338; (1939) 103 JP 393 HC KBD Not offence to drive with voidable insurance (even if knew was voidable).

Buck and another, Goodbarne v (1939-40) XXXI Cox CC 380; [1940] 109 LJCL 837; (1940) 162 LTR 259; (1939-40) LVI TLR 433 CA Not causing/permitting another to drive uninsured where provide money for voidable insurance policy.

Buckland, Petherick v (1955) 119 JP 82; (1955) 99 SJ 78; [1955] 1 WLR 48 HC QBD Disqualification can be limited to particular vehicle.

Buckley (Karl Peter), R v (1994) 15 Cr App R (S) 695; [1994] Crim LR 387 CA Life disqualification for driver guilty here and previously of numerous driving offences.

Buckley (Nicholas), R v (1988) 10 Cr App R (S) 477; [1989] Crim LR 386 CA Driving disqualification appropriate only where competency as driver in issue.

Buckoke and others v Greater London Council (1970) 134 JP 465; (1970) 120 NLJ 337t; [1970] RTR 327; (1970) 114 SJ 269; [1970] 1 WLR 1092 HC ChD London Fire Brigade Order 144/8 not unlawful as did not require though did in essence permit violation of red traffic lights.

Buckoke and others v Greater London Council [1971] Ch 655; (1971) 135 JP 321; (1971) 121 NLJ 154t; [1971] RTR 131; (1971) 115 SJ 174; [1971] 2 WLR 760 CA London Fire Brigade Order 144/8 not unlawful as did not require though did in essence permit violation of red traffic lights — disciplinary action against persons disputing terms of lawful Order not halted.

Buffel v Cardox (Gt Britain), Ltd and another [1950] 2 All ER 878; (1950) 114 JP 564 CA On ignoring traffic signs.

Bugge v Taylor (1939-40) XXXI Cox CC 450; (1940) 104 JP 467; [1941] 1 KB 198; [1941] 110 LJ 710; (1941) 164 LTR 312; (1941) 85 SJ 82 HC KBD Valid conviction for keeping unlighted vehicle on road where 'road' was forecourt of hotel.

Bullen v Keay [1974] Crim LR 371; [1974] RTR 559 HC QBD Non-disqualification not merited by fact that accused had attempted to commit suicide by taking drugs and had not intended to drive after taking same.

Bullen v Picking and others [1973] Crim LR 765; (1973) 123 NLJ 1043t; (1973) 117 SJ 895; [1974] RTR 46 HC QBD Agricultural machine could be used on public road for carrying non-agricultural goods for use on/of farm without having to pay more in excise.

Bulman v Bennett [1974] Crim LR 121; (1973) 123 NLJ 1113t; [1974] RTR 1; (1973) 117 SJ 916 HC QBD Successful prosecution for failure to report accident as soon as was reasonably practicable.

Bulman v Godbold [1981] RTR 242 HC QBD Not unlawful waiting for frozen fish deliverer to halt on restricted street but was not attending van (with motor still running) where entered premises to re-load fish into hotel refrigerator.

Bulman v Lakin [1981] RTR 1 HC QBD Absent explanatory reason ten-hour delay in reporting accident was not reporting of same as soon as reasonably practicable.

Bundock, Rathbone v [1962] 2 All ER 257; [1962] Crim LR 327; (1962) 126 JP 328; (1962) 112 LJ 404; [1962] 2 QB 260; (1962) 106 SJ 245 HC QBD Requirement to give information under Road Traffic Act 1960.

175

Coghlan, Childs v [1968] Crim LR 225t; (1968) 118 NLJ 182t; (1968) 112 SJ 175 HC QBD Earth excavator was a motor vehicle for purposes of Road Traffic Act 1960.

Cogley v Sherwood; Car Hire Group (Skyport) Ltd v Sherwood; Howe v Kavanaugh; Howe v Kavanaugh; Car Hire Group (Skyport) Ltd v Kavanaugh [1959] 2 All ER 313; [1959] Crim LR 464t; (1959) 123 JP 377; (1959) 109 LJ 330; [1959] 2 QB 311; (1959) 103 SJ 433 HC QBD Hackney carriage not plying for hire unless shown for hire.

Cole Brothers v Harrop [1916] 85 LJKB 494; (1915-16) 113 LTR 1013 HC KBD On what constituted use of locomotive for agricultural purposes under Locomotives Act 1898 (and so obviated need for locomotive licence).

Cole v Director of Public Prosecutions [1988] RTR 224 HC QBD Doctor must specifically advise that driver's condition possibly resulting from drug-taking before blood specimen may be validly requested.

Cole v Young [1938] 4 All ER 39; (1938) 86 LJ 293 HC KBD That brakes fail once not prove that breaking system ineffective and hence unlawful.

Cole, Baker v [1971] 3 All ER 680; [1966] Crim LR 453; (1971) 135 JP 592; [1972] RTR 43; [1971] 1 WLR 1788 HC QBD Factors to be considered as mitigating when sentencing for third traffic conviction.

Cole, Parkes v (1922) 86 JP 122; (1922) 127 LTR 152; (1922) WN (I) 123 HC KBD Speeeding conviction quashed where had been no warning of intended prosecution as required under Motor Car Act 1903.

Cole, Smith v [1971] 1 All ER 200; [1970] Crim LR 701; (1971) 135 JP 97; [1970] RTR 459; (1970) 114 SJ 887 HC QBD On taking blood specimens.

Coleman and another, Dent v [1977] Crim LR 753; [1978] RTR 1 HC QBD Were guilty of carrying improperly secured load where straps securing load defective though did not know were so.

Coleman, R v [1974] RTR 359 CA Seven/eight minute wait (while not under arrest) for arrival of breathalyser from police station did not invalidate test.

Coles v Underwood (1984) 148 JP 178; [1983] TLR 644 HC QBD On admissibility/usefulness of evidence of driving short way away from scene of collision following which were charged with driving without due care and attention.

Collacott, Pine v (1985) 149 JP 8; [1985] RTR 282; [1984] TLR 496 HC QBD Need not show that constable (at station where no breathalyser) specifically considered alternative specimen could seek and plumped for one type of specimen rather than another.

Collier, R v [1973] Crim LR 188 CA Determination of appropriate factual basis on which to sentence person found guilty of dangerous driving.

Collins (George), R v [1994] RTR 216 CA Jury to be warned that joint owner of car may have loaned car to third party where description evidence indicates that car actually driven by other owner at relevant time.

Collins v Lucking [1983] Crim LR 264; (1983) 147 JP 307; [1983] RTR 312; [1982] TLR 545 HC QBD Drunk driving conviction quashed where rested on accused's being diabetic and was no evidence either way as to whether was/was not diabetic.

Collins, R v [1997] Crim LR 578 CA Objective test whether (as here) driving (whereby caused death) was dangerous.

Collinson (Alfred Charles), R v (1931-32) 23 Cr App R 49; (1931) 75 SJ 491 CCA Admission of public to field (not normally open to public) made field a public place under Road Traffic Act 1930, s 15(1).

Colombo-Ratnapura Omnibus Company, Limited, Kelani Valley Motor Transit Company, Limited v [1946] 115 LJ 76; (1946) 90 SJ 599; (1945-46) LXII TLR 459 PC 'Route' under Motor Car Ordinance 1938 not same as 'highway' under Omnibus Service Licensing Ordinance 1942: effect of same on awarding of road service licences.

Commissioner of Metropolitan Police, ex parte Holloway, R v (1911) 75 JP 490; (1911) 46 LJ 525; (1912) 81 LJKB 205; (1911-12) 105 LTR 532; (1910-11) 55 SJ 773; (1911) WN (I) 184 CA On licensing responsibilities of Metropolitan Police Commissioner under Metropolitan Public Carriage Act 1869 and Order as to Hackney and Stage Carriages 1907.

Commissioner of Metropolitan Police, ex parte Pearce, R v (1911) 75 JP 85; (1910) 45 LJ 809; (1911) 80 LJCL 223; (1911) 104 LTR 135 HC KBD On discretion of Metropolitan Police Commisisoner regarding granting of taxi licences pursuant to Metropolitan Public Carriage Act 1869.

Commissioner of Police of the Metropolis and another, Adams and another v; Aberdeen Park Maintenance Co Ltd (Third Party) [1980] RTR 289 HC QBD Failed application to have 'road' declared such (where police commissioner had been refusing to prosecute road users on basis that was not road).

Commissioner of Police of the Metropolis, ex parte Humphreys, R v; R v Commissioner of Police of the Metropolis, ex parte Randall (1911) 75 JP 486; (1910-11) 55 SJ 726 HC KBD Metropolitan Police Commissioner cannot refuse hackney-cab licences out of hand on basis of precept that are to be unavailable to those who enjoy cab on hire-purchase terms.

Commissioner of Police of the Metropolis, Stepniewski v [1985] Crim LR 675; [1985] RTR 330 HC QBD Was failure to give (breath) specimen where gave one specimen and then refused to provide second.

Computer Cab Co Ltd and others, Director of Public Prosecutions v [1996] RTR 130 HC QBD Persons prohibited from plying for hire in central London did not do so where accepted passengers within that area who had arranged all details of prospective trip with radio booking service.

Confer, Morton v [1963] 2 All ER 765; [1963] Crim LR 577; (1963) 127 JP 433; (1963) 113 LJ 530; (1963) 107 SJ 417; [1963] 1 WLR 763 HC QBD Accused to show (civil burden) that had no intention to drive until fit to do so; claim to be considered in light of all the facts.

Connop, Weatherson v [1975] Crim LR 239 HC QBD Spiking of drink did not here constitute special reason justifying non-disqualification: on when could do so.

Connor v Graham and another [1981] RTR 291 HC QBD Tyres (though must have capacity to bear maximum axle weight of vehicle) need only be inflated to extent required for actual use to which being put.

Connor v Paterson [1977] 3 All ER 516; [1977] Crim LR 428; (1978) 142 JP 20; (1977) 127 NLJ 639t; [1977] RTR 379; (1977) 121 SJ 392; [1977] 1 WLR 1450 HC QBD Was unlawful under 'Zebra' Pedestrian Crossings Regulations 1971/1524 to overtake car which had stopped at crossing to let people cross even though all people crossing had crossed at moment of overtaking.

Considine, R v (1965) 109 SJ 78 CCA Four months' imprisonment merited by dangerous driver who had appealed sentence on basis that fact that was drinking ought to have been mitigating factor.

Conway, R v [1988] 3 All ER 1025; (1989) 88 Cr App R159; [1989] Crim LR 74; (1988) 152 JP 649; [1989] QB 290; [1989] RTR 35; (1988) 132 SJ 1244; [1988] 3 WLR 1238 CA Necessity a defence to reckless driving if duress of circumstances present.

Cook (Arthur Paul), R v [1996] 1 Cr App R (S) 350 CA £500 for driving with excess alcohol.

Cook and another v Briddon [1975] RTR 505; (1975) 119 SJ 462 HC QBD Was improper use of articulated vehicle of exceptional length to carry load that could be carried on standard or shorter than standard articulated vehicle.

Cook and another, Hulin v (1977) 127 NLJ 791t; [1977] RTR 345; (1977) 121 SJ 493 HC QBD Person plying for hire on railway premises required local authority licence and nor special permission to do so on rail premises.

Cook v Alfred Plumpton, Ltd and another [1935] All ER 806; (1935) 99 JP 308; (1935) 80 LJ 27; (1935) 153 LTR 462; (1935) 79 SJ 504; (1934-35) LI TLR 513; (1935) WN (I) 131 HC KBD Sum of driving periods in one day contravened maximum driving times allowed under Road Traffic Act 1930, s 19(1).

Cook v Atchison [1968] Crim LR 266; (1968) 112 SJ 234 HC QBD Once credible defence raised in road traffic prosecution is for prosecutor to disprove the defence.

Cook v Henderson (1934-39) XXX Cox CC 270; (1935) 99 JP 308 HC KBD Driver calculated to have breached permitted working hours by adding of separate periods punctuated by breaks.

Cook v Hobbs (1910) 45 LJ 710; (1910) WN (I) 219 HC KBD Vehicle had been built for purpose of carrying goods/burden (appellant and family were burden) but as could be used for other purposes was breach of Customs and Inland Revenue Act 1888.

Cook v Lanyon [1972] Crim LR 570t; (1972) 122 NLJ 657t; [1972] RTR 496 HC QBD On fraudulent intent necessary to be guilty of fraudulent use of excise licence contrary to Vehicles (Excise) Act 1971, s 26(1).

Cook v Southend Borough Council (1990) 154 JP 145; [1990] 2 QB 1; [1990] 2 WLR 61 CA Council could appeal as aggrieved person against justices' order to pay damages to/costs of hackney driver whose licence it had withdrawn.

Cook v Southend Borough Council (1987) 151 JP 641 HC QBD Local authority could be aggrieved person and could therefore appeal against decision of justices to allow appeal against authority's revocation of hackney carriage licence.

Cook, R v [1964] Crim LR 56t; (1964) 114 LJ 59 CCA Judge to decide whether facts can constitute attempt, jury to decide whether (as here) facts did constitute attempt.

Cooke (Philip), R v [1971] Crim LR 44 Sessions On elements of offence of causing bodily harm by misconduct when in charge of a vehicle.

Cooke v McCann [1973] Crim LR 522; [1974] RTR 131 HC QBD On determining whether person before court is person properly charged: name and address from licence matching those of person in court.

Coomaraswamy (Subramaniam), R v (1976) 62 Cr App R 80; [1976] Crim LR 260t; (1975) 125 NLJ 1167t; [1976] RTR 21 CA Prosecution must show blood-alcohol concentration exceeded prescribed limit not that figure arrived at precisely correct.

Coombes v Cardiff County Council [1976] Crim LR 75; [1975] RTR 491 HC QBD Decelerometer meter evidence indicated that handbraking system on omnibus not an efficient braking system.

Coombs v Kehoe [1972] 2 All ER 55; [1972] Crim LR 560; (1972) 136 JP 387; (1972) 122 NLJ 153t; [1972] RTR 224; (1972) 116 SJ 486; [1972] 1 WLR 797 HC QBD Driving merely to park not a special reason justifying non-disqualification.

Coombs, Hunter v [1962] 1 All ER 904; (1962) 112 LJ 321; (1962) 106 SJ 287 HC QBD Conviction quashed where bad information (despite prosecution being so notified).

Cooney, Cotgrove v [1987] Crim LR 272; (1987) 151 JP 736; [1987] RTR 124 HC QBD On finding there to be reasonable cause why specimen not provided.

Cooper v Hall [1968] 1 All ER 185; [1968] Crim LR 116t; (1968) 132 JP 152; (1967) 117 NLJ 1243t; (1967) 111 SJ 928; [1968] 1 WLR 360 HC QBD Not defence to parking in restricted area that was no yellow line.

Cooper v Hawkins [1904] 2 KB 164; (1902-03) 47 SJ 691; (1903-04) 52 WR 233 HC KBD Local authority restrictions on locomotive speed under Locomotives Act 1865 could not apply to Crown locomotive driven by Crown servant on Crown business.

Cooper v Rowlands [1972] Crim LR 53; [1971] RTR 291 HC QBD Could presume from circumstances/absent alternative evidence that police constable in uniform/approved breathalyser used.

Cooper, Benjamin v [1951] 2 All ER 907; (1951) 115 JP 632; [1951] 2 TLR 906; (1951) WN (I) 553 HC KBD Plying for hire in forecourt of tube station not offence as is private place.

Cooper, Cripps v [1936] 2 All ER 48 HC KBD Carriage of indivisible load contravened road traffic regulations as was danger to persons on trailer/highway.

Cooper, R v [1974] RTR 489 CA On whether/when handing over of car-keys to constable constituted ceasing driving.

Cooper, Thoms v (1986) 150 JP 53 HC QBD Not guilty of driving uninsured motor vehicle where vehicle at issue was completely incapable of being directed by person ostensibly controlling it.

Coote and another v Stone (1970) 120 NLJ 1205; [1971] RTR 66 CA Parked car on clearway not a common law nuisance.

Coote v Parkin [1977] Crim LR 172; [1977] RTR 61 HC QBD Tread pattern requirements related to that area of wheel normally touching road when car moving.

de Clifford, The Trial of Lord (1936) 81 LJ 60 HL Failed prosecution of peer for manslaughter (following motor accident).

De Meersman (Eddy Louis), R v [1997] 1 Cr App R (S) 106 CA Three years' imprisonment for causing death by dangerous driving (driver who fell asleep at the wheel).

De Munthe v Stewart [1982] RTR 27 HC QBD Person (with excess alcohol) requested by constable to re-park car just parked could not plead re-park request as special reason meriting non-disqualification.

Deacon v AT [1976] Crim LR 135; [1976] RTR 244 HC QBD Council housing estate road not a 'road' for purpose of Road Traffic Act 1972, s 196(1).

Deacon, Westover Garage Limited v (1931-34) XXIX Cox CC 327; (1931) 95 JP 155; (1931) 145 LTR 357; (1930-31) XLVII TLR 509 HC KBD General trade licence did not cover conveyances of prospective purchasers' goods.

Dean and another, Seeney v [1972] Crim LR 545; [1972] RTR 25 HC QBD On what constitutes a recovery vehicle (Land Rover with towing buoy attached a recovery vehicle).

Dean, Pilgram v [1974] 2 All ER 751; [1974] Crim LR 194; (1974) 138 JP 502; (1974) 124 NLJ 102t; [1974] RTR 299; (1974) 118 SJ 149; [1974] 1 WLR 601 HC QBD Conviction for not having and not displaying licence valid.

Dear v Director of Public Prosecutions [1988] Crim LR 316; [1988] RTR 148 HC QBD Division of blood single blood sample (one portion for police, one for defendant) invalid where latter portion placed in jar containing few droplets of blood already taken from defendant's other arm.

Dear, Director of Public Prosecutions v (1988) 87 Cr App R 181 HC QBD Medical procedure for extracting analysis specimen must be rigorously observed.

DeFreitas v Director of Public Prosecutions [1992] Crim LR 894; (1993) 157 JP 413; [1993] RTR 98; [1992] TLR 337 HC QBD Genuine (though baseless) fear that would contract AIDS by using breathalyser a reasonable excuse for refusing breath specimen.

Delaroy-Hall v Tadman; Watson v Last; Earl v Lloyd (1969) 53 Cr App R 143; [1969] Crim LR 93t; (1969) 133 JP 127; (1968) 118 NLJ 1173t; [1969] 2 QB 208; (1968) 112 SJ 987; [1969] 2 WLR 92 HC QBD Amount by which exceeded blood-alcohol limit not special reason justifying non-disqualification.

Dell, Evans v [1937] 1 All ER 349; (1934-39) XXX Cox CC 558; (1937) 101 JP 149; (1937) 156 LTR 240; (1937) 81 SJ 100; (1936-37) LIII TLR 310 HC KBD Owner's innocent unawareness that vehicle used as unlicensed stage carriage precluded liability therefor.

Dennis v Leonard (1926-30) XXVIII Cox CC 621; (1929) 141 LTR 94 HC KBD Petrol-driven tractor was motor car for purposes of Motor Cars (Use and Construction) Order 1904.

Dennis v Miles (1921-25) XXVII Cox CC 649; (1924) 88 JP 105; [1924] 2 KB 399; [1924] 93 LJCL 1115; (1924) 131 LTR 146; (1923-24) 68 SJ 755; (1923-24) XL TLR 643; (1924) WN (I) 173 HC KBD Properly convicted for overloaded motor omnibus under Railway Passenger Duty Act 1842.

Dennis v Tame [1954] 1 WLR 1338 HC QBD Unless there are special reasons justifying non-disqualification (here for driving uninsured motor-vehicle) disqualification must be ordered.

Denny v Director of Public Prosecutions (1990) 154 JP 460; [1990] RTR 417 HC QBD Not improper for police officer to seek breath specimens on properly working brethalyser after failed attempts on defective breathalyser.

Denscombe, Lyons v [1949] 1 All ER 977; (1949) 99 LJ 259; (1949) 113 JP 305; (1949) WN (I) 257 HC KBD Weekly lump sum payment from fellow employees for conveying home not use as express carriage.

Dent Transport (Spennymoor) Ltd's Appeal (No X15 of 1961) (1961) 111 LJ 486 TrTb On basis on which 'A' licence could be revoked.

Dent v Coleman and another [1977] Crim LR 753; [1978] RTR 1 HC QBD Were guilty of carrying improperly secured load where straps securing load defective though did not know were so.

Denton (Stanley Arthur), R v (1987) 85 Cr App R 246; [1987] RTR 129; (1987) 131 SJ 476 CA Defence of necessity unavailable where person charged with reckless driving claimed had driven carefully at relevant time.

Director of Public Prosecutions v Doyle [1993] RTR 369; [1992] TLR 651 HC QBD Could not find special reasons meriting non-disqualification where offender had wilfully elected to drink and drive.

Director of Public Prosecutions v Drury (1989) 153 JP 417; [1989] RTR 165 HC QBD If person unaware of car accident becomes aware of same within twenty-four hours is under duty to report it to police.

Director of Public Prosecutions v Eddowes [1990] Crim LR 428; [1991] RTR 35 HC QBD Person willing to and who tried but failed to give second breath specimen guilty of failure to provide same.

Director of Public Prosecutions v Elstob (1993) 157 JP 229 HC QBD On when specimen must be divided to have been "divided at . . . time . . . was provided' (Road Traffic Offenders Act 1988, s 15(5)).

Director of Public Prosecutions v Elstob [1992] Crim LR 518; [1992] RTR 45; [1991] TLR 526 HC QBD Two-minute delay in obtaining and dividing specimen permissible; division did not have to take place before defendant.

Director of Public Prosecutions v Enston [1996] RTR 324; [1995] TLR 89 HC QBD Woman's threat to accuse man who had been drinking of raping her unless he drove her to cash-point was special reason justifying non-disqualification.

Director of Public Prosecutions v Evans and another [1988] RTR 409 HC QBD Mobile crane in two parts could be one vehicle (not two as was charged) but could still nonetheless be in violation of motor vehicle construction and use regulations.

Director of Public Prosecutions v Feeney (1989) 89 Cr App R 173 HC QBD Disqualification justified: were special reasons justifying conveyance of passenger to her home but not justifying continuing driving on to own home.

Director of Public Prosecutions v Frost [1989] Crim LR 154; (1989) 153 JP 405; [1989] RTR 11 HC QBD On nature of offences of being in charge of vehicle while unfit to drive (matter which ordinary person could decide)/in charge with excess alcohol (provable by reference to expert evidence).

Director of Public Prosecutions v Fruer; Director of Public Prosecutions v Siba; Director of Public Prosecutions v Ward [1989] RTR 29 HC QBD Where motorway not in ordinary use could be that there were special reasons justifying driving on central reservation.

Director of Public Prosecutions v Garrett (1995) 159 JP 561; [1995] RTR 302; [1995] TLR 55 HC QBD On procedure as regards taking of specimens: that was procedural error as regards taking of blood specimen does not mean urine specimen taken in accordance with procedure is inadmissible.

Director of Public Prosecutions v Godwin (1992) 156 JP 643; [1991] RTR 303 HC QBD Evidence as to excess alcohol in defendant's breath validly excluded on basis that obtained on foot of unlawful arrest.

Director of Public Prosecutions v Gordon; Director of Public Prosecutions v Griggs [1990] RTR 71 HC QBD On need to inform defendant of alternative specimens that may be required and of consequences of not giving required specimen to police officer.

Director of Public Prosecutions v Guy [1997] TLR 354 HC QBD Person guilty of tachograph offence though had finished work and was driving home at time when challenged.

Director of Public Prosecutions v H [1997] TLR 238 HC QBD Insanity not a defence to strict liability offence (here driving with excess alcohol).

Director of Public Prosecutions v Harris [1995] 1 Cr App R 170; [1995] Crim LR 73; (1994) 158 JP 896; [1995] RTR 100; [1994] TLR 151 HC QBD Emergency vehicle driver could not plead necessity to charge of driving without due care and atttention.

Director of Public Prosecutions v Hastings (1994) 158 JP 118; [1993] RTR 205 HC QBD Not reckless driving for passenger to grab steering wheel so as to swerve towards friend (whom inadvertently hit).

Director of Public Prosecutions v Hawkins [1996] RTR 160 CA Not offence to use emergency vehicle for alternative purpose when blue light not lit.

Director of Public Prosecutions v Hill (1992) 156 JP 197; [1991] RTR 351 HC QBD Can challenge presumption that breath-alcohol concentration not lower than that shown by breathalyser but was inadequate evidence here that latter not functioning properly.

Director of Public Prosecutions v Hill-Brookes [1996] RTR 279 HC QBD Must tell person consequences of providing/not providing optional blood/urine specimen so that may properly exercise option.

Director of Public Prosecutions v Holtham [1990] Crim LR 600; (1990) 154 JP 647; [1991] RTR 5; [1990] TLR 132 HC QBD On what constitutes a 'motor tractor'.

Director of Public Prosecutions v Howard (1991) 155 JP 198; [1991] RTR 49 HC QBD User of cement-mixing lorry did not require operator's licence (Transport Act 1968, s 60(1)).

Director of Public Prosecutions v Hutchings (1992) 156 JP 702; [1991] RTR 380 HC QBD Duplicate print-outs from breathalyser admissible where original print-outs lost/police officer could have given oral evidence as to original readings by reference to duplicate print-outs.

Director of Public Prosecutions v Johnson [1995] 4 All ER 53; [1994] Crim LR 601; (1994) 158 JP 891; [1994] TLR 144; [1995] RTR 9; [1995] 1 WLR 728 HC QBD Prescribed injection of painkiller containing alcohol was consumption of alcohol for purposes of Road Traffic Act 1988.

Director of Public Prosecutions v Jones [1990] RTR 33 HC QBD Person using car to escape attack by another could not succeed on defence of necessity insofar as entire journey was concerned.

Director of Public Prosecutions v Khan [1997] RTR 82 HC QBD Acquittal of inconsiderate driving not autrefois acquit to charge of dangerous driving.

Director of Public Prosecutions v Kinnersley [1993] RTR 105; (1993) 137 SJ LB 12; [1992] TLR 651 HC QBD Fear of contracting HIV (though not mentioned at time refused to give breath specimen) could be special reason meriting non-disqualification: police not under duty to explain breathalyser is sterile.

Director of Public Prosecutions v Kinnersley (1993) 14 Cr App R (S) 516 CA Could (as here) find special reasons justifying non-disqualification though did not accept plea of reasonable excuse for non-provision of breath specimen.

Director of Public Prosecutions v Knight [1994] RTR 374 HC QBD Bona fide fear of attack on baby-sitter and baby (plus fact that driving not notably bad/were only slightly over limit) meant had special reasons meriting non-disqualification.

Director of Public Prosecutions v Lowden [1993] RTR 349; [1992] TLR 185 HC QBD Successful defence to excess alcohol charge whereby established it was extra alcohol consumed after finished driving which pushed driver over prescibed limit.

Director of Public Prosecutions v Magill [1989] Crim LR 155; [1988] RTR 337 HC QBD Upon breath reading indicating less than 50mg alcohol in 100ml of breath police officer to specifically inform defendant that may substitute either breath or urine sample instead.

Director of Public Prosecutions v Mansfield [1997] RTR 96 HC QBD On proving identity of person charged with driving while disqualified.

Director of Public Prosecutions v Marshall and Bell (1990) 154 JP 508; [1990] RTR 384 HC QBD Construction of Road Vehicles (Construction and Use) Regulations 1986, regs 80(1) and 80(2).

Director of Public Prosecutions v McGladrigan [1991] Crim LR 851; (1991) 155 JP 785; [1991] RTR 297; [1991] TLR 216 HC QBD On effect of unlawful arrest on admissibility of breath test evidence.

Director of Public Prosecutions v McKeown; Director of Public Prosecutions v Jones [1997] 1 All ER 737; [1997] 2 Cr App R 155; [1997] 2 Cr App R (S) 289; [1997] Crim LR 522; (1997) 161 JP 356; (1997) 147 NLJ 289; [1997] RTR 162; [1997] TLR 88; [1997] 1 WLR 295 HL Malfunction of clock on otherwise properly functioning breathalyser did not render print-out consequent upon breath test inadmissible.

Director of Public Prosecutions v O'Connor (Michael Dennis) (1992) 13 Cr App R (S) 188 HC QBD On non-disqualification for drink driving offences arising from 'spiking' of drinks unbeknownst to driver.

Docherty v South Tyneside Borough [1982] TLR 358 CA Council decision to increase number of Hackney carriage licences was valid.

Doctorine v Watt [1970] Crim LR 648; [1970] RTR 305 HC QBD Person given adequate portion of specimen need not also be instructed how to store it properly.

Dolan, R v [1969] 3 All ER 683; (1969) 53 Cr App R 556; [1970] Crim LR 43; (1969) 133 JP 696; (1969) 119 NLJ 996t; [1970] RTR 43; (1969) 113 SJ 818; [1969] 1 WLR 1479 CA Issue for jury if person alleges was not warned what failure to comply with specimen request entailed.

Dolbey, Kelly v [1984] RTR 67 HC QBD Arrest of driver after failed to inflate bag but crystals afforded negative reading was valid (though crystals later found at station to give positive reading): appellant properly convicted inter alia of failure to give specimen.

Donahue v Director of Public Prosecutions [1993] RTR 156 HC QBD Disqualification valid despite finding of special reasons which would justify non-disqualification.

Donegani v Ward [1969] 3 All ER 636; [1969] Crim LR 493t; (1969) 133 JP 693; (1969) 119 NLJ 649t; (1969) 113 SJ 588; [1969] 1 WLR 1502 HC QBD Issue of fact for justices whether test in a place 'nearby'.

Donnelly, R v [1975] 1 All ER 785; (1975) 60 Cr App R 250; [1975] Crim LR 178T; (1975) 139 JP 293; (1975) 125 NLJ 111; [1975] RTR 243; (1975) 119 SJ 138; [1975] 1 WLR 390 CA Re-taking of driving test unjustified unless solid ground for believing incompetent driver.

Dooley, R v [1964] 1 All ER 178; [1964] Crim LR 315; (1964) 128 JP 119; (1964) 114 LJ 106; (1964) 108 SJ 384; [1964] 1 WLR 648 CrCt Analyst's certificate admissible/inadmissible where sample requested by doctor/police officer and portion offered/not offered to accused.

Dormer, Police v [1955] Crim LR 252 Magistrates Electrician carrying testing tools and equipment in his van was not carrying 'goods'.

Dowker, Wurzal v [1953] 2 All ER 88; (1953) 117 JP 336; (1953) 103 LJ 349; [1954] 1 QB 52; (1953) 97 SJ 390; [1953] 2 WLR 1196 HC QBD 'Special occasion' means occasion special to place where made not to persons on trip.

Downes, R v [1991] Crim LR 715; [1991] RTR 395 CA On admissibility of expert evidence on excess alcohol charge.

Downey, R v [1970] RTR 257; [1970] Crim LR 287 CA May refuse breath test where believe no circumstances arose meriting test but do so at own risk.

Dowswell, Walker v [1977] RTR 215 HC QBD Was dangerous driving where failed to comply with traffic sign.

Doyle v Leroux [1981] Crim LR 631; [1981] RTR 438 HC QBD On when abuse of process merits halting of prosecution; person's destroying own specimen after informed that would not be prosecuted (and then was) did not have special reason meriting non-disqualification.

Doyle, Director of Public Prosecutions v [1993] RTR 369; [1992] TLR 651 HC QBD Could not find special reasons meriting non-disqualification where offender had wilfully elected to drink and drive.

Drake v Director of Public Prosecutions [1994] Crim LR 855; (1994) 158 JP 828; [1994] RTR 411; [1994] TLR 192 HC QBD Not guilty of being in charge of car with excess alcohol where car had been wheel-clamped.

Drew v Dingle [1933] All ER 518; (1934) 98 JP 1; [1934] 1 KB 187; (1933) 76 LJ 304; (1933) 77 SJ 799; (1933-34) L TLR 101; (1933) WN (I) 255; (1934-39) XXX Cox CC 53; (1934) 103 LJCL 97; (1934) 150 LTR 219 HC KBD Payment of fares meant vehicle unlawfully used as express carriage (even though passengers each carrying produce).

Driscoll, Groome v [1969] 3 All ER 1638; [1970] Crim LR 47t; (1970) 134 JP 83; [1970] RTR 105; (1969) 113 SJ 905 HC QBD On notices of prosecution.

Driver (Jabez), R v (1925-26) 19 Cr App R 86 CCA Four months in second division for bodily harm by wanton driving of trap conviction of middle-aged farmer of reasonably good character.

Drury, Director of Public Prosecutions v (1989) 153 JP 417; [1989] RTR 165 HC QBD If person unaware of car accident becomes aware of same within twenty-four hours is under duty to report it to police.

Edge v Director of Public Prosecutions [1993] RTR 146 HC QBD Conviction quashed where motorist not asked whether any reason why doctor could/should not take specimen.

Edkins v Knowles [1973] 2 All ER 503; (1973) 57 Cr App R 751; [1973] Crim LR 446; (1973) 137 JP 550; [1973] RTR 257; (1973) 123 NLJ 469t; [1973] QB 748; (1973) 117 SJ 395; [1973] 2 WLR 977 HC QBD Must form opinion that person driving in excess of limit when person is driving, not after.

Edmonds, Hosein v [1970] RTR 51; (1969) 113 SJ 759 HC QBD Request to speak with solicitor at same moment that refused specimen not special reason justifying non-disqualification.

Edmonton Justices, ex parte Brooks, R v [1960] 2 All ER 475; [1960] Crim LR 564; (1960) 124 JP 409; (1960) 110 LJ 462; (1960) 104 SJ 547 HC QBD Need not contend as preliminary point that no notice of intended prosecution: can wait until later.

Edward Ash, Limited, Sparks v [1943] KB 223; [1943] 112 LJ 289; (1943) 168 LTR 118; (1943) WN (I) 21 CA Despite black-out were still required to approach pedestrian crossing at speed that could safely stop if pedestrian on it.

Edward Ash, Ltd, Sparks v [1942] 2 All ER 214; (1942) 106 JP 239; [1942] 111 LJ 587; (1942) 167 LTR 64; (1941-42) LVIII TLR 324; (1942) WN (I) 152 HC KBD Despite black-out were still required to approach pedestrian crossing at speed that could safely stop if pedestrian on it but pedestrian could be contributorily negligent.

Edwards v Davies [1982] RTR 279 HC QBD Requirement of specimen valid where properly requested same of hospital patient though received it after patient discharged self.

Edwards v Griffiths [1953] 2 All ER 874 HC QBD '[D]isqualified' means court disqualification not mere refusal of licence.

Edwards v Rigby [1980] RTR 353 HC QBD Bus driver who had girl take actions which resulted in her being dragged by bus did not fail to take all reasonable steps to ensure passenger safety.

Edwards v Wood [1981] Crim LR 414 HC QBD Valid finding that there had not been refusal to provide specimen where specimen not provided but did not refuse to provide it.

Edwards, Hawkins v (1899-1901) XIX Cox CC 692; [1901] 70 LJK/QB 597; (1901) LXXXIV LTR 532; (1900-01) 45 SJ 447; (1901) WN (I) 88; (1900-01) 49 WR 487 HC KBD Any carriage occasionally used to stand/ply for hire is a hackney carriage under Town Police Clauses Act 1847, s 38.

Edwards, Rogerson v (1951) 95 SJ 172; (1951) WN (I) 101 HC KBD Inappropriate for justices to dismiss informations outright, without even hearing them just because notice of intended prosecution mis-addressed.

Eldridge v British Airports Authority [1970] 2 All ER 92; [1970] Crim LR 284; (1970) 134 JP 414; [1970] 2 QB 387; [1970] RTR 270; (1970) 114 SJ 247; [1970] 2 WLR 968 HC QBD Taxi can 'stand' in such a way that is not obliged to take passengers.

Elieson v Parker (1917) 81 JP 265; (1917-18) 117 LTR 276; (1916-17) 61 SJ 559; (1916-17) XXXIII TLR 380 HC KBD Electrically propelled one-quarter horse-power bath chair capable of travel at about 2mph is a motor car under Motor Car Act 1903.

Elkins v Cartlidge [1947] 1 All ER 829; (1947) 177 LTR 519; (1947) 91 SJ 573 HC KBD Parking enclosure a public place.

Elliot, R v [1976] RTR 308 CA Jury could prefer conflicting evidence of one expert to that of another so long as believed evidence correct beyond reasonable doubt.

Elliott v Grey [1960] Crim LR 63; (1960) 124 JP 58; (1959) 109 LJ 720; [1960] 1 QB 367; (1959) 103 SJ 921; [1959] 3 WLR 956 HC QBD Guilty of use of uninsured motor vehicle where car could be moved though not driven.

Elliott v Loake [1983] Crim LR 36 HC QBD Proper inference that owner of car was driver of car when offence committed.

Ellis (John T), Ltd v Hinds [1947] 1 All ER 337; [1947] KB 475 HC KBD Motor vehicle deemed insured though driven by unlicensed driver.

Ellis v Smith [1962] 3 All ER 954; [1963] Crim LR 128; (1963) 127 JP 51; (1963) 113 LJ 26; (1962) 106 SJ 1069; [1962] 1 WLR 1486 HC QBD Until relief (bus) driver assumes charge last (bus) driver is person in charge of vehicle.

Ellis, Brooks v [1972] 2 All ER 1204; [1972] Crim LR 439; (1972) 136 JP 627; [1972] RTR 361; (1972) 116 SJ 509 HC QBD Request for specimen must be within immediate time of driving.

Ellis, McKoen v [1987] Crim LR 54; (1987) 151 JP 60; [1987] RTR 26 HC QBD On what constitutes 'driving' (here on motorcycle).

Ellis, Yorkshire (Woollen District) Electric Tramways Limited v (1901-07) XX Cox CC 795; (1904-05) 53 WR 303 HC KBD Light railway carriage not an omnibus nor a hackney carriage.

Ellison, Dyson v [1975] 1 All ER 276; (1975) 60 Cr App R 191; [1975] Crim LR 48; (1975) 139 JP 191; (1974) 124 NLJ 1132t; [1975] RTR 205; (1975) 119 SJ 66; [1975] 1 WLR 150 HC QBD Court to consider previous endorsements in deciding if disqualification necessary; appeal against obligatory endorsement is by way of re-hearing.

Elstob, Director of Public Prosecutions v [1992] Crim LR 518; (1993) 157 JP 229; [1992] RTR 45; [1991] TLR 526 HC QBD Two-minute delay in obtaining and dividing specimen permissible; division did not have to take place before defendant.

Elwes v Hopkins (1907-09) XXI Cox CC 133; (1906) 70 JP 262; [1906] 2 KB 1; [1906] 75 LJKB 450; (1906) XCIV LTR 547 HC KBD Evidence of customary traffic on highway admissible in prosecution for excessive speed 'in all the circumstances'. (Motor Car Act 1903, s 1).

Elwood-Wade (Roger David), R v (1990-91) 12 Cr App R (S) 51 CA £750 fine/twelve month disqualification for reckless driving.

Ely v Godfrey (1921-25) XXVII Cox CC 191; (1922) 86 JP 82; (1922) 126 LTR 664 HC KBD Conviction for taking greater fare than permitted quashed as relevant by-law deemed inapplicable.

Ely v Marle [1977] Crim LR 294; [1977] RTR 412 HC QBD Breathalyser must be made available to suspect in reasonable time but attempt to walk way from scene after ten minutes did render accused guilty of failure to provide specimen.

Emerson, Taylor v [1962] Crim LR 638t; (1962) 106 SJ 552 HC QBD Expired licence was a licence so use of same was improper use of licence.

Emery, R v [1985] RTR 415 CA Community service appropriate for young man of previously good character guilty of taking motor cycle without authority/gravely reckless driving.

Endean v Evens [1973] Crim LR 448 HC QBD Person did not still have to be driving when breath testing equipment being assembled for testing to be valid.

English, Jones v [1951] 2 All ER 853; (1951) 115 JP 609; (1951) 101 LJ 625; (1951) 95 SJ 712; [1951] 2 TLR 973; (1951) WN (I) 552 HC KBD Must be/hear evidence to back special reasons justifying non-disqualification.

Ennion, Morgan v (1920) 84 JP 205; (1920) 123 LTR 399; (1920) WN (I) 296 HC KBD Valid conviction for not having communication cord from waggons to locomotive (travelling on road).

Enston, Director of Public Prosecutions v [1996] RTR 324; [1995] TLR 89 HC QBD Woman's threat to accuse man who had been drinking of raping her unless he drove her to cash-point was special reason justifying non-disqualification.

Entwhistle, Swaits v (1929) 93 JP 232; [1929] 2 KB 171; (1929) 67 LJ 380; (1929) 98 LJCL 648; (1930) 142 LTR 22; (1929) 73 SJ 366; (1929) WN (I) 143 HC KBD Road Vehicle (Registration and Licensing) Amendment Regulations 1928, r 2 not ultra vires act of Minister of Transport.

Epping Justices, ex parte Quy, R v [1993] Crim LR 970 HC QBD Fear of needles could be medical reason for refusing blood sample.

Erskine v Hollin [1971] Crim LR 243t; (1971) 121 NLJ 154t; [1971] RTR 199; (1971) 115 SJ 207 HC QBD Improper to arrest person and then wait for breathalyser to be brought to scene.

Esau, Baker v [1972] Crim LR 559; [1973] RTR 49 HC QBD Ambulance trailer and car together constituted single four-wheeled trailer.

Essendon Engineering Co Ltd v Maile [1982] Crim LR 510; [1982] RTR 260 HC QBD Company not liable for employee's false issuing of test certificate as not proven that had fully delegated relevant responsibilites to employee.

Evans, R v [1962] 3 All ER 1086; (1963) 47 Cr App R 62; [1963] Crim LR 112; (1963) 127 JP 49; (1963) 113 LJ 9; [1963] 1 QB 412; (1962) 106 SJ 1013; [1962] 3 WLR 1457 CCA Degree of recklessness/carelessness not relevant in death by dangerous driving charge.

Evens v Lewis [1964] Crim LR 472t HC QBD Publican not insured where hired car and completed insurance proposal form under pretence that was printer from garage that did not rent cars to publicans.

Evens, Endean v [1973] Crim LR 448 HC QBD Person did not still have to be driving when breath testing equipment being assembled for testing to be valid.

Everall v Barnwell [1944] KB 333; (1944) WN (I) 64 HC KBD Motor fuel used for purposes specified in application for motor fuel coupons so no offence.

Everett, Pinner v [1969] 3 All ER 257; (1977) 64 Cr App R 160; [1969] Crim LR 607; (1969) 133 JP 653; [1970] RTR 3; (1969) 113 SJ 674; [1969] 1 WLR 1266 HL On when (unlike here) person may be said to be driving car.

Everett, Pinner v [1969] Crim LR 378t; (1969) 119 NLJ 438 HC QBD On when (unlike here) person may be said to be driving car.

Everitt v Trevorrow [1972] Crim LR 566 HC QBD Was reasonable cause to suspect had alcohol in body where driving not especially bad but was smell of alcohol from driver's breath when police stopped him.

Evesham Motors Ltd, Burgin and Heward, D'Arcy v [1971] RTR 35 Magistrates Sign on private hire vehicle contravened legislative requirement that such signs not indicate vehicle could presently be hired.

Ex parte Beecham (1913-14) XXIII Cox CC 571; [1913] 82 LJKB 905; (1913-14) 109 LTR 442; (1912-13) XXIX TLR 586 HC KBD On prosecution of car owner who refuses to divulge identity of driver who committed offence.

Ex parte Hepworth (1931) 75 SJ 408; (1930-31) XLVII TLR 453 HC KBD Failed appeal against unqualified right of licensing authority to refuse licence.

Ex parte JEB Stone (1909) 73 JP 444; (1908-09) XXV TLR 787 HC KBD Driving car at twenty-three miles per hour through village justified speeding conviction under Motor Car Act 1903, s 1.

Ex parte Newsham [1964] Crim LR 57t HC QBD Leave to seek certiorari order granted to appellant convicted of careless driving where charged with dangerous driving.

Ex parte Symes (1911-13) XXII Cox CC 346; (1911) 75 JP 33; (1910-11) 103 LTR 428; (1910-11) XXVII TLR 21; (1910) WN (I) 219 HC KBD Licence can be indorsed for failure to drive with light.

Eynon (Graham), R v (1988) 10 Cr App R (S) 437 CA One hundred and eighty hour community service order imposed on reckless driver.

F and H Croft (Yeadon), Isaac Swires and Sons, Ltd v (1959) 109 LJ 461 TrTb Applicants declaration as to intended normal user of vehicle cannot be changed without consent of applicant.

F, Chief Constable of Avon and Somerset v (1987) 84 Cr App R 345 HC QBD Scrambler motor bicycle not proved to be motor vehicle.

Fairbank, London County Council v (1911) WN (I) 96 HC KBD Are only guilty of unlicensed keeping of carriage if possess and use it (mere possession not enough).

Fairbanks (John), R v (1986) 83 Cr App R 251; [1986] RTR 309; (1970) 130 SJ 750; [1986] 1 WLR 1202 CA Was material irregularity not to have left alternative verdict of driving without due care and attention where person charged with causing death by reckless driving.

Fairclough (TM), and Sons Limited, Beer v (1934-39) XXX Cox CC 551; (1937) 101 JP 157; (1937) 156 LTR 238; (1937) 81 SJ 180; (1936-37) LIII TLR 345 HC KBD Employer to ensure worker-driver has rest periods, not to police how he spends them.

Fardy, R v [1973] Crim LR 316; [1973] RTR 268 CA On reasonableness of constable's suspicion that driver had alcohol in his body.

Farnborough Urban District Council, ex parte Aldershot District Traction Co, R v (1919) WN (I) 271 HC KBD Improper exercise by local authority of its discretion whether or not to renew omnibus licence.

Fisher and others, Shimmell v [1951] 2 All ER 672; (1951-52) 35 Cr App R 100; (1951) 115 JP 526; (1951) 101 LJ 483; (1951) 95 SJ 625; [1951] 2 TLR 753; (1951) WN (I) 484 HC KBD Need not drive car to commit offence of taking and driving away.

Fisher, Chief Constable of Norfolk v (1992) 156 JP 93 HC QBD Permission that car could be used by insured driver only ought to have been communicated by owner to prospective driver.

Fisher, Phesse v (1915) 79 JP 174; [1915] 1 KB 572; [1915] 84 LJKB 277; (1914) WN (I) 438 HC KBD Overcrowding of tramcar: what was 'inside' and what was 'outside' for purpose of London County Council (Tramways and Improvements) Act 1913, s 27.

Fisher, R v [1992] Crim LR 201; [1993] RTR 140 CA Best that R v Lawrence direction on recklessness be given ipissima verba.

Flatman v Poole; Flatman v Oatey [1937] 1 All ER 495 HC KBD No need to keep records in respect of agricultural lorry transporting mixture of non-/agricultural goods from one farm to another.

Fleming, Broadway Haulage and others v (1965) 115 LJ 108 TrTb 'B' licence granted in continuation of existing one (and not to expire later than) existing licence.

Fletcher, ex parte Ansonia, R v (1907-09) XXI Cox CC 578; (1908) 72 JP 249; (1908) XCVIII LTR 749 HC KBD Omnibus wrongfully standing/plying for hire.

Flewitt v Horvath [1972] Crim LR 103; [1972] RTR 121 HC QBD Case remitted to justices who accepted inadmissible hearsay evidence of 'spiking' of drink with alcohol and then (oddly) imposed £50 fine.

Flint, Young v [1987] RTR 300 HC QBD Defence ought to have been allowed cross-examine prosecution witness called to testify to alterations to breathalyser (which may have rendered it an unapproved device).

Flores v Scott [1984] Crim LR 296; [1984] RTR 363; (1984) 128 SJ 319; [1984] 1 WLR 690 HC QBD Foreign postgraduate was resident in, not temporarily in Great Britain and so required to comply with road traffic licensing laws.

Flower Freight Co, Ltd v Hammond [1962] 3 All ER 950; [1963] Crim LR 55; (1963) 127 JP 42; (1962) 112 LJ 817; [1963] 1 QB 275; (1962) 106 SJ 919; [1962] 3 WLR 1331 HC QBD If vehicle not within certain category if originally so built, does not come within it through later adaptation.

Floyd v Bush [1953] 1 All ER 265; [1953] 1 WLR 242 HC QBD Motorised pedal cycle a 'motor vehicle' requiring licence.

Forbes, Lucking v [1985] Crim LR 793; (1987) 151 JP 479; [1986] RTR 97 HC QBD Magistrates Decision on basis of expert evidence that breathalyser evidence defective not unreasonable and so valid; evidence as to excess alcohol in blood irrelevant where charged with driving with excess alcohol in breath.

Forbes, R v [1971] Crim LR 174; [1970] RTR 491 CA Valid request for blood specimen (though failed to mention urine option).

Ford, Gullen v; Prowse v Clarke [1975] 2 All ER 24; [1975] Crim LR 172; (1975) 139 JP 405; (1975) 125 NLJ 111t; [1975] RTR 302; (1975) 119 SJ 153; [1975] 1 WLR 335 HC QBD Overtaking of person stopped because sees person about to cross crossing is offence.

Ford, Marshall v (1907-09) XXI Cox CC 731; (1908) 72 JP 480; (1908-09) XCIX LTR 796 HC KBD Evidence of police officer on contents of driving licence not before court was admissible.

Ford, R v [1982] RTR 5 CA Three year disqualification merited where overtook another on hill-crest.

Forrow, Bryan v [1950] 1 All ER 294 HC KBD 'Paid driver' included person employed by another in turn reimbursed by owner: insured.

Forster, Gill v [1972] Crim LR 45; [1970] RTR 372 HC QBD On effect on validity of breath test of non-compliance with manufacturer's instructions regarding proper storage of breathalyser.

Fossett, Eccles and Supertents Ltd, Creek v [1986] Crim LR 256 HC QBD Application of operator licence, test certificate and tachograph legislation in circus vehicles context.

Frost, Director of Public Prosecutions v [1989] Crim LR 154; (1989) 153 JP 405; [1989] RTR 11 HC QBD On nature of offences of being in charge of vehicle while unfit to drive (matter which ordinary person could decide)/in charge with excess alcohol (provable by reference to expert evidence).

Frow (John), R v (1995) 16 Cr App R (S) 609 CA Nine months' imprisonment for dangerous driving.

Fruer, Director of Public Prosecutions v; Director of Public Prosecutions v Siba; Director of Public Prosecutions v Ward [1989] RTR 29 HC QBD Where motorway not in ordinary use could be that there were special reasons justifying driving on central reservation.

Fry, Adur District Council v [1997] RTR 257 HC QBD Licensed private hire firm operating in one district who received call from another district to carry out private hire journey in that district not guilty of 'operation' of private hire business in district where unlicensed.

Fuller, Jarvis v [1974] Crim LR 116; [1974] RTR 160 HC QBD Not driving without due care and attention where on dark, drizzly day stuck cyclist wearing dark clothes whom could only see from distance of 6-8 feet.

Funnell v Johnson [1962] Crim LR 488g HC QBD On burden of proof as regards proving that person had been waiting in restricted street.

Furness, R v [1973] Crim LR 759 CA Did not invalidate breath testing request that police officer suspected appellant had breath taken before saw appellant driving.

Fussell, R v [1951] 2 All ER 761; (1951-52) 35 Cr App R 135; (1951) 115 JP 562; (1951) 101 LJ 582 CCA Could try attempted (taking and driving away) indictable offence summarily.

G (a minor) v Jarrett [1980] Crim LR 652; [1981] RTR 186 HC QBD Missing pedal rest did not mean moped ceased to be moped.

Gabrielson v Richards [1975] Crim LR 722; [1976] RTR 223 HC QBD Third urine specimen ought not to have been relied upon by prosecution where second had been given within hour of request.

Gage v Jones [1983] RTR 508 HC QBD Reasonable to infer from circumstances that breath test taken by police officer in uniform.

Gaimster v Marlow [1985] 1 All ER 82; (1984) 78 CrAppR 156; [1984] Crim LR 176; (1984) 148 JP 624; [1984] QB 218; [1984] RTR 49; (1983) 127 SJ 842; [1983] TLR 736; [1984] 2 WLR 16 HC QBD Computer printout from breathalyser certificate, and police operator's statement stating results related to accused's specimen were admissible in evidence.

Gannon (Kevin), R v (1988) 87 Cr App R 254; [1988] RTR 49 CA Drunkenness precluded defence on basis of belief had lawful authority for unauthorised taking of conveyance.

Gant, Piridge v [1985] RTR 196 HC QBD On exercising discretion not to disqualify where defendant pleads as special reason that non-alcoholic drink was 'spiked' with alcohol.

Gardner v Director of Public Prosecutions (1989) 89 Cr App R 229; (1989) 153 JP 357; [1989] RTR 384 HC QBD Disqualification discretionary where person in charge of (ie, not driving) vehicle refused to provide breath specimen upon request.

Gardner, Richards v [1974] Crim LR 119; [1976] RTR 476 HC QBD That were driving vehicle dangerously inferred (absent other evidence) from peculiar behaviour of vehicle when brakes applied.

Garforth, R v [1970] Crim LR 704t; (1970) 120 NLJ 945t; (1970) 114 SJ 770 CA On what constitutes 'driving' a vehicle (locking it up to make it safe from theft did not).

Garland, George v [1980] RTR 77 HC QBD Police officer exempted from disabled parking space requirements where acting in course of duty.

Garland, Jacob v [1974] Crim LR 194; [1974] RTR 40; (1973) 117 SJ 915 HC QBD On ingredients of offence of non-provision of information by keeper of vehicle to police as to identity of driver when properly requested to do so.

Garner v Burr and others [1950] 2 All ER 683; (1950) 114 JP 484; [1951] 1 KB 31; (1950) 94 SJ 597; [1950] 66 (2) TLR 768; (1950) WN (I) 445 HC KBD Poultry shed with iron wheels a 'trailer'.

Great Western Rail Co v West Midland Traffic Area Licensing Authority [1935] All ER 396; (1935) 80 LJ 323; [1936] 105 LJCL 37; (1936) 154 LTR 39; (1935) 79 SJ 941; (1935-36) LII TLR 44; (1935) WN (I) 209 HL On application for/user under 'B' licence.

Great Yarmouth Borough Council, ex parte Sawyer, R v [1989] RTR 297 CA Failed action against decision of council de-regulating local taxi provision.

Greatbanks (David Patrick), R v (1986) 8 Cr App R (S) 478 CA Thirty months' youth custody for seventeen year old guilty of causing death by reckless driving in course of police chase of car taken without consent.

Greater London Council, Buckoke and others v (1970) 134 JP 465; (1970) 120 NLJ 337t; [1970] RTR 327; (1970) 114 SJ 269; [1970] 1 WLR 1092 HC ChD London Fire Brigade Order 144/8 not unlawful as did not require though did in essence permit violation of red traffic lights.

Greater London Council, Buckoke and others v [1971] Ch 655; (1971) 135 JP 321; (1971) 121 NLJ 154t; [1971] RTR 131; (1971) 115 SJ 174; [1971] 2 WLR 760 CA London Fire Brigade Order 144/8 not unlawful as did not require though did in essence permit violation of red traffic lights — disciplinary action against persons disputing terms of lawful Order not halted.

Green (Hugh), R v [1980] RTR 415 CA Application for leave to appeal against conviction (on basis of impossible disparity between blood and breath tests) refused.

Green v Dunn [1953] 1 All ER 550 HC QBD Person giving all details to other driver but not telling police within twenty-four hours not guilty of offence.

Green v Harrison [1979] Crim LR 395; (1979) 129 (2) NLJ 734; [1979] RTR 483 HC QBD Puncture did not cause unavoidable delay in completion of journey so were guilty of driving for excess hours.

Green v Turkington; Green v Cater; Craig v Cater [1975] Crim LR 242; [1975] RTR 322; (1975) 119 SJ 356 HC QBD Private hire car with sticker-signs on it reading 'Speedicars Ltd' did suggest car was used for carrying of passengers for hire/reward contrary to London Cab (No 2) Order 1973.

Green, Atkins v [1970] Crim LR 653; [1970] RTR 332 HC QBD Failed prosecution for advertising cabs without making clear were unlicensed (advertisement contained term 'non-hackney').

Green, Bassam v [1981] Crim LR 626; [1981] RTR 362 HC QBD Hackney carriage licensed and used as such could not be divested of that attribute; was offence for hackney carriage driver to seek/take booking fee where ride arranged via radio service.

Green, R v [1970] 1 All ER 408; [1970] Crim LR 289; (1970) 134 JP 208; [1970] RTR 193 Assizes On requiring blood/urine specimens of person in hospital.

Greenaway v Director of Public Prosecutions (1994) 158 JP 27; [1994] RTR 17 HC QBD Pre- and post-test calibration to be given in evidence (may be given orally).

Greenough, Hateley v [1962] Crim LR 329t HC QBD On who is owner of car bought on hire purchase.

Greenwood, Issatt v [1971] RTR 476 HC QBD Not required to stop at school crossing once children are no longer on crossing.

Greenwood, R v [1962] Crim LR 639t HC QBD No duty to stop for lollipop lady unless children present who are crossing/seeking to cross.

Gregson v Director of Public Prosecutions [1993] Crim LR 884 HC QBD Was adequate proof that blood sample taken from defendant was sample that was forensically examined.

Grew v Cubitt (1951) 95 SJ 452; [1951] 2 TLR 305 HC KBD Failed prosecution of driver involved in road traffic accident who did not stop as was inadequate evidence that accused's vehicle had been involved.

Grey Coaches Limited and another, Griffin v (1926-30) XXVIII Cox CC 576; (1929) 93 JP 61; (1929) 67 LJ 32; (1929) 98 LJCL 209; (1929) 140 LTR 194; (1928) 72 SJ 861; (1928-29) XLV TLR 109; (1928) WN (I) 313 HC KBD Motor coach owner plying for hire though no motor coach present.

Grey, Elliott v [1960] Crim LR 63; (1960) 124 JP 58; (1959) 109 LJ 720; [1960] 1 QB 367; (1959) 103 SJ 921; [1959] 3 WLR 956 HC QBD Guilty of use of uninsured motor vehicle where car could be moved though not driven.

Hastings, Director of Public Prosecutions v (1994) 158 JP 118; [1993] RTR 205 HC QBD Not reckless driving for passenger to grab steering wheel so as to swerve towards friend (whom inadvertently hit).

Hateley v Greenough [1962] Crim LR 329t HC QBD On who is owner of car bought on hire purchase.

Haughton, Rooney v [1970] 1 All ER 1001; [1970] Crim LR 236t; (1970) 134 JP 344; [1970] RTR 119; (1970) 114 SJ 93; [1970] 1 WLR 550 HC QBD Person arrested after first breath test cannot require second test at same station.

Havell v Director of Public Prosecutions [1993] Crim LR 621; (1994) 158 JP 680 HC QBD Private members' club car park not a public place.

Hawes v Director of Public Prosecutions [1993] RTR 116; [1992] TLR 217 HC QBD Breath test requirement of person arrested on private property was valid.

Hawkes (James Albert), R v (1930-31) Cr App R 172; (1931) 75 SJ 247 CCA Guilty of drunk driving under Road Traffic Act 1930, s 15(1) if were under influence of drink and were incapable of controlling vehicle.

Hawkins v Director of Public Prosecutions [1988] RTR 380 HC QBD Failed appeal where had sought to prove that had not received analysis evidence by mail at least seven days before hearing.

Hawkins v Ebbutt [1975] Crim LR 465; [1975] RTR 363 HC QBD Prosecution blood analysis inadmissible where portion supplied to defendant had been in inadequate container that resulted in blood being incapable of analysis.

Hawkins v Edwards (1899-1901) XIX Cox CC 692; [1901] 70 LJK/QB 597; (1901) LXXXIV LTR 532; (1900-01) 45 SJ 447; (1901) WN (I) 88; (1900-01) 49 WR 487 HC KBD Any carriage occasionally used to stand/ply for hire is a hackney carriage under Town Police Clauses Act 1847, s 38.

Hawkins v Harold A Russett Ltd [1983] Crim LR 116; (1983) 133 NLJ 154; [1983] RTR 406 HC QBD Detachable container attached to lorry was part of vehicle and so to be included as part of overhang.

Hawkins v Holmes [1974] Crim LR 370; [1974] RTR 436 HC QBD That took real care to ensure brakes were in working order not defence to their being defective.

Hawkins v Phillips and another [1980] Crim LR 184; [1980] RTR 197 HC QBD Filter lane/slip road was 'main carriageway'.

Hawkins v Roots; Hawkins v Smith [1975] Crim LR 521; [1976] RTR 49 HC QBD Non-endorsement not justified just because of slightness of incident prompting careless/inconsiderate driving charge.

Hawkins, Cooper v [1904] 2 KB 164; (1902-03) 47 SJ 691; (1903-04) 52 WR 233 HC KBD Local authority restrictions on locomotive speed under Locomotives Act 1865 could not apply to Crown locomotive driven by Crown servant on Crown business.

Hawkins, Devon County Council v [1967] 1 All ER 235; (1967) 131 JP 161; [1967] 2 QB 26; (1966) 110 SJ 893; [1967] 2 WLR 285 HC QBD Suffer from prescribed disease if taking drugs that suppress its effects.

Hawkins, Director of Public Prosecutions v [1996] RTR 160 CA Not offence to use emergency vehicle for alternative purpose when blue light not lit.

Hawkins, Johnstone v [1959] Crim LR 459t HC QBD Driver on main road who passed junction at speed of 85-90 mph was prima facie guilty of dangerous driving.

Hawkins, Johnstone v [1959] Crim LR 854t HC QBD Failed appeal against acquittal of dangerous driving of driver who drove past junction on main road at speed of 85-90 mph.

Hawkins, Walton v [1973] Crim LR 187; [1973] RTR 366 HC QBD Failed appeal against conviction for being unlawfully positioned on road (appeal on basis that double white lines not of form prescribed in traffic signs regulations).

Hay v Shepherd [1974] Crim LR 114; (1973) 123 NLJ 1137t; [1974] RTR 64 HC QBD Person stopped without reasonable suspicion on part of officer that have been drinking can validly be breathalysed if officer then immediately forms reasonable suspicion.

Hill, Director of Public Prosecutions v (1992) 156 JP 197; [1991] RTR 351 HC QBD Can challenge presumption that breath-alcohol concentration not lower than that shown by breathalyser but was inadequate evidence here that latter not functioning properly.

Hill-Brookes, Director of Public Prosecutions v [1996] RTR 279 HC QBD Must tell person consequences of providing/not providing optional blood/urine specimen so that may properly exercise option.

Hillman, R v [1977] Crim LR 752; [1978] RTR 124 CA On refusal to provide laboratory specimens as evidence that drove while unfit to drive through drink.

Hindhaugh, Lang v [1986] RTR 271 HC QBD Was driving on road to drive on footpath that was highway.

Hindle and Palmer v Noblett (1908) 72 JP 373; (1908-09) XCIX LTR 26 HC KBD Valid conviction for using motor engines on highway that did not (contrary to Locomotives on Highways Act 1896) so far as was practicable consume their own smoke.

Hinds, John T Ellis, Ltd v [1947] 1 All ER 337; [1947] KB 475 HC KBD Motor vehicle deemed insured though driven by unlicensed driver.

Hird, George Cohen 600 Group Ltd v [1970] 2 All ER 650; [1970] Crim LR 473t; (1970) 134 JP 598; (1970) 120 NLJ 550t; [1970] RTR 386; (1970) 114 SJ 552; [1970] 1 WLR 1226 HC QBD Notice to highway and bridge authority needed for use of heavy goods vehicle.

Hirst v Wilson [1969] 3 All ER 1566; [1970] Crim LR 106; [1970] RTR 67; (1969) 113 SJ 906; [1970] 1 WLR 47 HC QBD Arrest for failing to provide proper specimen — though medical reason — lawful: drunk driving conviction sustained.

Hitchins, McCormick v [1988] RTR 182 HC QBD Non-endorsement of licence (of owner of parked car — damaged by another — who refused breath test on basis that was not about to drive) was valid as were special reasons present.

Hives, R v [1991] RTR 27 CA Was reckless driving where struck woman crossing road (who was only a footstep away from the pavement) and thought had hit something like a lamp post.

Hobbs v Clark [1988] RTR 36 HC QBD On need to specifically inform defendant that can substitute blood/urine specimen for that of breath specimen which proves to have less than 50mg of alcohol to 100ml of breath.

Hobbs, Cook v (1910) 45 LJ 710; (1910) WN (I) 219 HC KBD Vehicle had been built for purpose of carrying goods/burden (appellant and family were burden) but as could be used for other purposes was breach of Customs and Inland Revenue Act 1888.

Hobday, Nugent v [1972] Crim LR 569; [1973] RTR 41 HC QBD Part specimen supplied must be capable of analysis using ordinary equipment/skill.

Hockin v Reed and Co (Torquay) Ltd [1962] Crim LR 400; (1962) 112 LJ 337; (1962) 106 SJ 198 HC QBD On what constituted a land implement for purposes of Motor Vehicles (Construction and Use) Regulations 1955.

Hockin v Weston [1972] Crim LR 541t; (1971) 121 NLJ 690t; [1972] RTR 136; (1971) 115 SJ 675 HC QBD Could not be reasonable excuse for failure to provide specimen on part of person pleading guilty of said failure.

Hoddell (P and A), v Parker (1910) 74 JP 315; (1910-11) 103 LTR 2; (1910) WN (I) 146 HC KBD Locomotive drawing corn to market not an 'agricultural' locomotive for purposes of Locomotives Act 1898, s 9.

Hodgkins, Walker v [1983] Crim LR 555; (1983) 147 JP 474; [1984] RTR 34; [1983] TLR 149 HC QBD Justices could take into account normal laboratory procedure when deciding whether to accept evidence regarding blood specimens.

Hodgson v Burn [1966] Crim LR 226; (1965-66) 116 NLJ 501t; (1966) 110 SJ 151 HC QBD Obligation to name driver of car fell on owner personally.

Hoffman v Thomas [1974] 2 All ER 233; [1974] Crim LR 122; (1974) 138 JP 414; (1974) 124 NLJ 36t; [1974] RTR 182; (1974) 118 SJ 186; [1974] 1 WLR 374 HC QBD Police may regulate traffic to protect life and property not for traffic census purposes.

Holt v Dyson [1950] 2 All ER 840; (1950) 100 LJ 595; (1950) 114 JP 558; [1951] 1 KB 364; (1950) 94 SJ 743; [1950] 66 (2) TLR 1009; (1950) WN (I) 498 HC KBD Improper notice of prosecution if send notice to address where know accused not going to receive it.

Holt, Bunting v [1977] RTR 373 HC QBD Point where curving kerb joined adjacent road, not nominal intersection point, was point from which distance of parked car from junction to be measured.

Holt, Crack v (1926-30) XXVIII Cox CC 319; (1927) 91 JP 36; (1927) 136 LTR 511; (1926-27) XLIII TLR 231 HC KBD Licence required when plying for hire wherever did so.

Holt, R v [1968] 3 All ER 802; [1969] Crim LR 27; (1969) 133 JP 49; (1968) 112 SJ 928; [1968] 1 WLR 1942 CA Written statement of Home Office assistant-secretary that breathalyser approved by Home Secretary is admissible evidence thereof.

Holt, R v [1962] Crim LR 565t CCA No special reasons meriting non-disqualification of driver for driving under influence of drink/drugs even though his doctor did not warn him of adverse effects of drinking small amount of alcohol in tandem with taking prescription drug.

Holt, Spicer v [1976] 3 All ER 71; [1977] AC 987; (1976) 63 Cr App R 270; [1977] Crim LR 364; (1976) 140 JP 545; (1976) 126 NLJ 937t; [1976] RTR 389; (1976) 120 SJ 572; [1976] 3 WLR 398 HL Unlawfully arrested person cannot be compelled to give specimen.

Holt, Spicer v [1976] Crim LR 139; (1976) 126 NLJ 44t; [1976] RTR 1 HC QBD Analysis of specimen given by unlawfully arrested person was not admissible in evidence.

Holtham, Director of Public Prosecutions v [1990] Crim LR 600; (1990) 154 JP 647; [1991] RTR 5; [1990] TLR 132 HC QBD On what constitutes a 'motor tractor'.

Hood v Lewis [1976] Crim LR 74; [1976] RTR 99 HC QBD Not defence to speeding charge that did not see speed signs as should have appreciated from distance between street-lights that had entered 30mph area.

Hooper, Garrett v [1973] Crim LR 61; [1973] RTR 1 HC QBD That partner drove motor vehicle in course of partnership work did not mean other party guilty of 'use' of vehicle.

Hooper, Rendell v [1970] 2 All ER 72; [1970] Crim LR 285t; (1970) 134 JP 441; [1970] RTR 252; (1970) 114 SJ 248; [1970] 1 WLR 747 HC QBD Reasonable bona fide departure from manufacturer's instructions in giving breath test does not invalidate it.

Hooper, Secretary of State for the Environment v [1981] RTR 169 HC QBD Crown not Secretary of State owned Government department vehicle so latter improperly convicted of loading offence in respect of vehicle.

Hooper, Thomas v [1986] Crim LR 191; [1986] RTR 1 HC QBD Inoperable car whose wheels did not turn as was towed along road was not in use on road.

Hope v Director of Public Prosecutions [1992] RTR 305 HC QBD Person who provided breath specimens as required and who agreed to provide blood could then decide not to provide blood.

Hopkins, Elwes v (1907-09) XXI Cox CC 133; (1906) 70 JP 262; [1906] 2 KB 1; [1906] 75 LJKB 450; (1906) XCIV LTR 547 HC KBD Evidence of customary traffic on highway admissible in prosecution for excessive speed 'in all the circumstances' (Motor Car Act 1903, s 1).

Hopper v Stansfield (1950) 114 JP 368 HC KBD That car was stopped/battery flat not special reasons justifying non-disqualification following drunk driving conviction.

Hornsby, Hudson v [1972] Crim LR 505; [1973] RTR 4 HC QBD Where defence pleading without notice to prosecution that part specimen supplied could not be analysed prosecution ought to have been allowed call rebutting evidence.

Horrocks v Binns [1986] RTR 202 HC QBD Failure to provide blood specimen charge dismissed where had not properly sought breath specimen.

Horsfield, McDermott Movements Ltd v [1982] Crim LR 693; [1983] RTR 42 HC QBD Owners of insecurely loaded lorry which did not spill any goods wrongfully prosecuted under provision of motor construction regulations concerning spilling of loads.

Horton v Twells [1983] TLR 738 HC QBD Justices should have considered whether police officer's non-compliance with instructions of breathalyser manufacturer was bona fide — if was, then ought to have continued with drunk driving trial.

James v Audigier (1932) 74 LJ 407; (1932-33) XLIX TLR 36; (1932) WN (I) 250 CA Dismissal of appeal over questions that were asked in running-down case over earlier accident in which defendant involved.

James v Audigier (1931-32) XLVIII TLR 600; (1932) WN (I) 181 HC KBD On appropriate questions which may be put to driver in 'running down' case.

James v Cavey [1967] 1 All ER 1048; [1967] Crim LR 245t; (1967) 131 JP 306; [1967] 2 QB 676; (1967) 111 SJ 318; [1967] 2 WLR 1239 HC QBD Inadequate signing by local authority vitiated offence.

James v Davies [1952] 2 All ER 758; (1952) 116 JP 603; (1952) 102 LJ 625; [1953] 1 QB 8; (1952) 96 SJ 729; [1952] 2 TLR 662; (1952) WN (I) 480 HC QBD Unloaded vehicle pulling loaded trailer was a goods vehicle.

James v Evans Motors (County Garages), Ltd [1963] 1 All ER 7; [1963] Crim LR 58t; (1963) 127 JP 104; (1963) 113 LJ 26; (1963) 107 SJ 650; [1963] 1 WLR 685 HC QBD Use by garage owners of lorry to carry rubble not use as manufacturers/repairers/dealers.

James v Morgan [1988] RTR 85 HC QBD Successful appeal by prosecutor against successful 'spiking of drinks' claim by defendant before trial court as special reason justifying non-disqualification.

James, Patterson v [1962] Crim LR 406t HC QBD On deciding whether person to be fined/ imprisoned for driving while disqualified.

Jamison, Ambrose v [1967] Crim LR 114 CrCt Disqualified driver not disqualified where (acting under instructions from police officer) he drove while disqualified.

Janes, R v [1985] Crim LR 684 CA Two years' imprisonment plus five years' disqualification for person guilty of causing death by reckless driving.

Jarrett, G (a minor) v [1980] Crim LR 652; [1981] RTR 186 HC QBD Missing pedal rest did not mean moped ceased to be moped.

Jarvis v Director of Public Prosecutions [1996] RTR 192 HC QBD Not allowing doctor to take blood after breathalyser failed was failure to provide specimen; no judicial notice as to number of breathalysers in particular police station.

Jarvis v Fuller [1974] Crim LR 116; [1974] RTR 160 HC QBD Not driving without due care and attention where on dark, drizzly day stuck cyclist wearing dark clothes whom could only see from distance of 6-8 feet.

Jarvis v Norris [1980] RTR 424 HC QBD Vengefulness not a defence to/could afford evidence of requisite intent for reckless driving.

Jarvis v Williams [1979] RTR 497 HC QBD Absent other information could infer from car overturning after sharp bend that driver was driving without due care and attention.

Jeavons (Phillip), R v (1990) 91 Cr App R 307; [1990] RTR 263 CA Circumstances did not merit jury being given opportunity of convicting accused of careless driving where was charged with reckless driving.

Jefferies, Brooks v [1936] 3 All ER 232; (1936) 80 SJ 856; (1936-37) LIII TLR 34 HC KBD Innocent failure to observe road traffic sign deemed offence.

Jefford, Cannon v (1915) 79 JP 478; [1915] 3 KB 477; [1915] 84 LJKB 1897; (1915-16) 113 LTR 701 HC KBD Conviction for improper brakes on motor car.

Jelf, Scott v [1974] Crim LR 191; (1974) 124 NLJ 148t; [1974] RTR 256 HC QBD Person breaching condition of provisional licence (validly obtained) while subject to disqualification order (in force until passed new driving test) was guilty of driving while disqualified.

Jelliff v Harrington [1951] 1 All ER 384; (1951) 115 JP 100; (1951) 95 SJ 108; [1951] 1 TLR 324; (1951) WN (I) 74 HC KBD Breach of general trade licence by using to tow non-mechanically propelled vehicles.

Jenkins, Anthony v [1972] Crim LR 596; [1971] RTR 19 HC QBD Person who had stopped car and run away from it was not driving when constable formed opinion that had been drinking alcohol.

Jenkins, Arrowsmith v [1963] Crim LR 353; (1963) 127 JP 289; (1963) 113 LJ 350; (1963) 107 SJ 215; [1963] 2 WLR 856 HC QBD Person who does act that leads to obstruction of highway which did not intend is nonetheless guilty of obstruction of highway.

Jenkins, Moss v [1974] Crim LR 715; [1975] RTR 25 HC QBD Absent explanation as to how station sergeant had reasonable suspicion that certain driver had been drinking, breath test request by him invalid.

Jenkins, R v [1978] RTR 104 CA Disqualification for drunk driving reduced from three years to eighteen months.

Jenkins, Watmore v [1962] 2 All ER 868; [1962] Crim LR 562t; (1962) 126 JP 432; (1962) 112 LJ 569; [1962] 2 QB 572; (1962) 106 SJ 492 HC QBD Diabetic driver's behaviour upon reaction to insulin not automatism; diabetic having injected insulin and driven dangerously acquitted of driving under influence of drug.

Jenks, Adams v (1937) 101 JP 393 HC KBD 'Stop, Road Traffic Officer' a proper traffic sign, failure to observe which an offence.

Jenks, Langley Cartage Company, Limited v; Adams v Same [1937] 2 All ER 525; (1934-39) XXX Cox CC 585; [1937] 2 KB 382; [1937] 106 LJCL 559; (1937) 156 LTR 529; (1937) 81 SJ 399; (1936-37) LIII TLR 654; (1937) WN (I) 194 HC KBD Board stating 'Stop, Road Traffic Officer, Bucks CC' a valid road sign.

Jenner, Smith v [1968] Crim LR 99t; (1967) 117 NLJ 1296t; (1968) 112 SJ 52 HC QBD Successful appeal by driving instructor against conviction for aiding and abetting learner driver in unlicensed driving of car.

Jennings v United States Government [1982] 3 All ER 104 (also HC QBD); [1983] 1 AC 624; (1982) 75 Cr App R 367; [1982] Crim LR 748; (1982) 146 JP 396; (1982) 132 NLJ 881; [1983] RTR 1 (also HC QBD); (1982) 126 SJ 659; [1982] TLR 424; [1982] 3 WLR 450 HL Causing death by reckless driving is manslaughter; character of offence for which sought determines if extradition possible.

Jennings, Dobson v (1919) 83 JP 259; [1920] 1 KB 243; [1920] 89 LJCL 281; (1919) WN (I) 272 HC KBD No need to specially license locomotive used/intended for use in special capacity contained in Locomotives Act 1898, s 9(1).

Jennings, R v [1956] 3 All ER 429; (1956) 40 Cr App R 147; [1956] Crim LR 698; (1956) 100 SJ 861; [1956] 1 WLR 1497 CMAC Could convict of dangerous driving though not notified of prosecution fourteen days beforehand.

Jerrum (Maurice Albert), R v (1967) 51 Cr App R 251 CA Eighteen months' imprisonment and eleven year disqualification appropriate for three death by dangerous driving convictions.

Jessop v Clarke (1908) 72 JP 358; (1908-09) XCIX LTR 28; (1907-08) XXIV TLR 672 HC KBD Person stopped and told officer thought he was speeding and would prosecute if another officer agreed was warning of intended prosecution.

Jest, Chief Constable of Avon and Somerset Constabulary v [1986] RTR 372 HC QBD Defendant's thumb print on internal mirror did not prove had taken car without consent for own (uninsured) use.

Jewell, R v [1982] Crim LR 52 CA Six months' imprisonment/three years' disqualification/licence endorsement/requirement that re-take driving test for person guilty of causing death by reckless driving.

Jobson, Burke v [1972] Crim LR 187 HC QBD Police officer in evidence-in-chief must give full account of notification to doctor that intended to seek specimen from patient in doctor's care.

John v Bentley [1961] Crim LR 552; (1961) 105 SJ 406 HC QBD Conviction for being drunk in charge of motor vehicle quashed where was so drunk that could not have driven same.

John v Humphreys (1955) 119 JP 309; (1955) 105 LJ 202; (1955) 99 SJ 222; [1955] 1 WLR 325 HC QBD Prosecution/defendant not required to/must prove that defendant possesses licence.

John, Allen v [1955] Crim LR 383 HC QBD Failed prosecution for uninsured driving.

John, R v [1974] 2 All ER 561; (1974) 59 Cr App R 75; [1974] Crim LR 670t; (1974) 138 JP 492; (1974) 124 NLJ 293t; [1974] RTR 332; (1974) 118 SJ 348; [1974] 1 WLR 624 CA Religious-philospohical beliefs not reasonable excuse for refusing to give specimen — excuse must relate to physical capacity.

John, Strowger v [1974] Crim LR 123; (1974) 124 NLJ 57t; [1974] RTR 124; (1974) 118 SJ 101 HC QBD Non-display of valid vehicle excise licence an absolute offence.

Johns, Saunders v [1965] Crim LR 49t HC QBD Was no case to answer where did not establish that person guilty of speeding offence was in fact the defendant.

Johnson v Finbow (1983) 5 Cr App R (S) 95; [1983] Crim LR 480; (1983) 147 JP 563; [1983] RTR 363; (1983) 127 SJ 411; [1983] TLR 203; [1983] 1 WLR 879 HC QBD Not stopping after accident and not reporting same though at different times were on same occasion for purposes of Transport Act 1981, s 19(1): on appropriate penalty points for same.

Johnson v Whitehouse [1984] RTR 38; [1983] TLR 222 HC QBD Must have reasonable basis for believing — not merely suspecting — that accused drove car involved in accident before can request breath specimen of same.

Johnson, Ashworth v; Charlesworth v Johnson [1959] Crim LR 735 Sessions Successful appeals against disqualifications with requirement that re-take driving test: on when such a form of disqualification merited.

Johnson, Director of Public Prosecutions v [1995] 4 All ER 53; [1994] Crim LR 601; (1994) 158 JP 891; [1994] TLR 144; [1995] RTR 9; [1995] 1 WLR 728 HC QBD Prescribed injection of painkiller containing alcohol was consumption of alcohol for purposes of Road Traffic Act 1988.

Johnson, Dryden v [1961] Crim LR 551t HC QBD Failed drunk driving prosecution sent back to justices with direction to convict as no reasonable bench of magistrates could have failed to convict.

Johnson, Funnell v [1962] Crim LR 488g HC QBD On burden of proof as regards proving that person had been waiting in restricted street.

Johnson, Langton and another v [1956] 3 All ER 474; (1956) 120 JP 561; (1956) 106 LJ 698; (1956) 100 SJ 802; [1956] 1 WLR 1322 HC QBD Handbrake operating on less than half of wheels of vehicle an offence.

Johnson, R v [1960] Crim LR 430 CCC On appropriate test when determining whether driving was dangerous.

Johnson, R v [1969] Crim LR 443t CA On multiple disqualification.

Johnson, R v [1995] Crim LR 250; (1994) 158 JP 788; [1995] RTR 15; [1994] TLR 104 CA Conviction for fraudulent use of vehicle excise licence quashed where vehicle was on private land and was inadequate proof of intention to use vehicle on public road.

Johnson, Sayer v [1970] RTR 286; [1970] Crim LR 589 HC QBD Unless doubt arises prosecutor is not required to prove breathalyser was stored in compliance with manufacturer's instructions.

Johnston (Malcolm Victor), R v (1972) 56 Cr App R 859 CA Unless is statutory provision allowing same disqualification order not to begin from future date.

Johnston v Over (1985) 149 JP 286; [1985] RTR 240; [1984] TLR 450 HC QBD On determining for purposes of Transport Act 1981 penalty points provisions whether offence committed on same occasion as another.

Johnston v Over (1984) 6 Cr App R (S) 420 CA Parking two uninsured vehicles on road (so as to remove parts from one to another) involved committing two offences on same occasion.

Johnston, Biddle v [1965] Crim LR 494; (1965) 109 SJ 395 HC QBD Certificate of insurance that was not a contract of no aid in deciding whether was valid insurance contract for purposes of the Road Traffic Act 1960.

Johnston, R v [1972] Crim LR 647; [1973] RTR 403 CA On ordering consecutive disqualification.

Johnstone v Hawkins [1959] Crim LR 459t HC QBD Driver on main road who passed junction at speed of 85-90 mph was prima facie guilty of dangerous driving.

Johnstone v Hawkins [1959] Crim LR 854t HC QBD Failed appeal against acquittal of dangerous driving of driver who drove past junction on main road at speed of 85-90 mph.

Jolliffe, R v [1970] Crim LR 50; [1972] RTR 188 CA On imprisonment for causing death by dangerous driving: here three months and twenty-five days substituted for nine months.

Jollye v Dale [1960] 2 All ER 369; [1960] Crim LR 565; (1960) 124 JP 333; (1960) 110 LJ 415; [1960] 2 QB 258; (1960) 104 SJ 467; [1960] 2 WLR 1027 HC QBD One and a half hour delay before notice of intended prosecution given was justified in circumstances.

Kellett v Daisy [1977] Crim LR 566; (1977) 127 NLJ 791t; [1977] RTR 396 HC QBD Person using Crown road could be liable under Road Traffic Act 1972, s 3, for careless driving thereon.

Kelliher, Chief Constable of Avon and Somerset v [1986] Crim LR 635; [1987] RTR 305 HC QBD That trained operator could have been got from another police station did not mean that obtaining blood at police station with device but without operator was not impracticable.

Kelly (HF), R v [1972] Crim LR 643; [1972] RTR 447 CA On what constitutes 'failure' to provide/'opportunity' to give breath specimen; on physical disability as reasonable excuse for non-provision of breath specimen.

Kelly (TD and C), Ltd, Wing v [1997] TLR 14 HC QBD On when emergency vehicle owners need not have operator's licence (here found that company ought to have had licence for 'emergency' vehicle).

Kelly v Dolbey [1984] RTR 67 HC QBD Arrest of driver after failed to inflate bag but cystals afforded negative reading was valid (though crystals later found at station to give positive reading): appellant properly convicted inter alia of failure to give specimen.

Kelly v Hogan [1982] Crim LR 507; [1982] RTR 352 HC QBD Person unfit to drive sitting into car and placing wrong key in ignition switch was attempting to drive.

Kelly v Shulman [1988] Crim LR 755; [1989] RTR 84; (1988) 132 SJ 1036; [1988] 1 WLR 1134 HC QBD On what constitutes a day for purposes of EEC driver-worktime legislation (Council Regulation 3820/85/EEC): successful prosecution for breach of UK driver-worktime legislation (Transport Act 1968, s 96(11A)).

Kelly v WRN Contracting, Ltd and another (Burke, third party) [1968] 1 All ER 369; (1968) 112 SJ 465; [1968] 1 WLR 921 Assizes Action for breach of statutory duty if car parked contrary to regulations (but not negligently) helped cause accident.

Kelly, Goodley v [1973] RTR 125; [1973] Crim LR 125 HC QBD Hospital procedure followed irregular course but contained essential steps so was valid.

Kelly, R v [1970] 2 All ER 198; (1970) 54 Cr App R 334; [1970] Crim LR 352t; (1970) 134 JP 482; (1970) 120 NLJ 384t; [1970] RTR 301; (1970) 114 SJ 357; [1970] 1 WLR 1050 CA Person not driving when voluntarily stopped car, got out to make telephone call and was questioned on different matter.

Kemp v Chief Constable of Kent [1987] RTR 66 HC QBD Failure (where was physically and medically possible) to provide blood specimen on basis that already gave same to hospital for other reason not a reasonable excuse to failure to provide blood.

Kemp v Lubbock (1919) 83 JP 270; [1920] 1 KB 253; [1920] 89 LJCL 239; (1920) 122 LTR 220; (1919) WN (I) 269 HC KBD Infant in arms need not be paid for as extra person under London Hackney Carriage Act 1853.

Kennet v Holding and Barnes Ltd and another; TL Harvey Ltd v Hall and another [1986] RTR 334 HC QBD Break-down/recovery vehicles did not breach goods vehicle testing and plating/ excise licensing provisons where each bore two disabled vehicles (one carried, one towed).

Kennett v British Airports Authority [1975] Crim LR 106; [1975] RTR 164; (1975) 119 SJ 137 HC QBD On application of regulations pertaining to car brakes (Motor Vehicles (Construction and Use) Regulations 1973).

Kennison, Musgrove v (1901-07) XX Cox CC 874; (1905) 69 JP 341; (1905) XCII LTR 865; (1904-05) 49 SJ 567; (1904-05) XXI TLR 600 HC KBD Royal Park notice pertaining to maximum driving speeds therein but not laid before Parliament was valid.

Kent (Kenneth Gordon), R v; R v Tanser (Herbert Paul) (1983) 5 Cr App R (S) 16; [1983] Crim LR 406 CA Court when disqualifying driver may take into account that offender makes his living by way of driving.

Kent (Michael Peter), R v (1983) 5 Cr App R (S) 171 CA On disqualification from driving.

Kent County Constabulary, Spittle v [1985] Crim LR 744; [1986] RTR 142 HC QBD Properly convicted of speeding though some of lamps on road (distance between which enabled person to gauge speed limit) were not lit.

Knapp, Lee v [1966] 3 All ER 961; [1967] Crim LR 182; (1967) 131 JP 110; (1965-66) 116 NLJ 1712; [1967] 2 QB 442; (1966) 110 SJ 981; [1967] 2 WLR 6 HC QBD Duty to stop after accident involves stopping for as long as is needed for necessary information to be recorded by authorities.

Knight v Baxter [1971] Crim LR 368; [1971] RTR 270; (1971) 115 SJ 350 HC QBD That had drunk alcohol on empty stomach not special reason justifying non-disqualification.

Knight v Sampson (1938) 82 SJ 524; (1937-38) LIV TLR 974 HC KBD Person at pedestrian crossing who stepped onto road at last moment when car approaching could not recover personal injury damages notwithstanding pedestrian crossing regulations.

Knight, Boldizsar v [1980] Crim LR 653; [1981] RTR 136 HC QBD Person accepting lift in uninsured car which discovered was being driven without owner's consent guilty/not guilty of allowing self to be carried in same/using same without insurance.

Knight, Director of Public Prosecutions v [1994] RTR 374 HC QBD Bona fide fear of attack on baby-sitter and baby (plus fact that driving not notably bad/were only slightly over limit) meant had special reasons meriting non-disqualification.

Knightley, R v [1971] 2 All ER 1041; (1971) 55 Cr App R 390; [1971] Crim LR 426t; (1971) 121 NLJ 409t; [1971] RTR 409; (1971) 115 SJ 448; [1971] 1 WLR 1073 CA Must tell medic of intention to give statutory warning when notifying of intent to seek specimen.

Knights, Thomson v [1947] 1 All ER 112; (1947) 111 JP 43; [1947] KB 336; [1947] 116 LJR 445; (1947) 176 LTR 367; (1947) 91 SJ 68; (1947) LXIII TLR 38; (1947) WN (I) 37 HC KBD Conviction for driving while 'under the influence of drink or a drug' not void for vagueness.

Knipe, Pugh v [1972] Crim LR 247; [1972] RTR 286 HC QBD Was not driving in public place to drive on land which belonged to private club/public did not use.

Knowler v Rennison [1947] 1 All ER 302; [1947] KB 488 HC KBD Whether special reasons justifying non-disqualification a question of law; hardship/that would not re-offend not a special reason.

Knowles (Christopher), R v (1993) 14 Cr App R (S) 224 CA Eighteen months' imprisonment for serious case of reckless driving (at high speed).

Knowles (JM), Ltd v Rand [1962] 2 All ER 926; [1962] Crim LR 561t; (1962) 126 JP 442; (1962) 112 LJ 650; (1962) 106 SJ 513 HC QBD Van carrying hatching eggs for business purposes a farmer's goods/not a carriers' vehicle.

Knowles Transport Ltd v Russell [1974] Crim LR 717; [1975] RTR 87 HC QBD No grounds on which to infer that company knew/could have known of excess hours worked by lorry driver.

Knowles, Edkins v [1973] 2 All ER 503; (1973) 57 Cr App R 751; [1973] Crim LR 446; (1973) 137 JP 550; [1973] RTR 257; (1973) 123 NLJ 469t; [1973] QB 748; (1973) 117 SJ 395; [1973] 2 WLR 977 HC QBD Must form opinion that person driving in excess of limit when person is driving, not after.

Koumourou, Smith v [1979] Crim LR 116; [1979] RTR 355 HC QBD Was obtaining pecuniary advantage by deception where driver without excise licence displayed undated police receipt for earlier expired licence to avoid non-payment of duty being discovered and so avoid paying excise duty.

Kozimor v Adey and another [1962] Crim LR 564; (1962) 106 SJ 431 HC QBD On duty of driver on approaching pedestrian crossing.

Krebs, R v [1977] RTR 406 CA Person unwittingly drinking alcohol stronger than that which had led to believe by another was consuming had special reason meriting non-disqualification.

Kreft v Rawcliffe [1984] TLR 306 HC QBD On determining whether road a road to which public had access for purposes of the Road Traffic Act 1972, s 196(1).

Kwame (Emmanuele), R v (1975) 60 Cr App R 65; [1974] Crim LR 676; [1975] RTR 106 CA Condition of bail that party not to drive before trial was valid and was not special reason later justifying non-disqualification.

LA and A Pinch Ltd, Thurrock District Council v [1974] Crim LR 425; [1974] RTR 269 HC QBD On pleading defence that overweight vehicle had been en route to weighbridge for weighing.

Lackenby v Browns of Wem Ltd [1980] RTR 363 HC QBD Employer not required to issue unused driver's record book to employee drivers.

Lafferty v Director of Public Prosecutions [1995] Crim LR 429 HC QBD Roadside breath test evidence validly admitted (though two police station breath samples unchallenged) as was pertinent to issue raised by driver of how much he had drunk.

Lake (Brian Edward) and others, R v (1986) 8 Cr App R (S) 69 CA Counsel to be told by sentencer that has disqualification in mind so that counsel may make submissions on that point.

Lake and others, Sales v [1922] All ER 689; (1921-25) XXVII Cox CC 170; (1922) 86 JP 80; [1922] 1 KB 553; [1922] 91 LJCL 563; (1922) 126 LTR 636; (1921-22) 66 SJ 453; (1922) WN (I) 66 HC KBD Tour bus picking up clients at pre-arranged points not 'plying for hire' contrary to statute.

Lakin, Bulman v [1981] RTR 1 HC QBD Absent explanatory reason ten-hour delay in reporting accident was not reporting of same as soon as reasonably practicable.

Lamb (Charles Roland), R v (1990) 91 Cr App R 181; [1990] RTR 284; [1990] TLR 94 CA On what constitutes reckless driving; on adequacy of R v Lawrence direction.

Lamb's, Ltd, Badham v [1945] 2 All ER 295; [1946] KB 45; [1946] 115 LJ 180; (1945) 173 LTR 139; (1944-45) LXI TLR 569 HC KBD No action for damages for sale of car with defective brakes: is punishable as offence.

Lambert (Attorney General's Reference (No 15 of 1990)), R v (1991) 92 Cr App R 194; [1991] Crim LR 312; [1991] RTR 195 CA Two years' imprisonment/four year disqualification for speeding drunk driver whose recklessness resulted in death of another.

Lambert v Roberts [1981] 2 All ER 15; (1981) 72 Cr App R 223; [1981] Crim LR 256; (1981) 145 JP 256; (1981) 131 NLJ 448; [1981] RTR 113 HC QBD Where police officer's licence to enter withdrawn has no legal entitlement to administer breath test.

Lambert, Greyhound Motors Limited v (1926-30) XXVIII Cox CC 469; (1927) 91 JP 198; [1928] 1 KB 322; (1927) 64 LJ 358; (1928) 97 LJCL 122; (1928) 138 LTR 269; (1927) 71 SJ 881; (1927) WN (I) 271 HC KBD Conviction possible for plying for hire without notice where possible that passengers collected not all pre-paid passengers.

Lambert, R v [1962] Crim LR 645 CCA Two years' imprisonment for wanton driving, five years' consecutive for receiving and a further three years' concurrent for more receiving.

Lambeth London Borough Council v Saunders Transport Ltd [1974] Crim LR 311 HC QBD Person who hired out skip not liable for its being unlighted at night, having agreed with person hiring that responsibility for lighting skip would fall on the latter.

Lambeth Metropolitan Magistrate, ex parte Everett, R v [1967] 3 All ER 648; (1967) 51 Cr App R 425; [1967] Crim LR 543t; (1968) 132 JP 6; (1967) 117 NLJ 730t; [1968] 1 QB 446; (1967) 111 SJ 545; [1967] 3 WLR 1027 HC QBD New disqualification begins when earlier disqualification ends.

Lambie v Woodage [1972] 2 All ER 462; [1972] Crim LR 442t; (1972) 136 JP 554; (1972) 122 NLJ 426t; [1972] RTR 396; (1972) 116 SJ 376; [1972] 1 WLR 754 HL In considering reasonable grounds for not disqualifying judges may consider anything reasonable.

Lambie, Woodage v [1971] 3 All ER 674; (1971) 135 JP 595; (1971) 121 NLJ 665t; [1972] RTR 36; [1972] Crim LR 536t; (1971) 115 SJ 588; [1971] 1 WLR 1781 HC QBD That earlier offences not serious not mitigating circumstance when sentencing for third traffic offence.

Lambourne, Du Cros v (1907-09) XXI Cox CC 311; (1906) 70 JP 525; [1907] 1 KB 40; (1906) 41 LJ 701; [1907] 76 LJKB 50; (1906-07) XCV LTR 782; (1906-07) XXIII TLR 3 HC KBD Owner convicted as principal for excessively fast driving by driver over/with whom had control/ whom was in car.

Lane, Police v [1957] Crim LR 542 Magistrates Harpoon gun a gun for purposes of the Gun Licence Act 1870.

Lane, R v [1986] Crim LR 574 CA Judge minded to make disqualification (in light of car being used in commission of crime) ought to forewarn counsel so that latter might make representations on matter.

Lang v Hindhaugh [1986] RTR 271 HC QBD Was driving on road to drive on footpath that was highway.

Lewis, R v [1965] Crim LR 50t; (1965) 49 Cr App R 26; (1964) 108 SJ 863 CCA Doctor, unlike police constable, not required to offer defendant a portion of the urine specimen which the doctor requests from the defendant.

Lex Vehicle Leasing Ltd, West Yorkshire Trading Standards Service v [1996] RTR 70 HC QBD On what constitutes 'use' of vehicle on road.

Licensing Authority for Goods Vehicles for the Metropolitan Traffic Area, ex parte BE Barrett Ltd, R v (1949) 113 JP 202; [1949] 2 KB 17; (1949) 99 LJ 135; [1949] 118 LJR 1522; (1949) LXV TLR 309; (1949) WN (I) 126 HC KBD Refusal of mandamus order requiring licensing authority to allow change of operating centre for holder of public carrier's 'A' licence.

Liddon v Stringer [1967] Crim LR 371t HC QBD Bus driver guilty of driving without due care and attention where acting on signals from conductor reversed into woman and killed her.

Liggett, Chorlton v (1910) 74 JP 458 HC KBD Conviction for not driving horse and cart as close to kerb as possible quashed.

Light v Director of Public Prosecutions [1994] RTR 396 HC QBD Transport manager found to have permitted breaches of tachograph legislation where he had not done everything possible to end same.

Lindsell, Burningham v (1936) 80 SJ 367 HC KBD Vehicle with permanent apparatus a goods vehicle.

Ling, R v (1993) 157 JP 931 CA Sentences for various offences committed on same evening reduced as had been prompted by accused's diabetic attack after non-injection/eating chocolates/drinking alcohol.

Liskerrett Justices, ex parte Child, R v [1972] RTR 141 HC QBD Endorsement quashed as accused when asked if would like to plead guilty by post was not given all facts on which prosecutor could later seek to rely.

Littell, R v [1981] 3 All ER 1; [1981] Crim LR 642; (1981) 145 JP 451; [1981] RTR 449; (1981) 125 SJ 465; [1981] 1 WLR 1146 CA Provision of breath specimen in manner other than indicated a failure to provide breath specimen.

Littman, Gubby v [1976] Crim LR 386t; (1976) 126 NLJ 567t; [1976] RTR 470 HC QBD That drove with new tyres which had not 'broken in' not evidence of driving without due care and attention.

Liverpool City Council, Hough v [1980] Crim LR 443; [1981] RTR 67 HC QBD On what constituted 'reasonable cause' to fail to give authorised officer information pursuant to Liverpool Corporation Act 1972, s 36(1).

Liversidge and Featherstone, Police v [1956] Crim LR 59 Magistrates Two parties guilty of being drunk in charge of same motor vehicle.

Llewellyn, Tidswell v [1965] Crim LR 732t HC QBD ROAD Restoration of conviction for emission of oil from lorry where 'spilling' occurred on private land but dripping continued on public road.

Lloyd v E Lee, Ltd [1951] 1 All ER 589; (1951) 115 JP 189; [1951] 2 KB 121; (1951) 95 SJ 206; [1951] 1 TLR 624; (1951) WN (I) 116 HC KBD Licence prohibiting user of vehicle for carriage of goods for hire/reward applies to all; identity certificate required no matter who is using vehicle.

Lloyd v Morris [1985] Crim LR 742; [1986] RTR 299 HC QBD Unusually wide discrepancy between two breathalyser readings did not per se merit decision by justices that breathalyser defective.

Lloyd v Ross (1913-14) XXIII Cox CC 460; (1913) 48 LJ 229; [1913] 82 LJKB 578; (1913-14) 109 LTR 71; (1912-13) XXIX TLR 400; (1913) WN (I) 108 HC KBD Proper conviction for driving overweight car over bridge.

Lloyd v Singleton [1953] 1 All ER 291; [1953] 1 QB 357; [1953] 2 WLR 278 HC QBD Employee could be properly charged with driving unlicenced vehicle owned by employer.

Lloyd, Marshall v [1956] Crim LR 483 Magistrates Successful prosecution of individual for being drunk in the highway contrary to the Licensing Act 1872, s 12.

Loake, Elliott v [1983] Crim LR 36 HC QBD Proper inference that owner of car was driver of car when offence committed.

London Passenger Transport Board, Chisholm v [1938] 4 All ER 850; [1939] 1 KB 426; (1939) 87 LJ 27; [1939] 108 LJCL 239; (1939) 160 LTR 79; (1938) 82 SJ 1050; (1938-39) LV TLR 284; (1939) WN (I) 15 CA Driver not negligent in striking pedestrian already on crossing: case decided by reference to common law principles.

London Passenger Transport Board, Kayser v [1950] 1 All ER 231; (1950) 114 JP 122; [1950] 1 All ER 231 HC KBD Driver stopped for pedestrians may proceed once considers it safe but must be able to stop again should pedestrians act dangerously/negligently.

London Passenger Transport Board, Upson v [1947] KB 930; (1947) 97 LJ 431; [1947] 116 LJR 1382; (1947) 177 LTR 475; (1947) LXIII TLR 452; (1947) WN (I) 252 CA Driver's duties at crossing extant even though light green.

London Passenger Transport Board, Upson v (1947) 176 LTR 356 HC KBD Driver's duties at crossing extant even though light green.

London Transport Executive, Smith v [1951] 1 All ER 667; (1951) 115 JP 213; [1951] 1 TLR 683; (1951) WN (I) 157 HL On rôle of British Transport Commission.

London Transport Executive, Smith v [1948] 2 All ER 306; [1948] Ch 652; [1948] 117 LJR 1483; (1948) LXIV TLR 426; (1948) WN (I) 292 HC ChD On rôle of British Transport Commission.

London Transport Executive, Smith v [1949] Ch 685; (1949) 99 LJ 401; (1949) LXV TLR 538; (1949) WN (I) 315 CA On rôle of British Transport Commission.

Long, Bosley v [1970] 3 All ER 286; [1970] Crim LR 591; (1970) 134 JP 652; (1970) 120 NLJ 968; [1970] RTR 432; (1970) 114 SJ 571; [1970] 1 WLR 1410 HC QBD Specimen requested at hospital to be provided at hospital; medic may object anytime after notified of proposal to take specimen.

Lord v Wolley [1956] Crim LR 493 Magistrates £2 fine for failure to observe 'Yield' sign placed on road on trial basis.

Losexis Ltd v Clarke [1984] RTR 174; [1983] TLR 323 HC QBD Successful prosecution for sale of crash helmet that did not comply with British Standard specifications: no defence that helmet for off-road use only.

Loukes, R v [1996] 1 Cr App R 444; [1996] Crim LR 341; [1996] RTR 164; [1995] TLR 706 CA Could not be convicted of procuring the causing of death by dangerous driving where person charged with causing same was acquitted.

Lovell v Archer; Lovell v Ducket [1971] Crim LR 240; (1971) 121 NLJ 128t; [1971] RTR 237; (1971) 115 SJ 157 HC QBD Provisional licence to be endorsed where holder of same does not comply with conditions of licence.

Lovell, Walker v [1975] 3 All ER 107; (1978) 67 Cr App R 249; [1975] Crim LR 720; (1975) 139 JP 708; (1975) 125 NLJ 820t; [1975] RTR 377; (1975) 119 SJ 544; [1975] 1 WLR 1141 HL Where enough breath to provide specimen arrest for failure to provide specimen unlawful and this being so there can be no conviction for drunk driving.

Lovell, Walker v [1975] Crim LR 102; (1975) 125 NLJ 43; [1975] RTR 61; (1975) 119 SJ 258 HC QBD Where enough breath to provide specimen arrest for failure to provide specimen unlawful and this being so there can be no conviction for drunk driving.

Lovett v Payne [1979] Crim LR 729; (1979) 143 JP 756; [1980] RTR 103 HC QBD On what was ment by 'nearest' weighbridge in Road Traffic Act 1972, s 40(6).

Lowden, Director of Public Prosecutions v [1993] RTR 349; [1992] TLR 185 HC QBD Successful defence to excess alcohol charge whereby established it was extra alcohol consumed after finished driving which pushed driver over prescibed limit.

Lowe and another v Stone [1948] 2 All ER 1076; (1949) 113 JP 59; [1949] 118 LJR 797; (1948) WN (I) 487 HC KBD Detachable boards fitted to side of lorry meant was being used in altered condition.

Lowe v Lester [1986] Crim LR 339; [1987] RTR 30 HC QBD Person required by local authority to give identity of car-driver must do so immediately/within reasonable period.

Lowe, Watson v (1950) 114 JP 85; (1950) 100 LJ 20; (1950) 94 SJ 15; [1950] 66 (1) TLR 169; (1950) WN (I) 26 HC KBD Person in driving seat of stopped car negligently causing injury to passing cyclist by opening door guilty of interference with free passage of person on highway (Highway Act 1835, s 78).

Lowery (P), and Sons Ltd v Wark [1975] RTR 45 HC QBD Necessary to prove mens rea on part of company charged with permitting use on road of vehicle with inadequately secured load.

Lowry (Thomas Gordon), R v [1996] 2 Cr App R (S) 416 CA Eighteen months' imprisonment for causing death by dangerous driving (through overtaking).

Lowton, Oxford v [1978] Crim LR 295; [1978] RTR 237 HC QBD Not necessary to be outside patient's presence/hearing when informing doctor of intention to require specimen of patient.

Lubbock, Kemp v (1919) 83 JP 270; [1920] 1 KB 253; [1920] 89 LJCL 239; (1920) 122 LTR 220; (1919) WN (I) 269 HC KBD Infant in arms need not be paid for as extra person under London Hackney Carriage Act 1853.

Lucking v Forbes [1985] Crim LR 793; (1987) 151 JP 479; [1986] RTR 97 HC QBD Magistrates Decision on basis of expert evidence that breathalyser evidence defective not unreasonable and so valid; evidence as to excess alcohol in blood irrelevant where charged with driving with excess alcohol in breath.

Lucking, Collins v [1983] Crim LR 264; (1983) 147 JP 307; [1983] RTR 312; [1982] TLR 545 HC QBD Drunk driving conviction quashed where rested on accused's being diabetic and was no evidence either way as to whether was/was not diabetic.

Lummis, Urey v [1962] 2 All ER 463; (1962) 126 JP 346; (1962) 112 LJ 473; (1962) 106 SJ 430 HC QBD Visiting force driving permit does not entitle one to supervise provisional driver.

Lund v Thompson [1958] 3 All ER 356; (1959) 43 Cr App R 9; [1958] Crim LR 816t; (1958) 122 JP 489; (1958) 108 LJ 793; [1959] 1 QB 283; (1958) 102 SJ 811 HC QBD Post-notice of prosecution indication by police that would not prosecute did not make notice inoperative.

Lundt-Smith, R v [1964] 3 All ER 225; [1964] Crim LR 543; (1964) 128 JP 534; [1964] 2 QB 167; (1964) 108 SJ 424; [1964] 2 WLR 1063 HC QBD Ambulance driver involved in fatal collision after crashed red light en route to hospital in emergency had special reasons justifying non-disqualification upon conviction of causing death by dangerous driving.

Lunn, Dawson v (1985) 149 JP 491; [1986] RTR 234; [1984] TLR 720 HC QBD Justices ought not to have dismissed drunk driving charge purely on basis of medical journal report produced by defence.

Luongo (Marcello), King (David John) v (1985) 149 JP 84; [1985] RTR 186; [1984] TLR 143 HC QBD On licence endorsement provisions of Transport Act 1981.

Luxton, Scutt v (1950) 94 SJ 33 HC KBD Use of towing lorry as regular carrier's lorry was improper even though it occurred in the course of business.

Lyndon and another, ex parte Moffat, R v (1908) 72 JP 227 HC KBD Causing obstruction by leaving unattended car on highway not offence in connection with driving for which licence can be endorsed.

Lynn (Frederick John), R v (1971) 55 Cr App R 423; [1971] RTR 369 CA Disqualification order approved where person driving while disqualified did not actually know was disqualified.

Lyon v Oxford [1983] RTR 257 HC QBD Conviction quashed as speeding legislation vaguely worded so defendant given benefit of higher of two speeding limits (in excess of which had not driven).

Lyons (Terence Patrick), R v (1971) 55 Cr App R 565 CA On sentencing under-twenty-one year old of previously good character for causing death by dangerous driving.

Lyons v Denscombe [1949] 1 All ER 977; (1949) 99 LJ 259; (1949) 113 JP 305; (1949) WN (I) 257 HC KBD Weekly lump sum payment from fellow employees for conveying home not use as express carriage.

Lyons v May [1948] 2 All ER 1062 HC KBD Permitting vehicle to be used whilst uninsured.

Lythgoe, Anderton v [1985] Crim LR 158; [1985] RTR 395; (1984) 128 SJ 856; [1984] TLR 596; [1985] 1 WLR 222 HC QBD Police could not rely on breath specimen as definitive evidence of accused's guilt where failed to inform accused when arrested him of his right to give blood/urine sample.

MacDonagh, R v [1974] 2 All ER 257; (1974) 59 Cr App R 55; [1974] Crim LR 317; (1974) 138 JP 488; (1974) 124 NLJ 222t; [1974] QB 448; [1974] RTR 372; (1974) 118 SJ 222; [1974] 2 WLR 529 CA 'Driving' must be driving within common understanding.

Mallone (Patrick), R v [1996] 1 Cr App R (S) 221 CA Four (not five) years' imprisonment for causing death by dangerous driving — reduction inter alia because of grave injuries offender suffered.

Mallows v Harris [1979] Crim LR 320; [1979] RTR 404 HC QBD Where defence does not raise matter of reasonable excuse prosecution have discharged burden in respect of same; excuse not pertaining to capacity to provide specimen could not be reasonable excuse.

Manchester City Justices, ex parte McHugh, R v; Manchester City Council (Intervening); R v Manchester City Council, ex parte Reid [1989] RTR 285 HC QBD Council could require new hackney carriage licence holders to make provision for wheelchair passengers whilst not imposing same requirement on old licence holders.

Manchester Justices, ex parte Gaynor, R v [1956] 1 All ER 610; (1956) 100 SJ 210; [1956] 1 WLR 280 HC QBD Justices having ordered disqualification removed from deferred date could hold another hearing to see if immediate removal justified.

Manders (Richard John), R v (1989) 11 Cr App R (S) 442 CA Eighteen months in young offender institution/five years' disqualification for eighteen year old guilty of causing three deaths by dangerous driving (speeding).

Manley v Dabson; Manley v Same [1949] 2 All ER 578; (1949) 113 JP 501; [1950] 1 KB 100; (1949) 99 LJ 455; [1949] 118 LJR 1427; (1949) LXV TLR 491; (1949) WN (I) 376 HC KBD Farmer selling goods to wholesaler/retailer within twenty-five miles of farm not required to maintain record of trip under Goods Vehicles (Keeping of Records) regulations 1935, reg 6(3).

Mann, Hunter v [1974] 2 All ER 414; (1974) 59 Cr App R 37; [1974] Crim LR 260; (1974) 138 JP 473; (1974) 124 NLJ 202t; [1974] QB 767; [1974] RTR 328; [1974] 2 WLR 742 HC QBD Doctor's professional duty superseded by duty to give information identifying driver in accident.

Manners-Astley, R v [1967] 3 All ER 899; (1968) 52 Cr App R 5; [1967] Crim LR 658t; (1968) 132 JP 39; (1967) 111 SJ 853; [1967] 1 WLR 1505 CA Must be intent to defraud for conviction for fraudulent use of licence.

Manning v Hammond [1951] 2 All ER 815; (1951) 115 JP 600; (1951) 101 LJ 611; (1951) WN (I) 525 HC KBD Goods vehicles to which thirty mile maximum speed applied.

Manning, Houghton v (1904-05) 49 SJ 446 HC KBD Was not criminally reckless/negligent driving for driver to drive car along road with toll-keeper who had sought toll hanging onto car until fell off.

Manning, Troughton v (1901-07) XX Cox CC 861; (1905) 69 JP 207; (1905) XCII LTR 855; (1904-05) XXI TLR 408; (1904-05) 53 WR 493 HC KBD Not reckless driving to drive at reasonable speed with person clinging to side of car.

Mansfield, Director of Public Prosecutions v [1997] RTR 96 HC QBD On proving identity of person charged with driving while disqualified.

Marchant (Stephen), R v; R v McCallister (Stephen) (1985) 80 Cr App R 361 CA Was attempt to take conveyance without authority even though did not use conveyance.

Marison, R v [1996] Crim LR 909; (1996) TLR 16/7/96 CA Diabetic who suffered hypoglycaemic attack while driving was legitimately found guilty of causing death by dangerous driving of other driver with whom collided following attack.

Markham v Stacey [1968] 3 All ER 758; [1969] Crim LR 35t; (1969) 133 JP 63; (1968) 118 NLJ 1006t; (1968) 112 SJ 866; [1968] 1 WLR 1881 HC QBD Meaning of 'land implement' in Motor Vehicles (Construction and Use) Regulations 1966, reg 3.

Marks v West Midlands Police [1981] RTR 471 HC QBD Slightness of speeding/driver's concern for unwell elderly passenger were special reasons justifying non-endorsement of licence upon speeding conviction.

Marks, Plancq v (1907-09) XXI Cox CC 157; (1906) 70 JP 216; (1906) XCIV LTR 577; (1905-06) 50 SJ 377; (1905-06) XXII TLR 432 HC KBD Could be convicted of excessive speed on evidence of constable using stop watch.

Marle, Ely v [1977] Crim LR 294; [1977] RTR 412 HC QBD Breathalyser must be made available to suspect in reasonable time but attempt to walk way from scene after ten minutes did render accused guilty of failure to provide specimen.

Marlow, Gaimster v [1985] 1 All ER 82; (1984) 78 CrAppR 156; [1984] Crim LR 176; (1984) 148 JP 624; [1984] QB 218; [1984] RTR 49; (1983) 127 SJ 842; [1983] TLR 736; [1984] 2 WLR 16 HC QBD Computer printout from breathalyser certificate, and police operator's statement stating results related to accused's specimen were admissible in evidence.

Marr, R v [1977] RTR 168 CA Person validly convicted of driving with excess alcohol despite massive discrepancy between prosecution/defence analyses of specimen provided.

Marriott (AT), Ltd v British Transport Commission (1958) 108 LJ 634 TrTb Sucessful appeal against refusal to allow substitution of new diesel articulated vehicle for old one already held under 'A' licence.

Marron (Paul James), R v (1993) 14 Cr App R (S) 615 CA Six months' imprisonment for aggravated vehicle-taking.

Marsh (William), R v [1997] 1 Cr App R 67; [1997] Crim LR 205; [1997] RTR 195; (1996) 140 SJ LB 157 CA Was aggravated vehicle taking where accident arising in connection with car taken without authority was not driver's responsibility.

Marshall (Gordon Kane), R v (1988) 10 Cr App R (S) 246; [1991] RTR 201 CA Two and a half years' imprisonment/seven years' disqualification for off-duty police officer (with drink taken) guilty of causing death by reckless driving.

Marshall (John), R v [1976] RTR 483 CA Five year disqualification appropriate for young man driving too fast so as to impress others.

Marshall and Bell, Director of Public Prosecutions v (1990) 154 JP 508; [1990] RTR 384 HC QBD Construction of Road Vehicles (Construction and Use) Regulations 1986, regs 80(1) and 80(2).

Marshall v Ford (1907-09) XXI Cox CC 731; (1908) 72 JP 480; (1908-09) XCIX LTR 796 HC KBD Evidence of police officer on contents of driving licence not before court was admissible.

Marshall v Lloyd [1956] Crim LR 483 Magistrates Successful prosecution of individual for being drunk in the highway contrary to the Licensing Act 1872, s 12.

Marshall, R v [1954] Crim LR 386 CCA Twenty-seven months' disqualification for person guilty of using motor car in commission of crime: disqualification generally merited in such cases.

Marsham, ex parte Chamberlain, R v (1907-09) XXI Cox CC 510; (1907) 71 JP 445; [1907] 2 KB 638; (1907) 42 LJ 429; [1907] 76 LJKB 1036; (1907-08) XCVII LTR 396; (1906-07) 51 SJ 592; (1906-07) XXIII TLR 629; (1907) WN (I) 163 HC KBD Need not endorse licence upon first conviction where is for speeding through Royal Park.

Marson v Thompson [1955] Crim LR 319t HC QBD HC QBD review of dangerous driving acquittal justified as decision of justices that defendant had not been driving dangerously was unsupportable.

Martin, Absalom v [1973] Crim LR 752; (1973) 123 NLJ 946t HC QBD Billposter's part-parking of van on pavement while putting up bill did not merit conviction for wilful obstruction of highway.

Martin, Hougham v [1964] Crim LR 414; (1964) 108 SJ 138 HC QBD Direction to convict for careless driving where had pleaded that latent defect in mass-produced motor vehicle led to driving for which charged.

Martin, R v [1989] 1 All ER 652; (1989) 88 Cr App R 343; [1989] Crim LR 284; (1989) 153 JP 231; [1989] RTR 63; [1988] 1 WLR 655 CA Defence of necessity arises if objectively viewed person acted reasonably/proportionately to avoid death/serious injury threat.

Martin, Sandy v [1974] Crim LR 258; (1975) 139 JP 241; [1974] RTR 263 HC QBD Inn car park a public place during hours that persons invited to use it but absent contrary evidence not thereafter (so were not drunk and in charge of motor vehicle in public place).

Martin, Tremelling v [1972] Crim LR 596; [1971] RTR 196 HC QBD Was not production of licence at police station to produce same but withdraw it before constable could examine it.

Marx, Bryant v (1931-34) XXIX Cox CC 545; (1932) 96 JP 383; (1932) 147 LTR 499; (1932) 76 SJ 577; (1931-32) XLVIII TLR 624 HC KBD Obstructing road under Motor Vehicles (Construction and Use) Regulations 1931, r 74(1) includes obstructing footpath.

McAleer, Middlemas v [1979] RTR 345 HC QBD Unlicensed use of public service vehicle (coach) as contract carriage.

McAllister (William Joseph), R v (1974) 59 Cr App R 7; [1974] Crim LR 716; [1974] RTR 408 CA On what constitutes refusal to give blood/urine specimen.

McBride, R v [1961] 3 All ER 6; (1961) 45 Cr App R 262; [1961] Crim LR 625; (1961) 125 JP 544; (1961) 111 LJ 518; [1962] 2 QB 167; (1961) 105 SJ 572; [1961] 3 WLR 625 CCA Evidence of drinking admissible in death by dangerous driving prosecution; death by dangerous driving may be coupled with reckless/dangerous driving but not drunk/drugged driving charge.

McCabe (James), R v (1989) 11 Cr App R (S) 154 CA Six months' suspended sentence/£750 fine for false tachograph entries.

McCabe, R v [1995] RTR 197 CA Three years' imprisonment/five year disqualification merited for person of generally good character with learned referees where alcohol levels over prescribed limit and had caused three deaths.

McCann, Cooke v [1973] Crim LR 522; [1974] RTR 131 HC QBD On determining whether person before court is person properly charged: name and address from licence matching those of person in court.

McCaul and another, Swain v [1997] RTR 102; (1996) 140 SJ LB 142; (1996) TLR 10/7/96 HC QBD Commercial skip firm was required to comply with tachograph legislation requirements.

McCausland, R v; R v Quaile [1968] Crim LR 49; (1967) 111 SJ 998 CA On sentencing under twenty-one year olds: eighteen months' imprisonment for shopbreaking with intent and various road traffic offences.

McCluskie (Alexander), R v (1992) 13 Cr App R (S) 334; [1983] Crim LR 273 CA Ten years' (not life) disqualification for repeated road traffic offender guilty of careless driving.

McConnell, Hughes v [1986] 1 All ER 268; [1985] RTR 244 HC QBD Challenge to validity of breathometer reading must establish device in itself defective.

McCormick v Hitchins [1988] RTR 182 HC QBD Non-endorsement of licence (of owner of parked car — damaged by another — who refused breath test on basis that was not about to drive) was valid as were special reasons present.

McCormick, Phipps v [1972] Crim LR 540; (1971) 115 SJ 710 HC QBD Last known address for correspondence (rather than hospital address) was appropriate place to which to send notice of intended prosecution.

McCrone v J and L Rigby (Wigan) Ltd; Same v Same [1951] 2 TLR 911 HC KBD Road dumper a mechanically propelled vehicle on which duty payable as road construction vehicle.

McCrone v Riding [1938] 1 All ER 157; (1934-39) XXX Cox CC 670; (1938) 102 JP 109; (1938) 85 LJ 107; (1938) 158 LTR 253; (1938) 82 SJ 175; (1937-38) LIV TLR 328; (1938) WN (I) 60 HC KBD Same standard of 'due care and attention' expected of all drivers irrespective of experience.

McCrory v Director of Public Prosecutions (1990) 154 JP 520; [1991] RTR 187; [1990] TLR 119 HC QBD On what consttitues 'heavy motor car' for purposes of Motorways Traffic (England and Wales) Regulations 1982.

McDermott Movements Ltd v Horsfield [1982] Crim LR 693; [1983] RTR 42 HC QBD Owners of insecurely loaded lorry which did not spill any goods wrongfully prosecuted under provision of motor construction regulations concerning spilling of loads.

McDermott v Director of Public Prosecutions (1997) 161 JP 244; (1996) TLR 27/11/96 HC QBD Issue of fact whether requirement that stop after accident has been complied with.

McDonnell, Rosenbloom v [1957] Crim LR 809t HC QBD Absolute discharge for cab-driver guilty of unlawfully demanding more than the proper fare.

McEachran v Hurst [1978] Crim LR 499; [1978] RTR 462 HC QBD Immobile moped still mechanically propelled vehicle where being brought to repairers to be fixed.

McGall, R v [1974] Crim LR 482; [1974] RTR 216 CA Slow driving on empty road and mistaken use of indicators could have prompted reasonable suspicion that driver had excess alcohol.

Morgan, James v [1988] RTR 85 HC QBD Successful appeal by prosecutor against successful 'spiking of drinks' claim by defendant before trial court as special reason justifying non-disqualification.

Morgan, Owen v [1986] RTR 151 HC QBD Could repeat request to person (who had refused to supply breath specimen) to supply breath specimen.

Morgan, Stubbs v [1972] Crim LR 443; [1972] RTR 459 HC QBD No requirement that traffic sign must be fit for illumination at night so as to be valid by day.

Morgan, Venn v [1949] 2 All ER 562; (1949) 113 JP 504; (1949) LXV TLR 571; (1949) WN (I) 353 HC KBD On format of charge of careless driving.

Morgan, Williams v (1921-25) XXVII Cox CC 37; (1921) 85 JP 191 HC KBD Locomotive used by non-farmer without licence/permit was permissible as used for agricultural purpose.

Moriarty (Terence Patrick), R v (1993) 14 Cr App R (S) 575 CA Reckless driving (of which police chase formed part) punishable only by custodial sentence.

Morris v Beardmore [1980] 2 All ER 753; [1981] AC 446; (1980) 71 Cr App R 256 (also HC QBD); [1979] Crim LR 394; (1980) 144 JP 331; (1980) 130 NLJ 707; [1980] RTR 321; (1980) 124 SJ 512; [1980] 3 WLR 283 HL Request to take breath test invalid where police officer trespassing on property of person requested at time of request.

Morris v Beardmore [1979] 3 All ER 290; (1980) 71 Cr App R 256 (also HL); (1980) 144 JP 30; [1980] QB 105; [1979] RTR 393; (1979) 123 SJ 300; [1979] 3 WLR 93 HC QBD Request to take breath test valid though police officer trespassing on property of person requested at time of request.

Morris v Tolman (1921-25) XXVII Cox CC 345; (1922) 86 JP 221; [1923] 92 LJCL 215; (1923) 128 LTR 118 HC KBD Aider and abettor could not be convicted as principal where was excluded by statute.

Morris, Abercromby v [1932] All ER 676; (1931-34) XXIX Cox CC 553; (1932) 96 JP 392; (1932) 147 LTR 529; (1932) 76 SJ 560; (1931-32) XLVIII TLR 635; (1932) WN (I) 201 HC KBD Owner not liable for act of person (not his agent or servant) in use of car.

Morris, Geraghty v [1939] 2 All ER 269; (1939-40) XXXI Cox CC 249; (1939) 103 JP 175; (1939) 87 LJ 288; (1939) 160 LTR 397; (1939) 83 SJ 359; (1938-39) LV TLR 599; (1939) WN (I) 153 HC KBD No review of driving test results.

Morris, Lloyd v [1985] Crim LR 742; [1986] RTR 299 HC QBD Unusually wide discrepancy between two breathalyser readings did not per se merit decision by justices that breathalyser defective.

Morris, Melhuish v [1938] 4 All ER 98; (1938) 86 LJ 311; (1938) 82 SJ 854 HC KBD Evidence of two police officers based on one speedometer not enough to secure speeding conviction.

Morris, R v [1972] 1 All ER 384; (1972) 56 Cr App R 175; [1972] Crim LR 116; (1972) 136 JP 194; (1971) 121 NLJ 1074t; [1972] RTR 201; (1972) 116 SJ 17; [1972] 1 WLR 228 CA 'Accident' in Road Safety Act 1967, s 2(2), to be given ordinary meaning.

Morris, Roberts v [1965] Crim LR 46g HC QBD Need only possess licence for carriage of goods for hire/reward when are actually physcally transporting the goods.

Morton (Bernard Charles), R v (1992) 13 Cr App R (S) 315; [1992] Crim LR 70 CA Nine months' imprisonment for reckless driving.

Morton v Confer [1963] 2 All ER 765; [1963] Crim LR 577; (1963) 127 JP 433; (1963) 113 LJ 530; (1963) 107 SJ 417; [1963] 1 WLR 763 HC QBD Accused to show (civil burden) that had no intention to drive until fit to do so; claim to be considered in light of all the facts.

Moscrop and Wills v Blair [1962] Crim LR 323; (1961) 105 SJ 950 HC QBD Two ton car attached to one ton trailer validly found to be one entity and so a heavy motor car.

Moses v Winder [1980] Crim LR 232; [1981] RTR 37 HC QBD Failed attempt by diabetic to establish automatism defence to driving without due care and attention charge.

Moss v Jenkins [1974] Crim LR 715; [1975] RTR 25 HC QBD Absent explanation as to how station sergeant had reasonable suspicion that certain driver had been drinking, breath test request by him invalid.

Motor Insurers' Bureau, Buchanan v [1955] 1 All ER 607; [1955] 1 WLR 488 HC QBD 'Road' not road under Road Traffic Act 1930, s 121(1) as no public access by right/tolerance.

Mould, Clarke v [1945] 2 All ER 551; (1945) 109 JP 175; (1945) 173 LTR 370; (1945) 89 SJ 370; (1945) WN (I) 125 HC KBD In adequate notice of prosecution as served on firm rather than individual member thereof.

Moulder v Neville [1974] Crim LR 126; [1974] RTR 53; (1974) 118 SJ 185 HC QBD Driver must yield to person stepping onto stripes at zebra crossing before driver enters onto that area.

Moule, R v [1964] Crim LR 303t; (1964) 108 SJ 100 CCA Valid conviction for causing unlawful obstruction of highway of person who incited others to sit down and block highway.

Mounsey v Campbell [1983] RTR 36 HC QBD Bumper to bumper parking by van driver occasioned unnecessary obstruction.

Mowe v Perraton [1952] 1 All ER 423; (1951-52) 35 Cr App R 194; (1952) 116 JP 139; (1952) 96 SJ 182; (1952) WN (I) 96 HC QBD Not taking and driving away if vehicle at outset in one's lawful possession.

Moxon-Tritsch (Leona), R v [1988] Crim LR 46 CrCt Ought not on basis of same facts to be prosecuted for causing death by reckless driving where had already pleaded guilty to careless driving/driving with excess alcohol.

Moynes, R v; R v Dawson [1976] RTR 303 CA Nine months' imprisonment for each of two offenders who conspired to defeat justice by agreeing to one impersonating other if stopped by police.

Mulcaster v Wheatstone [1979] Crim LR 728; [1980] RTR 190 HC QBD Valid breath test requirement where driving did not prompt policeman's first suspicion of excess alcohol but immediately formed that opinion on approaching driver.

Mullarkey v Prescott [1970] RTR 296 HC QBD That had artificial legs/did not drive far along quiet road in Winter/might need state assistance if were disqualified not special reasons justifying non-disqualifiction for driving with excess alcohol.

Mullarkey, R v [1970] Crim LR 406dt HC QBD Disqualification of legless driver from using invalid carriage (after drunk driving episode) would have harsh consequences but was merited.

Mullen, Bentley v [1986] RTR 7 HC QBD Supervisor of learner driver convicted of aiding and abetting latter in not stopping after accident occurred.

Mulroy, R v [1979] RTR 214 CA Disqualification merited where sentencing for taking conveyance without lawful authority/robbery but reduced to fifteen months given difficulties offender might otherwise face upon twelve months' imprisonment imposed for same offences.

Muncaster, Keene v [1980] Crim LR 587; (1980) 130 NLJ 807; [1980] RTR 377; (1980) 124 SJ 496 HC QBD Police officer could not give himself permission not to leave left/near side of parked vehicle turned to carriageway at night (Motor Vehicles (Construction and Use Regulations) 1973, reg 115 interpreted).

Muncaster, R v [1974] Crim LR 320t; (1974) 124 NLJ 103t CA Justices exhorted to reduce unmerited twenty year disqualification which had led to offender being brought before court on several occasions.

Munning, R v [1961] Crim LR 555 Magistrates Pushing scooter not driving of same so not guilty of driving while disqualified.

Munson v British Railways Board and others [1965] 3 All ER 41; (1965) 115 LJ 709; [1966] 1 QB 813; (1965) 109 SJ 597; [1965] 3 WLR 781 CA Answer in previous licence application on normal user relevant in late licence application.

Murphy (Anthony John), R v (1989) 89 Cr App R 176; [1989] RTR 236 CA Six years' disqualification and requirement that sit new driving test substituted for eight years' disqualification for reckless driving resulting in serious injury/death.

Murphy v Brown [1970] Crim LR 234; [1970] RTR 190; (1969) 113 SJ 983 HC QBD Motor repairer/dealer not licensed to transport pony (for which had given motor vehicle) to place of sale.

Murphy v Griffiths [1967] 1 All ER 424; [1967] Crim LR 181t; (1967) 131 JP 204; (1965-66) 116 NLJ 1600t; (1967) 111 SJ 76; [1967] 1 WLR 333 HC QBD Back-dating of test certificate makes it false in material particular.

Norris, Jarvis v [1980] RTR 424 HC QBD Vengefulness not a defence to/could afford evidence of requisite intent for reckless driving.

Norris, Willingale v (1907-09) XXI Cox CC 737; (1908) 72 JP 495; [1909] 1 KB 57; [1909] 78 LJKB 69; (1908-09) XCIX LTR 830 HC KBD Breach of regulations under London Hackney Carriage Act 1850, s 4 punishable in way prescribed in London Hackney Carriage Act 1853, s 19.

North v Gerrish [1959] Crim LR 462t; (1959) 123 JP 313 HC QBD Must both stop and give name and address after motor accident if are not to contravene Road Traffic Act 1930, s 22(1).

North West Traffic Area Licensing Authority v Brady [1981] Crim LR 407; [1981] RTR 265 CA Licence application 'made' where completed/mailed in December 1976 though not received by licensing authority until January 1977.

North West Traffic Area Licensing Authority v Brady [1979] Crim LR 397; (1979) 129 (2) NLJ 712; [1979] RTR 500 HC QBD Licence application not 'made' when completed/mailed in December 1976 but when received by licensing authority in January 1977.

North West Traffic Area Licensing Authority v Post Office [1982] RTR 304 HC QBD Post Office guilty of unlicensed use of goods vehicle used in part to transport items incidental to but not necessary for operation of machine on vehicle.

North-Western Traffic Area Licensing Authority, Smith and another v [1974] Crim LR 193; [1974] RTR 236 HC QBD Vehicle carrying twelve separate beams not carrying abnormal 'indivisible' load but vehicle carrying same found to breach vehicle construction/use regulations.

Northfield v Pinder [1968] 3 All ER 854; (1969) 53 Cr App R 72; [1969] Crim LR 96t; (1969) 133 JP 107; (1968) 118 NLJ 1053t; [1969] 2 QB 7; (1968) 112 SJ 884; [1969] 2 WLR 50 HC QBD Accused in charge of motor vehicle while drunk must show no chance of driving car while over blood-alcohol limit.

Norton v Hayward [1969] Crim LR 36; (1968) 112 SJ 767 HC QBD That part of land to which local authority prohibited waiting order extended was private did not mean order did not apply in respect of same.

Nottingham City Council v Woodings [1994] RTR 72; [1993] TLR 115 HC QBD Driver of obvious minicab not plying for hire when parked and got out of same to enter toilet nor when was coming back but was plying for hire when sat into car and told enquirers was free for hire.

Nowell, R v [1948] 1 All ER 794; (1946-48) 32 Cr App R 173; (1948) 112 JP 255; (1948) 98 LJ 245; (1948) LXIV TLR 277; (1948) WN (I) 154 CCA Police doctor's persuading person to give sample as in their own interest did not make doctor's evidence inadmissible.

Nugent v Hobday [1972] Crim LR 569; [1973] RTR 41 HC QBD Part specimen supplied must be capable of analysis using ordinary equipment/skill.

Nugent v Phillips [1939] 4 All ER 57; (1939-40) XXXI Cox CC 358; (1939) 103 JP 367; (1939) 88 LJ 253; (1939) 161 LTR 386; (1939) 83 SJ 978; (1939) WN (I) 359 HC KBD That person paid to carry horse in motor horse-box meant vehicle used for hire or reward.

Nugent v Ridley [1987] Crim LR 640; [1987] RTR 412 HC QBD Third specimen could be sent for laboratory examination (so long as procedures in acquiring sample complied with).

Nunn (Adam John), R v [1996] 2 Cr App R (S) 136; [1996] Crim LR 210 CA On relevance of opinions of deceased's survivors to sentencing of person who caused deceased's death by dangerous driving.

Nutland v R [1991] Crim LR 630 HC QBD On what is meant by 'structural attachment' to goods vehicle that renders it a tanker (Dangerous Substances (Conveyance by Road in Road Tankers and Tank Containers) Regulations 1981, regulation 2).

Nuttall, R v [1972] Crim LR 485; [1971] RTR 279; (1971) 115 SJ 489 Assizes Removal of five-year disqualification.

Nuttall, Wing v (1997) 141 SJ LB 98; [1997] TLR 225 HC QBD Employers are required to check tachograph charts: failure to do so is reckless blindness to workers' tachograph offences, so are guilty of permitting same.

Nutter, Gibson v [1984] RTR 8; [1983] TLR 261 HC QBD On what constitutes 'disabled' vehicle; goods vehicle test certificate necessary where using recovery vehicle in connection with more than one disabled vehicle.

O'Boyle, R v [1973] RTR 445 CA On what it means to 'require' a breath test of a driver.

O'Brien v Anderton [1979] RTR 388 HC QBD 'Italjet' a motor vehicle (had two wheels/22cc engine/seat/handle bars).

O'Brien v Trafalgar Insurance Company, Limited (1945) 109 JP 107; (1944-45) LXI TLR 225 CA Roads around factory deemed not to fall within definition of road in Road Traffic Act 1930, s 121(1).

O'Brien, Chief Constable of Avon and Somerset Constabulary v [1987] RTR 182 HC QBD That had sought to meet with doctor/solicitor before giving specimen not reasonable excuse defence to charge of failure to provide specimen.

O'Connell v Fraser [1963] Crim LR 289; (1963) 107 SJ 95 HC QBD Valid conviction of person who edged out slowly from behind parked cars but collided with motor cyclist.

O'Connell v Murphy [1981] Crim LR 256; [1981] RTR 163 HC QBD Breach of motor vehicle construction and use law where transported exhaust pipe part of which sticking from window exceeded permissible lateral projection limits.

O'Connor (David), R v [1979] RTR 467 CA Fine/three year disqualification merited by provisional driver who caused death by dangerous driving when ignored supervising driver's instructions.

O'Connor (Michael Dennis), Director of Public Prosecutions v (1992) 13 Cr App R (S) 188 HC QBD On non-disqualification for drink driving offences arising from 'spiking' of drinks unbeknownst to driver.

O'Connor, Director of Public Prosecutions v; Director of Public Prosecutions v Allatt; Director of Public Prosecutions v Connor; Director of Public Prosecutions v Chapman; R v Crown Court at Chichester, ex parte Moss; Director of Public Prosecutions v Allen (1992) 95 Cr App R 135; (1991) 141 NLJ 1004t; [1992] RTR 66; [1991] TLR 335 HC QBD Guidelines on non-disqualification from driving of those convicted of drunk driving.

O'Connor, The Police v [1957] Crim LR 478 Sessions That large vehicle properly parked on road did not constitute unreasonable user adequate to support conviction for obstruction of road.

O'Hagan, Chapman v; Chapman v Smith [1949] 2 All ER 690; (1949) 113 JP 518; (1949) 99 LJ 611; (1949) LXV TLR 657; (1949) WN (I) 399; (1949) 113 JP 464 HC KBD Need not seek to put analyst's certificate in evidence where respondent admitted had commercial petrol in tank.

O'Halloran v Director of Public Prosecutions (1990) 154 JP 837; [1990] RTR 62 HC QBD Driver must comply with arrow requiring that keep to left of double white line.

O'Keefe, Williamson v [1947] 1 All ER 307 HC KBD Any person driving uninsured car — whether or not are owner — commits offence.

O'Meara, Director of Public Prosecutions v [1989] RTR 24 HC QBD That did not know that drink from previous night could result in one being over prescribed level next morning not special reason meriting non-disqualification.

O'Meara, Director of Public Prosecutions v (1988) 10 Cr App R (S) 56 CA Not special reason justifying non-disqualification that excess alcohol result of previous evening's drinking (interrupted by sleep).

O'Neale, R v [1988] Crim LR 122; [1988] RTR 124 CA On relevancy of driver being in violation of provisional licence-holder legislation where charged with causing death by reckless driving.

O'Neill v Brown [1961] 1 All ER 571; [1961] Crim LR 317; (1961) 125 JP 225; (1961) 111 LJ 206 [1961] 1 QB 420; (1961) 105 SJ 208; [1961] 2 WLR 224 HC QBD Guilty of endangering with trailer where improperly joined — though car and trailer independently in good condition.

O'Neill v George [1969] Crim LR 202; (1969) 113 SJ 128 HC QBD Valid to convict for parking without payment in bay where was free parking on bank holidays but not on day when banks closed because of special royal proclamation.

O'Sullivan (Patrick John), R v; R v Burtoft (Duncan Paul) (1983) 5 Cr App R (S) 283; [1983] Crim LR 827 CA On appropriate senences for reckless driving/causing death by reckless driving.

O'Toole (Robert John), R v (1971) 55 Cr App R 206; [1971] Crim LR 294 CA On disqualification as a sentence; on privileges and responsibilities of driver of emergency vehicle answering emergency call.

Paterson, Anderson and Heeley Ltd v [1975] Crim LR 49; (1975) 139 JP 231; [1975] RTR 248; (1975) 119 SJ 115; [1975] 1 WLR 228 HC QBD On what constitutes a 'tower wagon'.

Paterson, Connor v [1977] 3 All ER 516; [1977] Crim LR 428; (1978) 142 JP 20; (1977) 127 NLJ 639t; [1977] RTR 379; (1977) 121 SJ 392; [1977] 1 WLR 1450 HC QBD Was unlawful under 'Zebra' Pedestrian Crossings Regulations 1971/1524 to overtake car which had stopped at crossing to let people cross even though all people crossing had crossed at moment of overtaking.

Patterson v Charlton [1985] Crim LR 449; (1986) 150 JP 29; [1986] RTR 18 HC QBD Admission that were driving is enough proof of same; burden is on person over the blood-alcohol limit to prove had not driven while so.

Patterson v Helling [1960] Crim LR 562 HC QBD Information for use of motor vehicle absent excise licence ought not to have been dismissed in light of mitigating factors — these should have only counted towards sentence.

Patterson v James [1962] Crim LR 406t HC QBD On deciding whether person to be fined/imprisoned for driving while disqualified.

Patterson v Redpath Brothers Ltd [1979] Crim LR 187; (1979) 129 (1) NLJ 193; [1979] RTR 431; (1979) 123 SJ 165; [1979] 1 WLR 553 HC QBD On what constituted 'indivisible load' for purposes of Motor Vehicles (Construction and Use) Regulations 1973, regs 3(1) and 9(1).

Paul, Gosling v (1961) 125 JP 389 HC QBD Shortness of remaining period of disqualification not reason justifying non-imprisonment for driving while disqualified.

Pawley v Whardall (1965) 115 LJ 529 HC QBD '[O]ther persons using the Road' (in the Road Traffic Act 1960, s) includes persons in vehicle being driven.

Payne v Allcock (1932) 101 LJCL 775; (1932) 76 SJ 308; (1932) WN (I) 111 HC KBD Offence under Finance Act 1922, s 14 to use vehicle licensed for private use as a goods vehicle.

Payne v Harland [1980] RTR 478 HC QBD Headlights and sidelights required on car when driven by day.

Payne, Lovett v [1979] Crim LR 729; (1979) 143 JP 756; [1980] RTR 103 HC QBD On what was ment by 'nearest' weighbridge in Road Traffic Act 1972, s 40(6).

Payne, R v [1963] 1 All ER 848; (1963) 47 Cr App R 122; [1963] Crim LR 288; (1963) 127 JP 230; (1963) 113 LJ 285; (1963) 107 SJ 97; [1963] 1 WLR 637 CCA Evidence of doctor though admissible ought not to be admitted where person submits to medical examination on basis of express assurance that results would not be admissible in evidence.

Peacock, Duck v [1949] 1 All ER 318; (1949) 113 JP 135; (1949) 99 LJ 49; (1949) LXV TLR 87; (1949) WN (I) 36 HC KBD Stopping car as soon as feel effects of drinking not special reason justifying non-disqualification.

Peak Trailer and Chassis, Ltd v Jackson [1967] 1 All ER 172; (1967) 131 JP 155; (1966) 110 SJ 927; [1967] 1 WLR 155 HC QBD Forty-six from 177 short-load trips in twelve months showed lorry not normally used for loads of indivisible length.

Pearce, R v [1961] Crim LR 122 CCA On whether taking and driving away motor vehicle without owner's consent a continuing offence.

Pearman, Director of Public Prosecutions v (1993) 157 JP 883; [1992] RTR 407; [1992] TLR 154 HC QBD Was open to justices to find that very shocked and distressed woman had been incapable of providing proper breath specimen.

Pearson (Donald), R v [1974] Crim LR 315; [1974] RTR 92 CA Whether breath test request made as soon as reasonably practicable after road traffic offence a question of fact for jury.

Pearson (E) and Son (Teesside) Ltd v Richardson [1972] 3 All ER 277; [1972] Crim LR 444t; (1972) 136 JP 758; [1972] RTR 552; (1972) 116 SJ 416; [1972] 1 WLR 1152 HC QBD Vehicle that can raise but not tow disabled vehicle not 'recovery vehicle'.

Pearson v Boyes [1953] 1 All ER 492; (1953) 117 JP 131; (1953) 97 SJ 134; [1953] 1 WLR 384 HC QBD Not using goods vehicle as goods vehicle when simply using to tow caravan.

Pearson v Rutterford and another [1982] RTR 54 HC QBD Period when driver acting as instructor not a 'daily rest period' but another period of work.

Pitt, Stimson v [1947] KB 668; [1948] 117 LJR 351; (1947) 177 LTR 187; (1947) 91 SJ 309; (1947) LXIII TLR 293; (1947) WN (I) 157 HC KBD Each of four cross-roads with traffic-lighted pedestrian crossings is 'controlled' by lights (Pedestrian Crossing Places (Traffic) Regulation 1941, r 5).

Pitts v Lewis (1989) 153 JP 220; [1989] RTR 71 HC QBD On defence under Local Government (Miscellaneous Provisions) Act 1976, s 75, that vehicle used under contract of hire for not less than seven days where are charged operating unlicensed private hire vehicle.

Plancq v Marks (1907-09) XXI Cox CC 157; (1906) 70 JP 216; (1906) XCIV LTR 577; (1905-06) 50 SJ 377; (1905-06) XXII TLR 432 HC KBD Could be convicted of excessive speed on evidence of constable using stop watch.

Plant (Barnett), R v (1987) 9 Cr App R (S) 241 CA Eight months' imprisonment for causing death by reckless driving (speeding).

Platten v Gowing [1983] RTR 352 HC QBD On appropriate fine for breach of temporary speed limit (where did not breach normal speed limit).

Platten v Gowing (1982) 4 Cr App R (S) 386; [1983] Crim LR 184; [1982] TLR 615 CA Maximum penalties for exceeding general speed limit inapplicable to situation where exceed temporary (but not general) speed limit.

Plowden, ex parte Braithwaite, R v (1911-13) XXII Cox CC 114; (1909) 73 JP 266; [1909] 2 KB 269; [1909] 78 LJKB 733; (1909) 100 LTR 856; (1908-09) XXV TLR 430; (1909) WN (I) 87 HC KBD Royal Park speeding regulations valid.

Plowman, Woodriffe v [1962] Crim LR 326; (1962) 106 SJ 198 HC QBD Absent contrary evidence traffic signals are presumed to be lawfully placed.

Plume v Suckling [1977] RTR 271 HC QBD Coach modified to carry passengers, kitchen items and stock car was a goods vehicle.

Police v Beaumont [1958] Crim LR 620 Magistrates Successful plea of automatism (consequent upon pneumonia) in careless driving prosecution.

Police v Dormer [1955] Crim LR 252 Magistrates Electrician carrying testing tools and equipment in his van was not carrying 'goods'.

Police v Hadelka [1963] Crim LR 706 Magistrates Absolute discharge for person who spent five minutes waiting in restricted area for load which had good reason to expect (but did not actually) receive.

Police v Lane [1957] Crim LR 542 Magistrates Harpoon gun a gun for purposes of the Gun Licence Act 1870.

Police v Liversidge and Featherstone [1956] Crim LR 59 Magistrates Two parties guilty of being drunk in charge of same motor vehicle.

Police v O'Connor [1957] Crim LR 478 Sessions That large vehicle properly parked on road did not constitute unreasonable user adequate to support conviction for obstruction of road.

Police v Okoukwo [1954] Crim LR 869 Magistrates Person may at own risk board stationary bus at any point along route.

Police v Wright [1955] Crim LR 714 Magistrates Person riding motorised pedal cycle on which motor did not work acquitted of driving an uninsured vehicle.

Pond, Simmonds v (1918-21) XXVI Cox CC 365; (1919) 83 JP 56; [1919] 88 LJCL 857; (1919) 120 LTR 124; (1918-19) XXXV TLR 187; (1919) WN (I) 25 HC KBD Conviction for unauthorised use of petrol was one connected with use of motor car.

Pontin v Price (1934-39) XXX Cox CC 44; (1933) 97 JP 315; (1934) 150 LTR 177 HC KBD Conviction for failing to keep to line of traffic into which directed by police officer acting in course of duty.

Poole (Anthony), Rumbles (Brian) v (1980) 2 Cr App R (S) 50 CA Cannot endorse licence upon conviction for failure to stop when told to do so by traffic warden.

Poole Corporation, Brownsea Haven Properties, Ltd v (1957) 121 JP 571; (1957) 107 LJ 554 HC ChD Improper to use powers under Town Police Clauses Act 1847 to test prospective one-way street system under Road Traffic Act 1930: order doing so was void.

Poole v Ibbotson (1949) 113 JP 466; (1949) LXV TLR 701 HC KBD Successful action for causing motor vehicle to be used as express carriage without road service licence contrary to Road Traffic Act 1934 (club bringing club members to away game by coach).

Poole v Lockwood [1980] Crim LR 730; (1980) 130 NLJ 909; [1981] RTR 285 HC QBD Blood specimen inadmissible where second urine specimen provided over an hour after requested discarded by police officer genuinely believing specimen inadequate.

Poole, Director of Public Prosecutions v (1992) 156 JP 571; [1992] RTR 177; [1991] TLR 163 HC QBD Person must have option of giving alternative specimens presented to them but if impede officer from relaying option are not later entitled to acquittal on basis that option not given.

Poole, Flatman v; Flatman v Oatey [1937] 1 All ER 495 HC KBD No need to keep records in respect of agricultural lorry transporting mixture of non-/agricultural goods from one farm to another.

Poole, Series v [1967] 3 All ER 849; [1967] Crim LR 712; (1968) 132 JP 82; (1967) 117 NLJ 1140t; [1969] 1 QB 676; (1967) 111 SJ 871; [1968] 2 WLR 261 HC QBD Employer delegating task of seeing complies with road traffic legislation remains liable for non-compliance.

Pope v Clarke [1953] 2 All ER 704; (1953) 117 JP 429; (1953) 103 LJ 525; (1953) 97 SJ 542; [1953] 1 WLR 1060 HC QBD Notice of prosecution valid though time of offence charged incorrectly stated therein.

Pope, Blake v [1986] 3 All ER 185; [1986] Crim LR 749; [1987] RTR 77; (1970) 130 SJ 731; [1986] 1 WLR 1152 HC QBD Police need not see person driving before forming that driving with excess alcohol.

Popperwell v Cockerton [1968] 1 All ER 1038; [1968] Crim LR 336; (1968) 132 JP 231; (1968) 112 SJ 175; [1968] 1 WLR 438 HC QBD On excessive speed of dual-purpose vehicle.

Port of Manchester Insurance Company Limited and another, Richards v (1935) 152 LTR 261 HC KBD Owner of car but not insurance company guilty of permitting use of car by uninsured person contrary to Road Traffic Act 1930, s 35(2).

Portal, Darnell v [1972] Crim LR 511; [1972] RTR 483 HC QBD Non-compliance with manufacturer's directions inadequate per se to render breath test invalid (here was not invalid).

Porter (Edward Charles Thomas), R v (1973) 57 Cr App R 290; [1973] Crim LR 124t; (1972) 122 NLJ 1134t; (1973) 123 NLJ 612t; [1973] RTR 116; (1973) 117 SJ 36 CA On when 'hospital procedure'/'police station' procedure appropriate for taking of blood specimen.

Porter, Bourlet v [1973] 2 All ER 800; (1974) 58 Cr App R 1; [1974] Crim LR 53; (1973) 137 JP 649; [1973] RTR 293; (1973) 117 SJ 489; [1973] 1 WLR 866 HL Need not be given opportunity of second test at police station; specimen requested from hospital patient need not be provided while patient.

Post Office and Mr A Harris v London Borough of Richmond; Post Office v London Borough of Richmond [1994] Crim LR 940; (1994) 158 JP 919; [1995] RTR 28; [1994] TLR 276 HC QBD Breach of permit condition (permit allowing certain lorries travel in certain areas at restricted hours) did not give rise to criminal liability.

Post Office, North West Traffic Area Licensing Authority v [1982] RTR 304 HC QBD Post Office guilty of unlicensed use of goods vehicle used in part to transport items incidental to but not necessary for operation of machine on vehicle.

Potter v Gorbould and another [1969] 3 All ER 828; [1970] Crim LR 46; (1969) 133 JP 717; [1970] 1 QB 238; (1969) 113 SJ 673; [1969] 3 WLR 810 HC QBD Non-driving overtime work not part of rest period.

Powell (James Thomas), R v; R v Elliott (Jeffrey Terence); R v Daley (Frederick); R v Rafferty (Robert Andrew) (1984) 79 Cr App R 277 CA On sentencing for death by reckless driving; when disqualifying should have reference to accused's driving record.

Powell v Gliha [1979] Crim LR 188; [1979] RTR 126 HC QBD Disqualification merited where emergency requiring person to drive after had been drinking arose in part through her own positive actions.

Price, Pontin v (1934-39) XXX Cox CC 44; (1933) 97 JP 315; (1934) 150 LTR 177 HC KBD Conviction for failing to keep to line of traffic into which directed by police officer acting in course of duty.

Price, R v [1963] 3 All ER 938; (1964) 48 Cr App R 65; [1964] Crim LR 60; (1964) 128 JP 92; (1964) 114 LJ 41; [1964] 2 QB 76; (1963) 107 SJ 933; [1963] 3 WLR 1027 CCA Must offer to supply part of specimen to suspect when request it/very shortly before.

Price, R v [1968] 3 All ER 814; (1969) 53 Cr App R 25; (1969) 133 JP 47; [1968] 1 WLR 1853 CA Lawful breath test request if at time any driving offence being committed; 'any person driving' can include person in stopped car/who has alighted from car.

Prime v Hosking (1995) 159 JP 755; [1995] RTR 189; [1994] TLR 686 HC QBD Driver working overtime in non-driving capacity ought to have entered same on tachograph record sheet.

Prince v Director of Public Prosecutions [1996] Crim LR 343 HC QBD Unnecessary to prove calibration of breathalyser correct where prosecution only seek to rely on blood specimen evidence.

Pring (Marcus Anthony), R v; R v Pring (Harold William John) [1996] 2 Cr App R (S) 53 CA Eight months' imprisonment for causing causing death by dangerous driving in light of medical status.

Pringle, Gouldie v [1981] RTR 525 HC QBD Justices' finding that person did not breach prohibition on right-hand turns when did U-turn unassailable on facts.

Printz v Sewell (1913-14) XXIII Cox CC 23; (1912) 76 JP 295; [1912] 2 KB 511; (1912) 81 LJKB 905; (1912) 106 LTR 880; (1911-12) XXVIII TLR 396 HC KBD Can plead have taken all reasonable steps not to obscure illumination of number plate.

Pritchard (Terry), R v (1995) 16 Cr App R (S) 666 CA Thirty months' imprisonment for motorcyclist who caused death by dangerous driving.

Pritchard v Dyke (1934-39) XXX Cox CC 1; (1933) 97 JP 179; (1933) 149 LTR 493; (1932-33) XLIX TLR 473 HC KBD No evidence on speed necessary to sustain speeding conviction where weight of vehicle over twelve tons.

Pritchard v Jones [1985] Crim LR 52 CA Need not be afforded second opportunity to give breath specimen where first opportunity was given in police station.

Probert, Another v [1968] Crim LR 564 HC QBD Giving misleading indications when driving was careless driving.

Prosser v Dickeson [1982] RTR 96 HC QBD Apparent provision of two urine specimens inside two minutes was actualy provision of one specimen.

Prosser v Richings and another [1936] 2 All ER 1627; (1934-39) XXX Cox CC 457; (1936) 100 JP 390; (1936) 82 LJ 175; (1936) 155 LTR 284; (1935-36) LII TLR 677; (1936) 80 SJ 794; (1936) WN (I) 260 HC KBD On proper testing of weight of heavy motor car.

Prosser, Phillips v [1976] Crim LR 262; [1976] RTR 300 HC QBD Private hire car not a vehicle servicing premises.

Prothero, Jones v [1952] 1 All ER 434; (1952) 116 JP 141; (1952) 102 LJ 121 HC QBD Driver is driver from taking of vehicle on road to completion of trip.

Proudlock, White and another v [1988] RTR 163 HC QBD Objective test as to whether breath specimen cannot be provided/should not be required for medical reasons: constable need not specifically consider this matter same before moving on to request blood specimen.

Provincial Motor Cab Company v Dunning (Parker's case); The Same v The Same (Kynaston's case) (1911-13) XXII Cox CC 159; [1909] 2 KB 599; (1909-10) 101 LTR 231; (1908-09) XXV TLR 646; [1909] 78 LJKB 822; (1909) 73 JP 387 HC KBD Cab company properly convicted of aiding and abetting identification plate illumination offence by employee.

Pugh v Knipe [1972] Crim LR 247; [1972] RTR 286 HC QBD Was not driving in public place to drive on land which belonged to private club/public did not use.

Pugh, Solesbury v [1969] 2 All ER 1171; (1969) 53 Cr App R 326; [1969] Crim LR 381t; (1969) 133 JP 544; (1969) 119 NLJ 438t; (1969) 113 SJ 429; [1969] 1 WLR 1114 HC QBD Failing to provide specimen without reasonable excuse an offence.

Reg Morris (Transport) Ltd, Cassady v [1975] Crim LR 398; [1975] RTR 470 HC QBD Employer's not punishing non-delivery/encouraging delivery by driver of completed record sheets within seven days did not in instant circumstances render employer guilty as aider and abettor to employee's offence.

Regan v Anderton [1980] Crim LR 245; (1980) 144 JP 82; [1980] RTR 126 HC QBD Person could still be deemed to be driving car (for purposes of taking breath test) though car not actually moving.

Regan v Director of Public Prosecutions [1989] Crim LR 832; [1990] RTR 102 HC QBD Person properly informed of right to give either blood or urine where this choice was apparent from information given by police officer.

Reid (John Joseph), R v (1990) 91 Cr App R 263; [1990] RTR 276 CA On necessary mens rea for reckless driving; on adequacy of R v Lawrence direction.

Reid (Philip), R v [1973] 3 All ER 1020; (1973) 57 Cr App R 807; [1973] Crim LR 760; (1974) 138 JP 51; [1973] RTR 536; (1972) 116 SJ 565; (1973) 117 SJ 681; [1973] 1 WLR 1283 CA That get out of car need not mean not driving/attempting to drive; mistaken belief that not obliged to give specimen not failure for not doing so; whether was refusal a factual question for jury.

Reid, R v [1992] 3 All ER 673; (1992) 95 Cr App R 391; [1992] Crim LR 814; (1994) 158 JP 517; [1992] RTR 341; (1992) 136 SJ LB 253; [1992] 1 WLR 793 HL Reckless driving occurs where accused disregards serious risk of injury to another/does not give that prospect consideration.

Reid, R v [1991] Crim LR 269 CA On test for recklessness (Lawrence formula applied).

Reigate Justices, ex parte Holland, R v [1956] 2 All ER 289; [1956] Crim LR 492; (1956) 120 JP 355; (1956) 100 SJ 436; [1956] 1 WLR 638 HC QBD No limit on county council's right to prosecute for excise duties offences.

Rendell v Hooper [1970] 2 All ER 72; [1970] Crim LR 285t; (1970) 134 JP 441; [1970] RTR 252; (1970) 114 SJ 248; [1970] 1 WLR 747 HC QBD Reasonable bona fide departure from manufacturer's instructions in giving breath test does not invalidate it.

Rennie, R v [1978] RTR 109 CA Improper that had activated suspended sentence (relating to different type of offence) when sentencing person for drunk driving.

Rennison, Knowler v [1947] 1 All ER 302; [1947] KB 488 HC KBD Whether special reasons justifying non-disqualification a question of law; hardship/that would not re-offend not a special reason.

Renouf v Franklin [1972] Crim LR 115; [1971] RTR 469 HC QBD Ply/cord structure of tyre not exposed where had to lift tear on tyre to view same.

Renouf, R v [1986] 2 All ER 449; (1986) 82 Cr App R 344; [1986] Crim LR 408; [1986] RTR 191; (1986) 130 SJ 265; [1986] 1 WLR 522 CA That was part of reasonable use of force in making arrest could excuse reckless driving.

Renshaw v Director of Public Prosecutions [1992] RTR 186 HC QBD On procedure as egards informing person of choice as to provision of sample (whether blood or urine).

Revel v Jordan; Hillis v Nicholson (1983) 147 JP 111; [1983] RTR 497; [1982] TLR 504 HC QBD Police can request second breath test.

Revill, Cantwell v [1940] LJNCCR 240 CyCt Driver who drove onto crossing while lights just went in his favour not liable in damages to pedestrian who stepped onto crossing when lights just went against her.

Rey (Frederick Brian), R v (1978) 67 Cr App R 244; [1978] RTR 413 CA Valid arrest for failure to provide specimen where had provided same but in inadequate amount.

Reynolds and Warren v Metropolitan Police [1982] Crim LR 831 CrCt On elements of 'interference' with motor vehicle (Criminal Attempts Act 1981, s 9).

Reynolds v GH Austin and Sons, Ltd [1951] 1 All ER 606; (1951) 115 JP 192; [1951] 2 KB 135; (1951) 101 LJ 135; (1951) 95 SJ 173; [1951] 1 TLR 614; (1951) WN (I) 135 HC KBD Not guilty of offence by doing act not unlawful in itself made unlawful by actions of third party (non-servant/agent) unknown to self.

Robert Millar (Contractors) Ltd and Robert Millar, R v [1970] 1 All ER 577; (1970) 54 Cr App R 158; (1970) 134 JP 240; (1969) 119 NLJ 1164t; [1970] 2 QB 54; [1970] RTR 147; (1970) 114 SJ 16; [1970] 2 WLR 541 CA Scottish employers of lorry driver causing death by dangerous driving in England with unsafe lorry liable for trial in England.

Roberts (Gwylim Ian), R v (1964) 48 Cr App R 300; [1964] Crim LR 531g CCA On police officer's duty to offer portion of suspect's specimen to suspect.

Roberts and another, Brown v [1963] 2 All ER 263; [1963] Crim LR 435; (1963) 107 SJ 666; [1965] 1 QB 1; [1963] 3 WLR 75 HC QBD On what constitutes use of motor vehicle on road (Road Traffic Act 1930, s 36(1)).

Roberts v Director of Public Prosecutions [1994] Crim LR 926; [1994] RTR 31; [1993] TLR 307 HC QBD Could not take judicial notice of approval by Secretary of State of 'Kustom Falcon' radar gun just because police officer had used same on many occasions.

Roberts v Jones [1969] Crim LR 90; (1968) 112 SJ 884 HC QBD Arrest of person in charge of motor car justified subsequent request that arrestee take breath test.

Roberts v Morris [1965] Crim LR 46g HC QBD Need only possess licence for carriage of goods for hire/reward when are actually physcally transporting the goods.

Roberts v Powell [1966] Crim LR 225; (1965-66) 116 NLJ 445g; (1966) 110 SJ 113 HC QBD Successful appeal against conviction for unlawful parking where parked in parking bay (albeit one where meter had been removed).

Roberts, Haines v [1953] 1 All ER 344; (1953) 117 JP 123; (1953) 97 SJ 117; [1953] 1 WLR 309 HC QBD Person in charge of vehicle until places in charge of another.

Roberts, Jones v [1973] Crim LR 123; [1973] RTR 26 HC QBD Person refusing specimen need not be given an hour within which to change mind.

Roberts, Lambert v [1981] 2 All ER 15; (1981) 72 Cr App R 223; [1981] Crim LR 256; (1981) 145 JP 256; (1981) 131 NLJ 448; [1981] RTR 113 HC QBD Where police officer's licence to enter withdrawn has no legal entitlement to administer breath test.

Roberts, R v (1964) 48 Cr App R 296; [1964] Crim LR 472; (1964) 128 JP 395; (1964) 114 LJ 489; [1965] 1 QB 85; (1964) 108 SJ 383; [1964] 3 WLR 180 CCA Not taking and driving away to release handbrake, set lorry going, then jump free and allow freewheel downhill.

Robertson and Son and others v Highland Haulage, Ltd (1958) 108 LJ 651 TrTb On having standing to challenge application for substitution of vehicles under 'A' licence.

Robertson v Bannister [1973] Crim LR 46; [1973] RTR 109 HC QBD Onus of proof on taxi-driver offering services at Heathrow to prove authorised to do so by British Airports Authority.

Robertson v Crew [1977] Crim LR 228; [1977] RTR 141 HC QBD For vehicle to be 'disabled' must have done more than simply taking out rotor arm.

Robertson, Standen v [1975] Crim LR 395; [1975] RTR 329 HC QBD Drunk driving prosecution could rely on second urine specimen provided over an hour after requested.

Robinson (WGA), (Express Haulage), Ltd, Wurzal v [1969] 2 All ER 1021; [1969] Crim LR 666; (1969) 113 SJ 408; [1969] 1 WLR 996 HC QBD Unless rigorously policed record-keeping system do not conform to Goods Vehicles (Keeping of Records) Regulations 1935.

Robinson (Dorothy), R v [1975] RTR 99 CA £50 fine/twelve months' disqualification for person guilty of death by dangerous driving as result of momentary inattention.

Robinson (Kristian Paul), R v; R v Scurry (Lee Patrick) (1994) 15 Cr App R (S) 452 CA Eighteen/twenty-one months' young offender detention for aggravated vehicle-taking.

Robinson v Director of Public Prosecutions [1989] RTR 42 HC QBD No special reasons meriting non-endorsement of licence of solicitor speeding to arrive at place and deal with matter that could be (and was) adequately dealt with by his clerk.

Robinson v Secretary of State for the Environment [1973] 3 All ER 1045; [1973] RTR 511; (1973) 117 SJ 603; [1973] 1 WLR 1139 HC QBD Suspension of public service vehicle licence for fixed period reasonable.

Robinson, Oakley-Moore v [1982] RTR 74 HC QBD Erroneous belief that did not have petrol not a defence to parking inside prohibited area of Pelican crossing.

Robinsons Limited v Richards (1926-30) XXVIII Cox CC 498; (1928) 92 JP 73; [1928] 2 KB 234; (1928) 65 LJ 401; (1928) 97 LJCL 483; (1928) 139 LTR 164; (1928) WN (I) 110 HC KBD Breach of Motor Cars (Use and Construction) Order 1904 where had to take both hands off wheel to apply tractor and trailer brakes.

Robson (James Keith), R v (1989) 11 Cr App R (S) 78 CA Three years' imprisonment for causing death by reckless driving (speeding after drinking).

Robson and others, Henderson v (1949) 113 JP 313 HC KBD Using tractor to haul pony not an agricultural purpose so were liable to higher rate of duty.

Robson, R v [1991] RTR 180 CA Three years' imprisonment appropriate for driver with alcohol guilty of prolonged speeding (despite passenger warnings) resulting in death of two passengers.

Roche v Willis [1934] All ER 613; (1934-39) XXX Cox CC 121; (1934) 98 JP 227; (1934) 151 LTR 154 HC KBD Agricultural tractor driven on farm exempt from licensing requirements of Road Traffic Act 1930, s 9(3).

Roche, Reynolds v [1972] RTR 282 HC QBD Rushing home late in case babysitter left child alone/that disqualification would adversely affect offender's family not special reasons justifying non-disqualification.

Rodenhurst (Ian Henry), R v (1989) 11 Cr App R (S) 219; [1989] RTR 333 CA Roughly three months in young offender institution for twenty year old guilty of causing three deaths by reckless driving through brief instant of recklessness.

Rodgers v Taylor [1987] RTR 86 HC QBD Exception to restricted waiting provisions applied to hackney carriages awaiting hire not to taxis left parked and unattended on street.

Roe, Nelms v [1969] 3 All ER 1379; (1970) 54 Cr App R 43; [1970] Crim LR 48t; (1970) 134 JP 88; (1969) 119 NLJ 997t; [1970] RTR 45; (1969) 113 SJ 942; [1970] 1 WLR 4 HC QBD Police inspector was acting under implied delegated responsibility of Police Commissioner when completed notice under Road Traffic Act 1960, s 232(2).

Rogers, Bryson v [1956] 2 All ER 826; [1956] Crim LR 570t; (1956) 120 JP 454; (1956) 106 LJ 507; [1956] 2 QB 404; (1956) 100 SJ 569; [1956] 3 WLR 495 HC QBD Dual-purpose vehicle covered by special speed limits whether or not conveying goods.

Rogers, Burditt v [1986] Crim LR 636; (1986) 150 JP 344; [1986] RTR 391 HC QBD Breathalyser not unreliable where printout merely contained spelling mistakes that were human in provenance.

Rogers, Maynard v (1970) 114 SJ 320 HC QBD Pedestrian (two thirds)/driver (one third) responsible for collision at pedestrian crossing; on driver's duty on approaching pedestrian crossing.

Rogers, Taylor v [1960] Crim LR 271t; (1960) 124 JP 217 HC QBD Is objective test whether person was driving with due care and attention.

Rogerson v Edwards (1951) 95 SJ 172; (1951) WN (I) 101 HC KBD Inappropriate for justices to dismiss informations outright, without even hearing them just because notice of intended prosecution mis-addressed.

Rogerson v Stephens [1950] 2 All ER 144 HC KBD No offence of uninsured driving of motor vehicle and trailer.

Rolfe, Bailey v [1976] Crim LR 77 HC QBD That problem with brakes probably pre-dated collision insufficient to support conviction for driving with defective brakes.

Roney v Matthews (1975) 61 Cr App R 195; [1975] Crim LR 394; [1975] RTR 273; (1975) 119 SJ 613 HC QBD That second specimen provided one minute over maximum period prescribed by statute did not render it inadmissible as specimen.

Rooney v Haughton [1970] 1 All ER 1001; [1970] Crim LR 236t; (1970) 134 JP 344; [1970] RTR 119; (1970) 114 SJ 93; [1970] 1 WLR 550 HC QBD Person arrested after first breath test cannot require second test at same station.

Roots, Hawkins v; Hawkins v Smith [1975] Crim LR 521; [1976] RTR 49 HC QBD Non-endorsement not justified just because of slightness of incident prompting careless/inconsiderate driving charge.

Sandland v Neale [1955] 3 All ER 571; (1955) 39 Cr App R 167; [1956] Crim LR 58; (1955) 119 JP 583; (1955) 105 LJ 761; [1956] 1 QB 241; (1955) 99 SJ 799; [1955] 3 WLR 689 HC QBD Notice of prosecution properly served when sent to place most likely to receive it even if know is elsewhere.

Sandwell (David Anthony), R v (1985) 80 Cr App R 78; [1985] RTR 45; [1984] TLR 481 CA Cannot impose consecutive disqualification periods — sentence varied in light of entirety of case.

Sandy v Martin [1974] Crim LR 258; (1975) 139 JP 241; [1974] RTR 263 HC QBD Inn car park a public place during hours that persons invited to use it but absent contrary evidence not thereafter (so were not drunk and in charge of motor vehicle in public place).

Sasson v Taverner [1970] 1 All ER 215; (1970) 134 JP 244; (1970) 120 NLJ 12t; [1970] RTR 63; (1970) 114 SJ 75; [1970] 1 WLR 338 HC QBD Specimen request off-road following pursuit after request on road a valid request.

Saunders Transport Ltd, Lambeth London Borough Council v [1974] Crim LR 311 HC QBD Person who hired out skip not liable for its being unlighted at night, having agreed with person hiring that responsibility for lighting skip would fall on the latter.

Saunders v Johns [1965] Crim LR 49t HC QBD Was no case to answer where did not establish that person guilty of speeding offence was in fact the defendant.

Saunders, R v [1978] Crim LR 98 CrCt Person properly convicted for non-compliance with signal of traffic warden (a non-endorseable offence).

Saycell v Bool [1948] 2 All ER 83; (1948) 112 JP 341; (1948) 98 LJ 315; (1948) LXIV TLR 421; (1948) WN (I) 232 HC KBD Freewheeling is 'driving'.

Sayer v Johnson [1970] RTR 286; [1970] Crim LR 589 HC QBD Unless doubt arises prosecutor is not required to prove breathalyser was stored in compliance with manufacturer's instructions.

Sayers, Stokes v [1988] RTR 89 HC QBD On what prosecution must (and here failed to) prove for seeking of blood specimen in light of defectively operating breathalyser to be valid.

Scammell, Beck v [1985] Crim LR 794; [1986] RTR 162 HC QBD Constable's written amendments to breathalyser printout/printout copy (that should read BST not GMT) did not render printout inadmissible.

Scammell, R v [1967] 3 All ER 97; (1967) 51 Cr App R 398; [1967] Crim LR 594; (1967) 131 JP 462; (1967) 117 NLJ 913; (1967) 111 SJ 620; [1967] 1 WLR 1167 CA On careless driving as defence to dangerous driving.

Scampion, Young and another v [1989] RTR 95 HC QBD On ingredients of offence of hackney carriage licensed in one area undertaking to stand/ply for hire/drive in another area in which unlicensed.

Scates, R v [1957] Crim LR 406 CCC Whether person driving dangerously a question of fact not one of intent.

Schofield, R v [1964] Crim LR 829; (1964) 108 SJ 802 CCA Five (not thirty) years' disqualification for person who drove while unfit to do so through drink or drugs.

Scholfield, Houghton and another v [1973] Crim LR 126; [1973] RTR 239 HC QBD Cul de sac validly found to be road to which public had access.

Scoble v Graham [1970] RTR 358 HC QBD Person stunned after collision might (but did not here) have reasonable excuse for refusing specimen.

Scott (Alexander), R v (1990-91) 12 Cr App R (S) 684 CA Suspended sentence quashed/two years' disqualification ordered/£500 fine approved for causing death by reckless driving (travelled with rear of long lorry jutting into fast lane of dual carriageway).

Scott (Stephen Anthony James), R v (1989) 11 Cr App R (S) 249; [1989] Crim LR 920 CA Judge minded to make life disqualification order to inform counsel of same so that counsel may make submissions on the matter.

Scott Greenham Ltd, Director of Public Prosecutions v [1988] RTR 426 HC QBD On determining whether trailer vehicle carrying mobile crane and hoist was a vehicle carrying goods necessary for running of vehicle or for use with fitted machine (was not).

Secretary of State for Transport, ex parte Cumbria County Council, R v [1983] RTR 88 HC QBD Minister could disagree wth factual findings of inspector and (as here) allow appeal against refusal of road service licence.

Secretary of State for Transport, ex parte Cumbria County Council, R v [1983] RTR 129 CA Minister's disagreeing with factual findings of inspector and allowing appeal against refusal of road service licence was improper.

Seeney v Dean and another [1972] Crim LR 545; [1972] RTR 25 HC QBD On what constitutes a recovery vehicle (Land Rover with towing buoy attached a recovery vehicle).

Segal, R v [1976] RTR 319 CA Person properly convicted of driving at dangerous speed but acquitted of driving in dangerous manner.

Selby v Chief Constable of Avon and Somerset [1988] RTR 216 HC QBD Once proved that person was involved in accident and did not report it to police, burden falls on person to prove did not know of accident.

Sellwood v Butt [1962] Crim LR 841; (1962) 106 SJ 835 HC QBD Car possessing device for measuring engine speed but not speedometer did not comply with requirement that every car have a speed measuring device.

Sergeant (Attorney General's Reference (No 37 of 1994)), R v [1995] RTR 309 CA Four years' imprisonment/ten year disqualification/ten penalty points merited for previously convicted road traffic offender guilty here of several traffic offences including causing death to another after lengthy piece of very bad driving after which sought to evade apprehension/frustrate investigation.

Series v Poole [1967] 3 All ER 849; [1967] Crim LR 712; (1968) 132 JP 82; (1967) 117 NLJ 1140t; [1969] 1 QB 676; (1967) 111 SJ 871; [1968] 2 WLR 261 HC QBD Employer delegating task of seeing complies with road traffic legislation remains liable for non-compliance.

Serle, Goodman v [1947] 2 All ER 318; (1947) 111 JP 492; [1947] KB 808; (1947) 97 LJ 389; [1948] 117 LJR 381; (1947) 177 LTR 521; (1947) 91 SJ 518; (1947) LXIII TLR 395; (1947) WN (I) 225 HC KBD Successful appeal against conviction for agreeing on bargained fare within area of City of London where stanadardised fares applicable.

Severn (Kevin), R v (1995) 16 Cr App R (S) 989 CA Eighteen months' imprisonment for causing death by dangerous driving (did not see person on crossing in darkness).

Seward, R v [1970] 1 All ER 329; (1970) 54 Cr App R 85; [1970] Crim LR 113t; (1970) 134 JP 195; (1969) 119 NLJ 1069t; [1970] RTR 102; (1969) 113 SJ 984; [1970] 1 WLR 323 CA Matter for jury whether presence of car on road prompted accident.

Sewell, Printz v (1913-14) XXIII Cox CC 23; (1912) 76 JP 295; [1912] 2 KB 511; (1912) 81 LJKB 905; (1912) 106 LTR 880; (1911-12) XXVIII TLR 396 HC KBD Can plead have taken all reasonable steps not to obscure illumination of number plate.

Sexton, Milstead v [1964] Crim LR 474g HC QBD Steering of uninsured car being towed from one place to another was causing unroadworthy/uninsured use of towed car.

Seymour (Edward John), R v (1983) 76 Cr App R 211; [1983] Crim LR 260; [1983] RTR 202 CA Direction to jury where manslaughter arising from reckless driving.

Seymour, R v [1983] 2 All ER 1058; [1983] 2 AC 493; (1983) 77 Cr App R 215; [1983] Crim LR 742; (1984) 148 JP 530; (1983) 133 NLJ 746; [1983] RTR 455; (1983) 127 SJ 522; [1983] TLR 525; [1983] 3 WLR 349 HL Direction to jury where manslaughter arising from reckless driving.

Sharkey (Bernard Lee), R v; R v Daniels (Andrew Anthony) (1995) 16 Cr App R (S) 257; [1994] Crim LR 866 CA Inappropriate that sixteen year old be sentenced to twelve months' young offender detention after guilty plea to aggravated vehicle-taking resulting in injury to another.

Sharman, Ferriby v [1971] Crim LR 288; [1971] RTR 163 HC QBD Relevant time for determining blood-alcohol concentration is time at which blood specimen is taken.

Sharman, R v [1974] Crim LR 129; [1974] RTR 213 CA Two year disqualification for driving with excess alcohol was not excessive.

Sharp v Spencer [1987] Crim LR 420 HC QBD Bona fide comment by police officer which led road traffic offender not to give blood specimen in lieu of breath specimens did not invalidate process.

Solihull Metropolitan Borough Council, Startin v (1978) 128 NLJ 1072t; [1979] RTR 228 HC QBD Charge of £10 for late-leaving of car in car park (reduced to £1.50 if paid in certain time) not unreasonable.

Solihull Metropolitan Borough Council, Young and Allen v (1989) 153 JP 321 HC QBD On what consttitues 'hackney carriage'/'street' for purposes of Town Police Clauses Act 1847.

Solman, Palastanga v [1962] Crim LR 334; (1962) 106 SJ 176 HC QBD Prosecution failure to produce Stationery Office copy of regulations in court and so prove same did not merit dismissal of summons.

Solomon v Durbridge (1956) 120 JP 231 HC QBD Car left in certain place for unduly long period could be (and here was) obstruction of highway.

Solomons, R v (1911-13) XXII Cox CC 178; (1909) 2 Cr App R 288; (1909) 73 JP 467; [1909] 2 KB 980; (1910) 79 LJKB 8; (1909-10) 101 LTR 496; (1908-09) XXV TLR 747 CCA Taxicab driver of cab owned by company received fares for/on behalf of company for purposes of Larceny Act 1901.

Somers, R v [1963] 3 All ER 808; (1964) 48 Cr App R 11; [1963] Crim LR 845; (1964) 128 JP 20; (1963) 113 LJ 821; (1963) 107 SJ 813; [1963] 1 WLR 1306 CCA Doctor's evidence based not on own tests but expert findings is admissible.

South Tyneside Borough, Docherty v [1982] TLR 358 CA Council decision to increase number of Hackney carriage licences was valid.

South Wales Traffic Licensing Authority, ex parte Ebbw Vale Urban District Council, R v [1951] 1 All ER 806; (1951) 115 JP 278; [1951] 2 KB 366; [1951] 1 TLR 742; (1951) WN (I) 192 CA Licensing authority could hear/decide application by company wholly owned by the British Transport Commission.

Southampton Justices, ex parte Tweedie, R v (1932) 96 JP 391; (1932) 74 LJ 168; (1933) 102 LJCL 11; (1932) 147 LTR 530; (1932) 76 SJ 545; (1931-32) XLVIII TLR 636 HC KBD Conviction quashed as charge of dangerous driving reduced to careless driving without adequate evidence of consent to reduction of charge.

Southend Borough Council, Cook v (1990) 154 JP 145; [1990] 2 QB 1; [1990] 2 WLR 61 CA Council could appeal as aggrieved person against justices' order to pay damages to/costs of hackney driver whose licence it had withdrawn.

Southend Borough Council, Cook v (1987) 151 JP 641 HC QBD Local authority could be aggrieved person and could therefore appeal against decision of justices to allow appeal against authority's revocation of hackney carriage licence.

Southend-on-Sea Justices, ex parte Sharp and another, R v [1980] RTR 25 HC QBD Person's licence wrongfully endorsed as his age/weight of vehicle brought him within exempting provisions of Road Traffic (Drivers' Ages and Hours of Work) Act 1976, Schedule 2.

Southwell, Wilmott v (1908) 72 JP 491; (1908-09) XCIX LTR 839; (1908-09) XXV TLR 22 HC KBD One brake plus being able to use engine to lock wheels did not fulfil requirement of having two independent brakes (Motor Car (Use and Construction) Order 1904, Article II).

Spalding, Rynsard v [1985] Crim LR 795; [1986] RTR 303 HC QBD Failed prosecution of person charged with driving after drinking where only rose above prescribed limit after ceased driving.

Spalding, Simpson v [1987] RTR 221 HC QBD Conviction for failure to provide blood specimen quashed where not warned that failure to provide same could result in prosecution.

Sparks v Edward Ash, Limited [1943] KB 223; [1943] 112 LJ 289; (1943) 168 LTR 118; (1943) WN (I) 21 CA Despite black-out were still required to approach pedestrian crossing at speed that could safely stop if pedestrian on it.

Sparks v Edward Ash, Ltd [1942] 2 All ER 214; (1942) 106 JP 239; [1942] 111 LJ 587; (1942) 167 LTR 64; (1941-42) LVIII TLR 324; (1942) WN (I) 152 HC KBD Despite black-out were still required to approach pedestrian crossing at speed that could safely stop if pedestrian on it but pedestrian could be contributorily negligent.

Sparrow v Bradley [1985] RTR 122 HC QBD Evidence obtained via second breath test at police station using different breathalyser was admissible.

Storer, Maddox v [1962] 1 All ER 831; [1962] Crim LR 328; (1962) 126 JP 263; (1962) 112 LJ 306; [1963] 1 QB 451; (1962) 106 SJ 372 HC QBD 'Adapted' does not mean 'altered so as to be apt' but 'fit and apt for'.

Storer, R v [1969] Crim LR 204 Sessions On correct procedure as regards proving that breathalyser used in case was an approved device.

Storey, R v [1973] Crim LR 189 CA Nine months' imprisonment plus three years' disqualification merited by first time road traffic offender guilty of seriously dangerous driving.

Stringer, Liddon v [1967] Crim LR 371t HC QBD Bus driver guilty of driving without due care and attention where acting on signals from conductor reversed into woman and killed her.

Stringer, R v (1931-34) XXIX Cox CC 605; (1933) 97 JP 99; [1933] 1 KB 704; (1933) 75 LJ 96; (1933) 102 LJCL 206; (1933) 148 LTR 503; (1933) 77 SJ 65; (1932-33) XLIX TLR 189; (1933) WN (I) 28 CCA Could be tried together for manslaughter/dangerous driving and convicted of latter.

Stripp (David Peter), R v (1979) 69 Cr App R 318 CA Once ground for automatism established for jury prosecution must prove accused acted wilfully.

Strong v Dawtry [1961] 1 All ER 926; [1961] Crim LR 319t; (1961) 125 JP 378; (1961) 111 LJ 240; (1961) 105 SJ 235 HC QBD Are guilty of not paying parking meter if park then dash for change and return immediately.

Strong, R v [1995] Crim LR 428 CA Failed conviction for dangerous driving where manner of driving arose from vehicle defect that would not have been obvious to careful/competent driver.

Strowger v John [1974] Crim LR 123; (1974) 124 NLJ 57t; [1974] RTR 124; (1974) 118 SJ 101 HC QBD Non-display of valid vehicle excise licence an absolute offence.

Strutt and another v Clift (1910) WN (I) 212 HC KBD Failed appeal against conviction for unlicensed use of carriage to convey burden.

Stubbs (Sydney), R v (1912-13) 8 Cr App R 238; (1912-13) XXIX TLR 421 CCA Twelve months' hard labour appropriate for motor car manslaughter; victim's negligence alleviated negligence of party convicted of manslaughter.

Stubbs v Chalmers [1974] Crim LR 257 HC QBD Person still driving when stopped on edge of road after signalled to do so by police who only then formed reasonable suspicion he had been drinking and validly required breath test.

Stubbs v Morgan [1972] Crim LR 443; [1972] RTR 459 HC QBD No requirement that traffic sign must be fit for illumination at night so as to be valid by day.

Studebakers, Limited, Griffiths v (1921-25) XXVII Cox CC 565; (1923) 87 JP 199; [1924] 1 KB 102; (1923) 58 LJ 484; [1924] 93 LJCL 50; (1924) 130 LTR 215; (1923-24) 68 SJ 118; (1923-24) XL TLR 26; (1923) WN (I) 278 HC KBD Company liable for road traffic offence of employee acting in course of duty — no mens rea necessary.

Such v Ball [1981] Crim LR 411; [1982] RTR 140 HC QBD Police stopping motorist without reasonable belief as to moving traffic offence could validly request breath test where reasonably suspected driver to have excess alcohol taken as was not random breath test.

Suckling, Plume v [1977] RTR 271 HC QBD Coach modified to carry passengers, kitchen items and stock car was a goods vehicle.

Sullivan, Roper v [1978] Crim LR 233; [1978] RTR 181 HC QBD Not giving name and address to other driver/not reporting accident to police are two separate offences; statement of facts not made by witness (being cross-examined) not to be put to same as being evidence disparate from court testimony.

Sulston v Hammond [1970] 2 All ER 830; [1970] Crim LR 473t; (1970) 134 JP 601; [1970] RTR 361; (1970) 114 SJ 533; [1970] 1 WLR 1164 HC QBD Notice of intended prosecution not needed for pelican crossing violations.

Sunter Brothers Ltd and another v Arlidge [1962] 1 All ER 510; [1962] Crim LR 320; (1962) 126 JP 159; (1962) 112 LJ 240; (1962) 106 SJ 154 HC QBD Load an 'abnormal indivisible load' if would be undue risk/damage in dividing.

Taylor v Ciecierski [1950] 1 All ER 319; (1950) 114 JP 162; (1950) 94 SJ 164; (1950) WN (I) 80 HC KBD Commercial petrol in private car.

Taylor v Emerson [1962] Crim LR 638t; (1962) 106 SJ 552 HC QBD Expired licence was a licence so use of same was improper use of licence.

Taylor v Kenyon [1952] 2 All ER 726; (1952) 116 JP 599; (1952) 96 SJ 749; (1952) WN (I) 478 HC QBD Driving while disqualified an absolute offence.

Taylor v Mead [1961] 1 All ER 626; [1961] Crim LR 411; (1961) 125 JP 286; (1961) 111 LJ 323; (1961) 105 SJ 159; [1961] 1 WLR 435 HC QBD 'Constructed or adapted' for use means later material alteration: here hanging rails inside car to transport samples not changing purpose.

Taylor v Rajan (1974) 59 Cr App R 11 HC QBD On special circumstances justifying non-disqualification.

Taylor v Rogers [1960] Crim LR 271t; (1960) 124 JP 217 HC QBD Is objective test whether person was driving with due care and attention.

Taylor v Thompson [1956] 1 All ER 352; [1956] Crim LR 206; (1956) 120 JP 124; (1956) 106 LJ 106; (1956) 100 SJ 133; [1956] 1 WLR 167 HC QBD Car a goods vehicle as constructed for and used as goods vehicle.

Taylor's Central Garages (Exeter), Limited, Roper v (1951) 115 JP 445; [1951] 2 TLR 284; (1951) WN (I) 383 HC KBD On degree of knowledge necessary to be guilty of operating return service without road service licence contrary to Road Traffic Act 1930, s 72(10).

Taylor, Bugge v (1939-40) XXXI Cox CC 450; (1940) 104 JP 467; [1941] 1 KB 198; [1941] 110 LJ 710; (1941) 164 LTR 312; (1941) 85 SJ 82 HC KBD Valid conviction for keeping unlighted vehicle on road where 'road' was forecourt of hotel.

Taylor, Langridge v [1972] RTR 157 HC QBD Container in which blood supplied to defendant clotted but remained capable of analysis was a 'suitable container'.

Taylor, Redman v [1975] Crim LR 348 HC QBD On when road traffic incident may be said to arise from 'the presence of a motor vehicle on the road' (Road Traffic Act 1972, s 8(2)).

Taylor, Rodgers v [1987] RTR 86 HC QBD Exception to restricted waiting provisions applied to hackney carriages awaiting hire not to taxis left parked and unattended on street.

Taylor, St Albans District Council v [1991] Crim LR 852; (1992) 156 JP 120; [1991] RTR 400 HC QBD That allowed unlicensed person to carry out complementary trips in unlicensed vehicle when short-staffed did not mean operator of private vehicle hire business had not contravened licensing legislation.

Taziker (Attorney General's Reference (No 36 of 1994)), R v [1995] RTR 413 CA Generally custodial sentences merited in death by careless driving cases: here two years' imprisonment imposed.

Teape v Godfrey [1986] RTR 213 HC QBD Failed 'reasonable excuse' plea where accused had not tried his hardest to give breath specimen/were no facts on which to ground plea.

Tebb, Fawcett v [1984] Crim LR 175; (1984) 148 JP 303; [1983] TLR 719 HC QBD Not failure to provide breath specimen where specimen provided was never tested to see if adequate to give reading.

Tee v Gough [1980] Crim LR 380; [1981] RTR 73 HC QBD Blood properly taken/analysed with driver's consent for excess alcohol charge admissible in driving while unfit through drink charge.

Tee, Yakhya v [1984] RTR 122; [1983] TLR 286 HC QBD Mere display of telephone number on car-roof did not contravene requirement that non-taxi not have sign on roof indicating is taxi.

Telford, R v [1954] Crim LR 137 Sessions On what is meant by 'wanton driving'/'wilful neglect'.

Templeton (Graham), R v [1996] 1 Cr App R (S) 380 CA Nine months' imprisonment for dangerous driving.

Terry (Neil William), R v (1983) 77 Cr App R 173; [1983] Crim LR 557; [1983] RTR 321; [1983] TLR 309 CA Must be intent not to pay licence fee when prosecuting for fraudulent use of licence.

Terry, R v [1984] 1 All ER 65; [1984] AC 374; (1984) 78 Cr App R 101; (1984) 148 JP 613; [1984] RTR 129; (1984) 128 SJ 34; [1983] TLR 771; [1984] 2 WLR 23 HL Crown need not show intent not to pay licence fee when prosecuting for fraudulent use of licence.

Venn v Morgan [1949] 2 All ER 562; (1949) 113 JP 504; (1949) LXV TLR 571; (1949) WN (I) 353 HC KBD On format of charge of careless driving.

Vickers (Attorney General's Reference (No 42 of 1994)), R v [1995] Crim LR 345; [1996] RTR 9 CA Five years' imprisonment for drunken driver whose wanton driving caused death of pedestrian.

Vickers v Bowman [1976] Crim LR 77; [1976] RTR 165 HC QBD Vehicle was used as express carriage carrying passengers at separate fares even though money collected weekly by one passenger from all and then handed to driver.

Vickers, Arthur Sanderson (Great Broughton), Ltd v [1964] Crim LR 474g; (1964) 108 SJ 425 HC QBD Licensing authority when issuing 'A' carrier's licence could specify which trailers might be used.

Vickers, Borthwick v [1973] Crim LR 317; [1973] RTR 390 HC QBD Justices could rely on own knowledge of local area when deciding if certain trip must have involved using public road.

Vickers, Dial Contracts Ltd v [1972] Crim LR 27; [1971] RTR 386 HC QBD Owner of vehicles not liable as aiders and abettors of breaches of plating and testing regulations by hirers of vehicles.

Victoria Motors (Scarborough), Ltd and another v Wurzal [1951] 1 All ER 1016; (1951) 115 JP 333; [1951] 2 KB 520; (1951) 101 LJ 247; (1951) 95 SJ 382; [1951] 1 TLR 837; (1951) WN (I) 233 HC KBD Special occasion must be special in itself, not just so to parties being carried.

Vincent v Whitehead [1966] 1 All ER 917; [1966] Crim LR 225; (1966) 130 JP 214; (1965-66) 116 NLJ 669; (1966) 110 SJ 112; [1966] 1 WLR 975 HC QBD Structure enabling second person to be carried meant vehicle constructed to carry more than one person.

Vines, Pumbien v [1996] Crim LR 124; [1996] RTR 37; [1995] TLR 337 HC QBD Vehicle on road that was irremovable unless repaired was being 'used' on road.

Vipond, Bason v; Same v Robson [1962] 1 All ER 520; [1962] Crim LR 320; (1962) 126 JP 178; (1962) 112 LJ 241; (1962) 106 SJ 221 HC QBD Test whether vehicle exceeds length restrictions is in its ordinary position as constructed.

Vogt, Gross Cash Registers, Ltd v [1965] 3 All ER 832; [1966] Crim LR 109; (1966) 130 JP 113; (1965-66) 116 NLJ 245; [1967] 2 QB 77; (1966) 110 SJ 174; [1966] 2 WLR 470 HC QBD Person not obliged to but using company van in course of employment a 'part-time driver'.

Wade v Director of Public Prosecutions (1995) 159 JP 555; [1996] RTR 177; [1995] TLR 77 HC QBD Blood test evidence inadmissible where correct procedure not observed in obtaining specimen.

Wade v Grange [1977] RTR 417 HC QBD Doctor blocking road by way in which parked car when answering emergency call was guilty of causing unnecessary obstruction.

Wagner, R v [1970] Crim LR 535; [1970] RTR 422; (1970) 114 SJ 669 CA Was refusal to provide breath specimen where requesting constable did not have necessary equipment and accused refused to wait.

Waite v Smith [1986] Crim LR 405 CA On determining reliability of breathalyser.

Wakefield (Paul), Darlington Borough Council v (1989) 153 JP 481 HC QBD Appeal to magistrates from refusal by local council of hackney carriage licence ought to have been by way of full rehearing.

Wakefield Metropolitan District Council, Ghafoor and others v [1990] RTR 389; [1990] TLR 597 HC QBD Failed appeal against refusal of hackney carriage licence: council not required to issue licences beyond number of carriages required in its area of authority.

Wakeley v Hyams [1987] Crim LR 342; [1987] RTR 49 HC QBD Blood and breath specimens rightly deemed inadmissible evidence where statutory procedure as regarded obtaining same not complied with.

Wakelin, Clark v (1965) 109 SJ 295 HC QBD Delay in road traffic accident actions unhelpful; overtaker not negligent where person overtaken behaved in manner that was not reasonably forseeable.

Wakeman v Catlow [1976] Crim LR 636; [1977] RTR 174 HC QBD On burden of proof on defendant seeking to prove that vehicle (here a Jeep) a land tractor.

Walden, Osgerby v [1967] Crim LR 307t; (1967) 111 SJ 259 HC QBD Successful prosecution for non-disclosure of identity of driver to police.

Walkden and another, Evans v [1956] 1 WLR 1019 HC QBD Father (giving son a driving lesson) and (unlicensed) son guilty of uninsured driving.

Walker (HL), Ltd v British Transport Commission and others (1961) 111 LJ 678 TrTb On allowable use of trailer pursuant to 'A' licence.

Walker v Dowswell [1977] RTR 215 HC QBD Was dangerous driving where failed to comply with traffic sign.

Walker v Hodgkins [1983] Crim LR 555; (1983) 147 JP 474; [1984] RTR 34; [1983] TLR 149 HC QBD Justices could take into account normal laboratory procedure when deciding whether to accept evidence regarding blood specimens.

Walker v Lovell [1975] 3 All ER 107; (1978) 67 Cr App R 249; [1975] Crim LR 720; (1975) 139 JP 708; (1975) 125 NLJ 820t; [1975] RTR 377; (1975) 119 SJ 544; [1975] 1 WLR 1141 HL Where enough breath to provide specimen arrest for failure to provide specimen unlawful and this being so there can be no conviction for drunk driving.

Walker v Lovell [1975] Crim LR 102; (1975) 125 NLJ 43; [1975] RTR 61; (1975) 119 SJ 258 HC QBD Where enough breath to provide specimen arrest for failure to provide specimen unlawful and this being so there can be no conviction for drunk driving.

Walker v Rawlinson [1975] Crim LR 523; [1976] RTR 94 HC QBD Not special reason justifying non-endorsement of licence of person guilty of speeding that did not appreciate that distance between street-lights on street meant had entered 30mph area.

Walker v Tolhurst [1976] Crim LR 261; [1976] RTR 513 HC QBD Essence of careless driving is whether acted without due care and here justices made reasonable decision on that basis.

Walker, Atkinson v [1976] Crim LR 138; (1976) 126 NLJ 64t; [1976] RTR 117 HC QBD Police officer to make clear that breath specimen request must be complied with but need not specify statutory basis of demand (test still valid if mistaken justification for it given but valid alternative reason exists).

Walker-Trowbridge Ltd and another v Director of Public Prosecutions [1992] RTR 182; [1992] TLR 92 HC QBD Load knocked from lorry had fallen from same for purposes of Road Vehicles (Construction and Use) Regulations 1986, reg 100(2); on determining adequacy of securing of load.

Walklett, Haime v (1983) 5 Cr App R (S) 165 CA Not special reason justifying non-disqualification that person only drove car from roadside to car park so as to leave it there overnight.

Wall (Harry), R v [1969] 1 All ER 968; (1969) 53 Cr App R 283; [1969] Crim LR 271; (1969) 133 JP 310; (1969) 119 NLJ 176t; (1969) 113 SJ 168; [1969] 1 WLR 400 CA Arresting officer's responsibilities upon arresting suspected drunk driver.

Wall v Walwyn [1973] Crim LR 376; [1974] RTR 24 HC QBD Must stop upon approaching school crossing at which 'Stop' sign being displayed and must continue to stop as long as sign displayed.

Wall v Williams; Wallwork v Giles [1966] Crim LR 50t; [1970] Crim LR 109t; (1969) 119 NLJ 1142t; [1970] RTR 117; (1970) 114 SJ 36 HC QBD Police officer wearing all of uniform save for headgear is a constable in uniform and can request person to take breath test.

Wall, Webster v [1980] Crim LR 186; [1980] RTR 284 HC QBD Could reasonably be found not guilty of careless driving where struck stationary car while motor-bicycling below restricted speed limit on dark, wet evening.

Waller, Director of Public Prosecutions v [1989] RTR 112 HC QBD Driving fiancee from restaurant where she faced threat of attack not special reason meriting non-disqualification as though began as emergency must have been other means of remedying situation.

Wallington, R v [1995] RTR 112 CA Conviction of police constable for causing death by dangerous driving quashed in light of harsh criticisms of particular police behaviour by trial judge when directing jury.

Wallwork v Rowland [1972] 1 All ER 53; [1972] Crim LR 52t; (1972) 136 JP 137; (1971) 121 NLJ 1025t; [1972] RTR 86; (1972) 116 SJ 17 HC QBD Motorway hard shoulder not part of motorway.

Wolley, Lord v [1956] Crim LR 493 Magistrates £2 fine for failure to observe 'Yield' sign placed on road on trial basis.

Wolverhampton Crown Court, Matto v [1987] RTR 337 HC QBD Bad faith, oppressive behaviour of police meant discretion ought to have been exercised and breath test evidence excluded.

Wood and others, ex parte Anderson and another, R v (1922) 86 JP 64; [1922] 91 LJCL 573; (1922) 126 LTR 522; (1921-22) 66 SJ 453; (1921-22) XXXVIII TLR 269; (1922) WN (I) 38 HC KBD Not liable to conviction under Roads Act 1920, s 8(3) where have general licence under Finance Act 1920 for lorry normally/sometimes used to carry goods/passengers.

Wood v Richards (1977) 65 Cr App R 300; [1977] Crim LR 295; (1977) 127 NLJ 467t; [1977] RTR 201 HC QBD Motor police officer called to emergency had to exercise same care as anyone else: absent evidence as to nature of emergency defence of necessity unavailable.

Wood, Beard v [1980] Crim LR 384; [1980] RTR 454 HC QBD Road Traffic Act 1972, s 159 empowered uniformed police constable to stop traffic independent of any common law powers.

Wood, Edwards v [1981] Crim LR 414 HC QBD Valid finding that there had not been refusal to provide specimen where specimen not provided but did not refuse to provide it.

Wood, R v (1971) 121 NLJ 749t CA On when life disqualification merited (not here).

Woodage v Jones [1975] Crim LR 47 HC QBD On when person may correctly be stated to be in charge of motor vehicle.

Woodage v Jones (No 2) (1975) 60 Cr App R 260; [1975] Crim LR 169; [1975] RTR 119; (1975) 119 SJ 304 HC QBD Offender properly convicted of being drunk and in charge of motor vehicle though was half-mile from vehicle when arrested.

Woodage v Lambie [1971] 3 All ER 674; (1971) 135 JP 595; (1971) 121 NLJ 665t; [1972] RTR 36; [1972] Crim LR 536t; (1971) 115 SJ 588; [1971] 1 WLR 1781 HC QBD That earlier offences not serious not mitigating circumstance when sentencing for third traffic offence.

Woodage, Lambie v [1972] 2 All ER 462; [1972] Crim LR 442t; (1972) 136 JP 554; (1972) 122 NLJ 426t; [1972] RTR 396; (1972) 116 SJ 376; [1972] 1 WLR 754 HL In considering reasonable grounds for not disqualifying judges may consider anything reasonable.

Woodard, Archer v [1959] Crim LR 461 HC QBD That drank whisky on empty stomach not special reason justifying reduced disqualification period.

Woodings, Nottingham City Council v [1994] RTR 72; [1993] TLR 115 HC QBD Driver of obvious minicab not plying for hire when parked and got out of same to enter toilet nor when was coming back but was plying for hire when sat into car and told enquirers was free for hire.

Woodriffe v Plowman [1962] Crim LR 326; (1962) 106 SJ 198 HC QBD Absent contrary evidence traffic signals are presumed to be lawfully placed.

Woods, Newton v (1987) 151 JP 436; [1987] RTR 41 HC QBD Difference in breathalyser readings peculiar but not such as to require decision that machine unreliable (and so evidence inadmissible).

Woodward v Dykes [1969] Crim LR 33; (1968) 112 SJ 787 HC QBD Valid conviction for knowingly making false statement: stated in driving licence application form that deaf person not suffering from illness/disability that might endanger public.

Woodward, Alker v [1962] Crim LR 313t HC QBD On what it means for unlicensed hackney cab to be 'plying for hire'.

Woodward, R v [1995] 3 All ER 79; [1995] 2 Cr App R 388; [1995] Crim LR 487; (1995) 159 JP 349; [1995] RTR 130; (1995) 139 SJ LB 18; [1994] TLR 632; [1995] 1 WLR 375 CA That a driver was adversely affected by alcohol is relevant to death by dangerous driving charge; that he merely consumed alcohol is not; failure to warn jury of this latter point a misdirection.

Woodward, Wells v [1956] Crim LR 207 HC QBD Justices could (unless contrary proven) infer that one traffic light red where other traffic light green.

Woolley v Moore [1952] 2 All ER 797; (1952) 116 JP 601; (1952) 102 LJ 639 [1953] 1 QB 43; (1952) 96 SJ 749; [1952] 2 TLR 673; (1952) WN (I) 480 HC QBD Unloaded goods vehicle not subject to special conditions of carrier's licence.

Woon v Maskell [1985] RTR 289 HC QBD Once positive reading showed police officer no longer required to keep 'READ' button depressed.

Worgan (TK), and Son, Ltd v Gloucestershire County Council; H Lancaster and Co, Ltd v Gloucestershire County Council [1961] 2 All ER 301; (1961) 125 JP 381; (1961) 111 LJ 304; [1961] 2 QB 123; (1961) 105 SJ 403 CA Tractors altered to carry felled timber were solely for haulage and not goods vehicles.

Worsman, Baldwin v [1963] 2 All ER 8; [1963] Crim LR 364; (1963) 127 JP 287; (1963) 113 LJ 349; (1963) 107 SJ 215; [1963] 1 WLR 326 HC QBD Vehicle capable of reversing does not cease to be such because means of reverse shut off.

Worthington, Bowers v [1982] RTR 400 HC QBD Interpretation and application of Road Vehicles (Registration and Licensing) Regulations 1971, reg 35(4).

Worthy v Gordon Plant (Services) Ltd; WR Anderson (Motors), Ltd v Hargreaves [1989] RTR 7; [1962] 1 All ER 129; [1962] Crim LR 115t; (1962) 126 JP 100; (1962) 112 LJ 154; [1962] 1 QB 425; (1961) 105 SJ 1127 HC QBD Not guilty of obstruction where park vehicles in parking place during operative hours, whatever intent.

Wright (Desmond Carl), R v (1979) 1 Cr App R (S) 82 CA Disqualification (for use of motor vehicle in course of offence) imroper if will impede person getting job after released from prison.

Wright (Ernest), R v [1979] RTR 15 CA Three years' imprisonment (concurrent) for each of three deaths caused by dangerous driving/twelve months' imprisonment (concurrent) for driving while unfit to drink justified even in light of offender's having been 'sent to Coventry' by small community in which lived.

Wright (John), R v (1975) 60 Cr App R 114; [1975] RTR 193 CA Portion of specimen supplied to accused to be capable of analysis at that moment and for reasonable time thereafter.

Wright v Brobyn [1971] Crim LR 241; [1971] RTR 204; (1971) 115 SJ 310 HC QBD Delay in taking second breath specimen did not mean driver lost quality of driving.

Wright v Howard [1972] Crim LR 710t; (1972) 122 NLJ 610t; [1973] RTR 12 HC QBD On meaning of phrase 'right-hand turn into' in City of Oxford traffic order.

Wright v Hunt [1984] TLR 309 HC QBD On offence of overtaking at zebra crossing.

Wright v Taplin [1986] RTR 388 HC QBD Evidence of defective breathalyser could be relied upon.

Wright v Wenlock [1972] Crim LR 49; [1971] RTR 228 HC QBD Could infer absent alternative reason being preferred that driving was careless but res ipsa loquitur doctrine per se inapplicable to criminal action.

Wright, British Car Auctions Ltd v [1972] 3 All ER 462; [1972] Crim LR 562t; (1972) 122 NLJ 680t; [1972] RTR 540; [1972] 1 WLR 1519 HC QBD Car auctioneer seeking bids not offering to sell car — conviction for selling unroadworthy vehicle quashed.

Wright, Caise v; Fox v Wright [1981] RTR 49 HC QBD Person steering towed vehicle was 'driving' so could be convicted of driving while disqualified.

Wright, Covington v [1963] 2 All ER 212; (1963) 113 LJ 366; (1963) 107 SJ 477; [1963] 2 WLR 1232 HC QBD Dishonesty not necessary for fare avoidance charge; full fare is that less any amount paid.

Wright, Gilligan v [1968] Crim LR 276 Sessions That were only slightly in excess of proscribed blood-alcohol concentration was not a basis for non-disqualification.

Wright, Police v [1955] Crim LR 714 Magistrates Person riding motorised pedal cycle on which motor did not work acquitted of driving an uninsured vehicle.

WRN Contracting, Ltd and another (Burke, third party), Kelly v [1968] 1 All ER 369; (1968) 112 SJ 465; [1968] 1 WLR 921 Assizes Action for breach of statutory duty if car parked contrary to regulations (but not negligently) helped cause accident.

Wurzal v Addison [1965] 1 All ER 20; [1965] Crim LR 116g; (1965) 129 JP 86; (1965) 115 LJ 42; [1965] 2 QB 131; (1964) 108 SJ 1046; [1965] 2 WLR 131 HC QBD Method of payment immaterial: if payment is hire for reward; 'adapted' in Road Traffic Act refers to capability.

Wurzal v Dowker [1953] 2 All ER 88; (1953) 117 JP 336; (1953) 103 LJ 349; [1954] 1 QB 52; (1953) 97 SJ 390; [1953] 2 WLR 1196 HC QBD 'Special occasion' means occasion special to place where made not to persons on trip.

UNIVERSITY OF WOLVERHAMPTON
LEARNING RESOURCES